Neurocultural Health and Wellbeing

Aim of this devoted book Series in neurology is to highlight the relationship between neuroscience and culture. Nowadays, there is more evidence of how our brain is influenced by the various artistic and cultural disciplines, both in the form of entertainment as well as – and above all – through the emotions and gratification which lead to wellbeing. The emotional mechanisms, the different cultural manifestations provided, are due to the activation of a perceptual and cognitive range that constitute the basis of the social behaviors. All this makes us aware of the benefits the arts have on personal and collective health.

This concept was already known to philosophers in ancient times, who were convinced that inner balance was influenced by culture and in particular by music.

During the centuries, men understood that the *cultivation of the spirit*, or *humanitas*, had a certain role on behavior and from the beginning of modern experimental science, in the fifteenth, the spread of the notions of neuroanatomy allowed artists to get closer to the knowledge of the brain's mechanisms and reveal the emotional and empathic responses at the basis of creativity, and indeed of their own psychophysical wellbeing. A dialogue between the science of the mind and artistic disciplines was born. The results of this meeting made it possible to better define which mental processes are involved when we come into contact with the various humanistic disciplines, and how they can be applied, for instance, to the treatment of mental disorders and neuropsychiatric diseases.

The goal of this Series is both to prove the role of biological brain mechanisms and the influence of various artistic forms on clinical practice, especially in neuropsychiatric disorders, as well as to trace different therapeutic and psycho-physical well-being applications based on scientific evidence from medical literature.

Volumes of the Series will be edited by experts under the supervision of an international editorial committee. Each book, focused on a specific discipline, will provide knowledge on relationship between the brain activity and different forms of language, communication, art. This close inter-relationship with specific focus on different forms of art will explain the effectiveness of this kind of approaches in neuro-psychiatric diseases.

This Series will allow to understand how the culture is one of the fundamental tools to improve general well-being, quality of life and motivation in neurological diseases.

This Series also find a correlation with SDG3 goal "ensuring healthy lives and promoting wellbeing for all at all ages" for the progress of the health and wellbeing considering that neurological diseases are on the rise worldwide including in the developing countries.

Thorsten Rudroff

AI for the Healthy Brain and Neurological Disorders

Neuroscience Innovation in an Anti-Science Era

Springer

Thorsten Rudroff
Turku PET Centre
Turku University Hospital
Turku, Finland

ISSN 2731-4464 ISSN 2731-4472 (electronic)
Neurocultural Health and Wellbeing
ISBN 978-3-032-11595-9 ISBN 978-3-032-11596-6 (eBook)
https://doi.org/10.1007/978-3-032-11596-6

This Springer imprint is published by the registered company Springer Nature Switzerland AG
The registered company address is: Gewerbestrasse 11, 6330 Cham, Switzerland

To my wife Nona, the love of my life, to my family across Germany and the United States, and to Jack and Luna, my beloved cats.

I dedicate this work to everyone who stands as guardians of scientific inquiry, free expression, and democratic values—those who understand that the pursuit of truth and the protection of liberty are inseparable endeavors.

Preface

In the summer of 2024, I witnessed two moments that crystallized the central tension explored in this book. In one laboratory, I watched an AI system decode a patient's imagined speech from neural signals with unprecedented accuracy, offering hope for those who had lost the ability to communicate. In another room, I observed a heated ethics committee debate about whether analyzing brain scans with machine learning violated patients' mental privacy in ways traditional medicine never could. These scenes were separated by mere hallways but worlds apart in their implications. They capture the extraordinary moment in which we find ourselves. We stand at an unprecedented convergence of artificial intelligence and neuroscience that promises to revolutionize how we understand, diagnose, and treat disorders of the brain and mind. Machine learning algorithms are beginning to show potential for predicting Alzheimer's disease years before symptoms appear. Early systems are demonstrating the ability to decode visual experiences from brain activity. Researchers are exploring how AI might optimize personalized treatments for depression with precision that would have seemed fantastical just decades ago. Brain-computer interfaces are enabling paralyzed patients to control computers with thought alone in experimental settings, while AI-powered analysis is revealing hidden patterns in neuroimaging data that often escape expert detection.

Yet these same capabilities emerge during an era when scientific institutions face unprecedented challenges. Anti-science movements systematically target neuroscience research, exploiting public fears about "mind reading" technologies and cognitive enhancement. This targeting reflects a broader pattern that Peter Hotez and Michael Mann documented in "Science Under Siege," where organized campaigns systematically undermine scientific expertise across multiple domains. As Hotez further detailed in "The Deadly Rise of Anti-Science," these movements have evolved beyond simple skepticism into coordinated efforts that have demonstrable, lethal consequences—with over 200,000 unvaccinated Americans dying unnecessarily during the COVID-19 pandemic despite widespread vaccine availability. Commercial pressures threaten research integrity as venture capital floods into neurotechnology, sometimes prioritizing profit over patient welfare. Academic freedom comes under strain as researchers navigate political sensitivities around brain research that touches fundamental questions of human nature, free will, and identity. Global collaboration fractures amid rising nationalism, just when the complexity of brain disorders demands international cooperation more than ever.

This book emerges from my recent immersion in the rapidly evolving intersection of AI and neuroscience. This convergence has accelerated dramatically over the past four years as artificial intelligence became a transformative force in brain research. As someone who has witnessed this revolution unfold in real-time, I have observed the remarkable potential of multimodal AI analysis to reveal neural signatures of complex conditions. These range from Long COVID fatigue to multiple sclerosis progression. I have studied how biomimetic approaches that draw inspiration from the brain's own learning mechanisms are creating more adaptive and efficient AI systems. I have seen firsthand how these technologies are beginning to reshape clinical practice. But I have also encountered the profound challenges that accompany these advances. I have observed institutional review board discussions that struggled to address consent for neural data analysis whose full implications we cannot yet foresee. I have observed how algorithmic bias can systematically disadvantage already marginalized populations, potentially exacerbating rather than reducing health disparities. I have watched promising research directions abandoned due to political pressure and witnessed how market forces can distort research priorities away from the most pressing human needs.

Most troublingly, I have experienced the growing disconnect between what AI neuroscience can actually accomplish and what the public believes—or fears—it can do. This pattern mirrors what Hotez experienced as a vaccine researcher, where his scientific expertise made him a target for harassment and death threats—a dynamic he chronicled in "The Deadly Rise of Anti-Science." His collaboration with climate scientist Michael Mann in "Science Under Siege" revealed how this disconnects between scientific reality and public perception create systematic vulnerabilities that anti-science movements exploit across disciplines. This gap, fueled by sensationalized media coverage and exploited by those with anti-science agendas, threatens to undermine beneficial applications while failing to address genuine risks. The result is a field caught between unrealistic expectations and unfounded fears, struggling to realize its potential for human benefit while navigating what Hotez and Mann termed "institutional vulnerability" to organized attacks on expertise itself. As Hotez's personal experience demonstrates, scientists advancing beneficial technologies increasingly face not just professional criticism but coordinated campaigns designed to discredit scientific authority altogether.

The 19 chapters that follow trace a comprehensive journey through this complex terrain. We begin with the technical foundations that make modern AI neuroscience possible. We explore how machine learning has transformed our ability to analyze brain structure and function, decode neural signals, and predict individual disease trajectories. The early chapters document remarkable advances in several areas. These include multimodal analysis that integrates diverse types of brain data, biomimetic AI systems that learn from the brain's own computational principles, and clinical applications that are already changing patient care in neurology and psychiatry.

The middle parts confront the ethical challenges these capabilities create. We examine how neural decoding technologies raise unprecedented questions about mental privacy and cognitive liberty. We explore how predictive algorithms complicate traditional approaches to informed consent and patient autonomy. We also address how systematic biases in AI development risk creating neurological technologies that serve some populations while systematically failing others. A particularly significant chapter addresses how AI alternatives to animal experimentation are transforming neuroscience research, creating both opportunities to reduce animal suffering and challenges for maintaining scientific rigor.

The latter parts investigate the institutional and social forces shaping how these technologies develop and deploy. We explore how anti-science movements target neuroscience research, how market pressures threaten scientific integrity, how academic freedom faces new constraints in politically sensitive domains, and how the inherently political nature of brain research intersects with democratic governance. These challenges are balanced against promising responses: AI systems that combat neuroscience misinformation, institutional models that protect scientific integrity while enabling innovation, strategies for maintaining international collaboration despite nationalist pressures, and approaches to meaningful public engagement that build trust through transparency rather than authority.

The book culminates in a comprehensive framework for responsible innovation that synthesizes insights from across this landscape. This framework addresses risk-proportionate governance that matches oversight to actual dangers rather than unfounded fears, privacy protection specifically designed for the unique challenges of neural data, justice considerations that ensure equitable access to beneficial innovations, and adaptive mechanisms that can evolve alongside rapidly advancing technologies.

Throughout this exploration, I have been guided by the conviction that AI neuroscience represents one of the most promising avenues for reducing human suffering available to our generation. The potential applications justify sustained investment and research effort. These include detecting neurological diseases before symptoms appear, developing personalized treatments based on individual brain characteristics, restoring lost cognitive functions through brain-computer interfaces, and understanding the neural basis of consciousness itself. But I am equally convinced that realizing this potential requires confronting honestly both the capabilities and limitations of current technologies, the values and interests that shape their development, and the institutional structures that determine whether they serve broad human benefit or narrow advantage.

The challenges we face extend far beyond technical considerations to encompass fundamental questions about human dignity, social justice, and the role of expertise in democratic societies. The anti-science resistance patterns documented throughout this book reflect deeper social tensions about authority, change, and the pace of technological development that cannot be resolved through better algorithms or more data alone. Addressing these challenges requires sustained engagement with the broader social contexts in which neuroscience operates, while maintaining the scientific rigor and ethical commitment essential for generating knowledge worthy of public trust.

This book is written for multiple audiences whose collaboration is essential for navigating these challenges successfully. It addresses neuroscientists and AI researchers pushing the boundaries of what computational analysis can reveal about brain function. It speaks to clinicians working to translate these insights into better patient care, and to bioethicists grappling with novel challenges that traditional frameworks were never designed to address. It engages with policymakers struggling to develop appropriate governance for technologies they may not fully understand, and with institutional leaders working to protect scientific integrity while enabling beneficial innovation.

Most fundamentally, this book is written for anyone who believes that our most powerful technologies should serve our deepest human values, and who is willing to engage with the difficult work of ensuring they actually do. The convergence of artificial intelligence and neuroscience will reshape how we understand ourselves and treat disorders of the brain and mind. Whether this transformation enhances human flourishing or undermines it remains an open question—one that will be answered not by the technologies themselves but by the choices we make about how to develop, govern, and deploy them.

The path forward requires multiple complementary elements. These include technical sophistication and ethical wisdom, innovative capability and institutional resilience, and global collaboration alongside culturally responsive implementation. Most fundamentally, it requires recognition that the extraordinary power of AI neuroscience comes with extraordinary responsibility to ensure this power serves the values that make human flourishing possible. This recognition becomes more urgent in light of what Hotez documented: that scientific passivity in the face of anti-science movements can have deadly consequences. His warning that "if we are passive and don't say something, then these new partisan factions could have their way" applies directly to neuroscience, where the stakes of public rejection could prevent life-saving treatments from reaching those who need them most. This book aims to provide both the understanding and the tools necessary for meeting that responsibility, while acknowledging that the commitment to use them wisely remains ultimately a human choice that each generation must make anew.

Turku, Finland Thorsten Rudroff

Competing Interests The author has no competing interests to declare that are relevant to the content of this manuscript.

Contents

Part IV Solutions and Resistance Models

About the Author

Thorsten Rudroff is a Professor (Docent) in Experimental Neuroimaging at the Turku PET Centre, University of Turku and Turku University Hospital, Finland. He previously held positions as Associate Professor at the University of Iowa and Assistant Professor at Colorado State University, USA. With over 20 years of experience in neurophysiology research, he has made significant contributions to understanding fatigue and cognitive dysfunction in neurological disorders, particularly in Multiple Sclerosis and Long COVID. His scientific work includes over 90 publications in prestigious journals and the Springer book *Long COVID Fatigue*, which addresses the neurological underpinnings of this prevalent post-viral syndrome. He received his Ph.D. in Neurophysiology from the University of Constance, Germany, and completed his postdoctoral training at the University of Colorado, Boulder. He is particularly passionate about leveraging artificial intelligence to advance medical research and improve patient outcomes. He serves on multiple editorial boards and has received several significant research grants from the National Institutes of Health. His current research focuses on the intersection of neuroimaging, artificial intelligence, and Long COVID, with particular emphasis on developing innovative approaches to understand and treat fatigue symptoms. Furthermore, his focus is on developing AI-driven approaches for early detection of Alzheimer's disease through neuroimaging biomarkers.

Part I

Foundations and Applications

Foundations of AI in Modern Neuroscience

1

The brain is a world consisting of a number of unexplored continents and great stretches of unknown territory.

—Santiago Ramón y Cajal

1.1 Introduction

In the century since Spanish neuroscientist Santiago Ramón y Cajal first sketched the intricate cellular architecture of the nervous system, our understanding of the brain has advanced dramatically. Yet despite these advances, the brain remains perhaps the most complex and enigmatic organ in the human body, comprising an estimated 86 billion neurons forming approximately 100 trillion connections [1]. This complexity has traditionally posed formidable challenges for researchers seeking to understand both normal brain function and the mechanisms underlying neurological disorders.

In recent decades, however, a powerful new set of tools has emerged to help navigate the "unexplored continents" of the brain: artificial intelligence (AI). The application of AI methodologies to neuroscience research represents not merely an incremental improvement in our analytical capabilities but a fundamental transformation in how we approach the study of the brain. Machine learning algorithms can detect patterns in vast datasets. Neural networks can model complex brain functions. Computer vision systems can automate the analysis of neuroimaging data. Together, these technologies are revolutionizing our ability to map neural circuits, understand brain-behavior associations, and diagnose and treat neurological disorders. This chapter explores the foundations of this revolutionary convergence between AI and neuroscience. We begin by tracing the historical development of AI applications in brain research, from early computational models to deep contemporary learning approaches. We then examine the current state of the field, highlighting how AI is transforming various domains of neuroscience research and clinical

T. Rudroff, *AI for the Healthy Brain and Neurological Disorders*, Neurocultural Health and Wellbeing, https://doi.org/10.1007/978-3-032-11596-6_1

practice. Next, we focus specifically on neuroimaging, exploring how machine learning and deep learning methods enhance our ability to visualize and interpret brain structure and function. We then consider how AI is enabling a shift from descriptive to predictive models in the study of neurological disorders, with profound implications for diagnosis, prognosis, and treatment. Finally, we present a case study on the early detection of neurodegenerative patterns using machine learning, illustrating the practical impact of these advances.

Throughout, we will see how AI is not merely automating traditional analytical approaches but enabling entirely new ways of understanding the brain—ways that may ultimately help us map those "unexplored continents" that Cajal identified over a century ago.

1.2 Historical Development of AI Applications in Neuroscience

The association between artificial intelligence and neuroscience has been bidirectional from its inception. Early AI pioneers drew inspiration from neural systems, while neuroscientists adopted computational methods to model brain function. This reciprocal association has evolved through several distinct phases, each characterized by different approaches and achievements.

Figure 1.1 illustrates the historical trajectory of AI applications in neuroscience from the 1940s to the present day. This timeline highlights key developments across four major eras. The early computational models period (1940s–1960s) established fundamental concepts. The cognitive revolution (1970s–1980s) saw the emergence of backpropagation and neural network architectures. The computational neuroscience era (1990s–2000s) marked when machine learning methods were first applied to neuroimaging. The deep learning revolution (2010s–present) has been

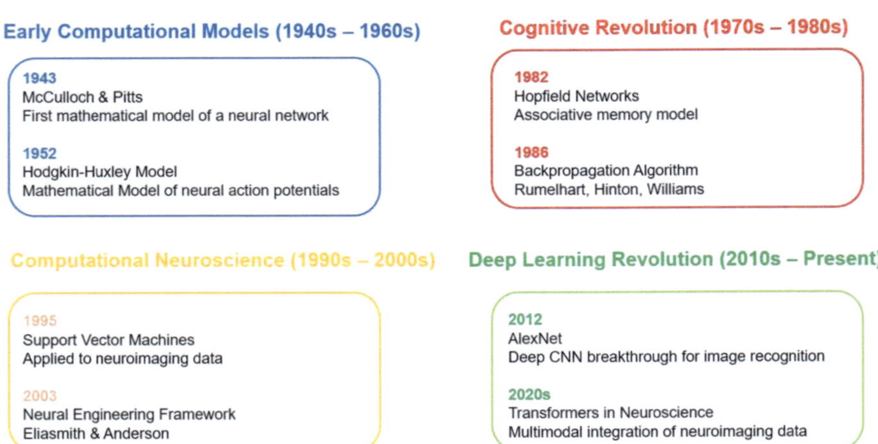

Fig. 1.1 Historical timeline of AI in neuroscience

characterized by breakthroughs in pattern recognition and predictive modeling. Each milestone represents a significant advancement in our ability to apply computational approaches to understanding the brain, demonstrating how the association between AI and neuroscience has strengthened and evolved over time.

1.2.1 Early Computational Models (1940s–1960s)

The modern history of AI in neuroscience begins with the pioneering work of Warren McCulloch and Walter Pitts, who in 1943 proposed the first mathematical model of a neural network [2]. Their model represented neurons as simple logical units that could be combined to compute complex functions. Although highly simplified compared to biological neurons, this model established the fundamental concept that neural processes could be understood in computational terms.

Building on this foundation, Frank Rosenblatt introduced the perceptron in 1958, an early neural network model capable of pattern recognition through a simple learning algorithm [3]. The perceptron generated significant excitement as a potential model of neural learning, although this enthusiasm was temporarily dampened by Minsky and Papert's 1969 demonstration of its limitations in handling nonlinearly separable problems [4].

During this same period, neurophysiologists began developing computational models of neural function. The Hodgkin-Huxley model, published in 1952, used differential equations to describe the ionic mechanisms underlying the action potential in neurons [5]. This groundbreaking work, which earned its authors the 1963 Nobel Prize in physiology or medicine, established mathematical modeling as a core approach in neuroscience.

1.2.2 The Cognitive Revolution and Neural Networks (1970s–1980s)

The 1970s and early 1980s saw the emergence of cognitive science as an interdisciplinary field combining neuroscience, psychology, linguistics, philosophy, and artificial intelligence [6]. This period witnessed significant advances in computational models of neural functions, exemplified by David Marr's influential work on vision [7], which proposed that visual processing occurs at multiple levels of analysis from computational theory to algorithm to neural implementation.

In parallel, new neural network architectures emerged that addressed some of the limitations identified by Minsky and Papert [4]. The development of backpropagation as an efficient learning algorithm for multilayer networks by Rumelhart et al. [8] was particularly significant, enabling neural networks to learn complex, nonlinear mappings between inputs and outputs. These advances renewed interest in neural network models as both practical tools for pattern recognition and theoretical frameworks for understanding brain function.

John Hopfield's work on recurrent neural networks in the early 1980s further strengthened the connection between AI and neuroscience [9]. Hopfield networks, with their ability to store and retrieve memories through attractor dynamics, provided a computational implementation of associative memory that resonated with emerging neuroscientific understanding of hippocampal function.

1.2.3 The Rise of Computational Neuroscience (1990s–2000s)

The 1990s saw the emergence of computational neuroscience as a distinct discipline characterized by increasingly sophisticated models of neural circuits and brain functions. Computational neuroscientists such as Terrence Sejnowski developed models that bridged single neurons and system-level phenomena [10]. The neural engineering framework, proposed by Chris Eliasmith and Charles Anderson [11], provided a systematic method for constructing neural models of cognitive functions, further tightening the connection between AI and neuroscience.

This period also witnessed the application of machine learning methods to neuroscientific data analysis. Support vector machines (SVMs), a class of supervised learning algorithms developed in the 1990s, became popular tools for classifying patterns in neuroimaging data [12]. These methods allowed researchers to distinguish between different cognitive states based on brain activity patterns and to identify neural correlations of various disorders.

In clinical neuroscience, early applications of AI focused primarily on computer-aided diagnosis systems. For example, neural networks were trained to assist in the interpretation of electroencephalogram (EEG) data for epilepsy diagnosis [13] and to classify brain tumors from magnetic resonance imaging (MRI) scans [14]. While promising, these early applications were limited by the computational resources and data available at the time.

1.2.4 The Deep Learning Revolution (2010s–Present)

The past decade has witnessed a dramatic transformation in AI applications in neuroscience, driven by breakthroughs in deep learning. The development of efficient training methods for deep neural networks, combined with increases in computational power and the availability of large datasets, has enabled unprecedented advances in pattern recognition, image analysis, and predictive modeling.

In 2012, AlexNet, a deep convolutional neural network, achieved breakthrough performance on the ImageNet challenge, demonstrating the power of deep learning for visual recognition tasks [15]. This development had immediate implications for neuroscience, particularly in neuroimaging analysis, where deep learning models began to outperform traditional methods in tasks such as brain segmentation, tumor detection, and classification of neurodegenerative disorders [16].

Beyond image analysis, deep learning approaches have been applied to model neural responses at multiple scales, from single neurons to entire brain regions. For example, Yamins et al. [17] demonstrated that deep convolutional neural networks could predict responses of neurons in the ventral stream visual pathway of primates better than any previous model. Similar approaches have been used to model auditory processing [18] and other sensory systems.

The emergence of generative models, particularly generative adversarial networks (GANs) and variational autoencoders (VAEs), has further expanded computational neuroscience. These models have been used to generate synthetic neuroimaging data for augmenting training datasets [19], to model the distribution of brain activity patterns associated with different cognitive states [20], and to explore potential disease trajectories in neurological disorders [21].

Most recently, transformer-based language models and their multimodal extensions have begun to impact neuroscience research. These models, characterized by attention mechanisms that enable them to process sequential data with awareness of context, have been applied to analyze time series of brain activity [22], to integrate multimodal neuroimaging data [23], and to develop more natural brain-computer interfaces [24].

This historical trajectory illustrates the increasingly close association between AI and neuroscience—an association that has accelerated dramatically in recent years and shows no signs of slowing. As we will see in the next section, the current state of the field represents a diverse ecosystem of approaches and applications, from basic research to clinical practice.

1.3 Current State of the Field: From Basic Applications to Breakthrough Technologies

Today, AI applications in neuroscience span a continuum from relatively straightforward applications of established methods to groundbreaking technologies that are transforming our understanding of the brain and creating new possibilities for treating neurological disorders. Here, we survey the current landscape, identifying key areas of innovation and impact.

1.3.1 Enhanced Neuroimaging Analysis

Perhaps the most widespread application of AI in contemporary neuroscience is in the analysis of neuroimaging data. Machine learning and deep learning approaches are now routinely used to process and interpret data from various neuroimaging modalities, including structural and functional MRI, positron emission tomography (PET), electroencephalography (EEG), and magnetoencephalography (MEG).

In structural MRI analysis, deep learning models have automated and improved the accuracy of brain segmentation, the process of delineating different brain structures and tissue types [25]. Convolutional neural networks (CNNs) can now segment brain structures with accuracy approaching that of human experts but with greater consistency and speed [25]. These advances are particularly significant for large-scale studies involving thousands of subjects, where manual segmentation would be prohibitively time-consuming.

In functional neuroimaging, machine learning methods are increasingly used for decoding mental states from brain activity patterns. For example, multivariate pattern analysis (MVPA) techniques can identify neural representations of specific cognitive processes, emotions, or perceptual experiences [27]. More recently, deep learning approaches have enhanced our ability to decode complex mental states from functional MRI data. Particularly notable is the work of Horikawa and Kamitani [28], who used deep neural networks to reconstruct visual experiences from brain activity, effectively "reading the mind's eye."

AI has also transformed the analysis of brain connectivity, a critical aspect of understanding neural function. Graph theoretical approaches combined with machine learning have enabled researchers to identify patterns of functional and structural connectivity associated with different cognitive states and neurological conditions [29]. For instance, Khosla et al. [30] used graph convolutional networks to model the association between brain connectivity and behavior, demonstrating how AI can help us understand the network properties of the brain.

Figure 1.2 demonstrates how AI enables the integration of multiple data modalities to create comprehensive models of brain function and dysfunction. AI algorithms combine information from multiple sources: structural and functional neuroimaging (MRI, fMRI, and PET), genetic data, clinical assessments, and digital biomarkers. This multimodal approach can identify patterns and associations that might not be apparent when analyzing any single data type in isolation. This multimodal approach is particularly valuable for understanding complex neurological conditions like Long COVID and multiple sclerosis, where the underlying mechanisms involve interactions across multiple biological systems. As shown in the

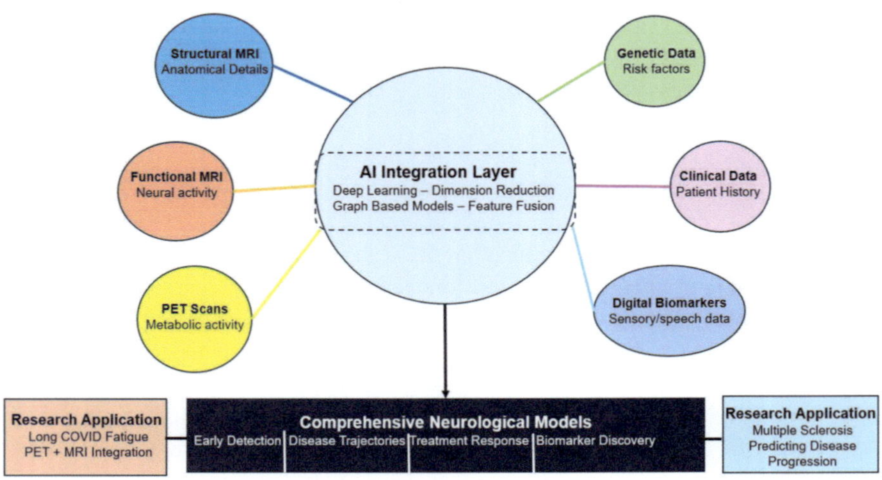

Fig. 1.2 Multimodal integration for brain analysis. AI-powered comprehensive models of neurological conditions

figure, these integrated models support various clinical applications, from early detection to treatment optimization, representing a significant advance over traditional single-modality analyses.

1.3.2 High-Throughput Analysis of Neural Data

Beyond neuroimaging, AI is transforming our ability to analyze other types of neural data, particularly those involving high-dimensional, temporally complex signals. For instance, machine learning algorithms have revolutionized the analysis of single-cell recordings, enabling researchers to identify functional cell types and to track the activity of large populations of neurons over time [31].

Particularly significant has been the application of dimensionality reduction techniques to neural population data. Methods such as principal component analysis (PCA), independent component analysis (ICA), and, more recently, manifold learning approaches have helped researchers identify low-dimensional structures in neural activity that correspond to behaviors or cognitive states [32].

AI has also transformed the analysis of microscopy data in neuroscience. Deep learning models can now automatically trace axons and dendrites in electron microscopy images, enabling the reconstruction of neural circuits at unprecedented scales [33]. These approaches are crucial for projects such as the Human Connectome Project, which aims to map the complete structural connectivity of the human brain [34].

1.3.3 Brain-Computer Interfaces

Brain-computer interfaces (BCIs) represent one of the most direct applications of AI in neuroscience, creating systems that translate neural activity into commands for external devices. Recent years have seen remarkable advances in this field, driven in part by improvements in machine learning algorithms for decoding neural signals.

Particularly notable is the work of the BrainGate consortium, which has developed BCIs that allow paralyzed individuals to control robotic limbs or computer cursors through neural activity [35]. These systems rely on machine learning algorithms to decode intended movements from patterns of neural firing, enabling increasingly naturalistic and precise control [36].

More recently, companies like Neuralink and Synchron have been developing minimally invasive BCI technologies with the goal of making these technologies more widely accessible [37, 38]. These systems use machine learning algorithms to interpret neural signals recorded from electrodes implanted in or near the brain, potentially opening new avenues for treating neurological disorders and enhancing human capabilities.

Noninvasive BCIs have also advanced significantly, with systems based on EEG or functional near-infrared spectroscopy (fNIRS) providing increasingly reliable control capabilities [39]. These developments have been enabled by improvements in signal processing and machine learning algorithms that can extract meaningful information from the noisy, low-resolution signals available through noninvasive recording methods.

1.3.4 Personalized Medicine in Neurology

AI is increasingly enabling personalized approaches to the diagnosis and treatment of neurological disorders. Machine learning models trained on large datasets of clinical, genetic, and neuroimaging data can identify patterns that predict individual responses to treatments, disease progression trajectories, and risk factors.

For example, in multiple sclerosis (MS), machine learning algorithms have been used to predict disease progression based on a combination of clinical features, MRI markers, and genetic data [40]. Similar approaches have been applied in Alzheimer's disease, where multimodal predictive models can identify individuals at risk of developing dementia years before symptoms appear [41].

In my own research on Long COVID-related fatigue, I have explored how AI-driven multimodal analysis of PET and MRI data can help identify the neurobiological mechanisms underlying this debilitating condition [42]. By integrating information from different imaging modalities, we can develop a more comprehensive understanding of the condition and potentially identify biomarkers that predict which patients will develop persistent symptoms.

Treatment optimization represents another promising application of AI in clinical neuroscience. Machine learning models can help identify which patients are likely to respond to specific interventions, enabling more targeted therapeutic approaches. For instance, AI algorithms have been used to predict response to antidepressant medications based on EEG patterns [43] and to optimize the parameters of deep brain stimulation for Parkinson's disease based on patient-specific characteristics [44].

1.3.5 Neuroscience-Inspired AI

While we have primarily focused on applications of AI in neuroscience, it is worth noting that the relationship between these fields remains bidirectional. Neuroscience continues to inspire developments in AI, particularly in the design of neural network architectures and learning algorithms.

Recent examples include spiking neural networks (SNNs), which more closely mimic the temporal dynamics of biological neurons [45]; neuromorphic computing systems that implement neural algorithms in specialized hardware [46]; and biologically inspired learning rules such as spike-timing-dependent plasticity (STDP) that provide alternatives to backpropagation [47].

These neuroscience-inspired approaches to AI have potential benefits in terms of energy efficiency, continual learning capabilities, and robustness to noise and adversarial attacks. As such, they represent a promising frontier in AI research, illustrating how insights from neuroscience continue to shape the development of artificial intelligence.

This overview of the current state of AI in neuroscience reveals a field that is both broad and deep, encompassing applications ranging from basic research tools to transformative clinical technologies. In the next section, we will focus specifically on neuroimaging, an area where AI has had particularly profound impacts.

1.4 Key AI Methodologies Transforming Neuroimaging

Neuroimaging technologies have revolutionized our ability to observe and understand the living brain, providing windows into both structure and function without the need for invasive procedures. However, the data generated by these technologies—whether structural MRI, functional MRI, PET, or other modalities—are complex, high-dimensional, and challenging to analyze using traditional methods. Artificial intelligence, particularly machine learning and deep learning approaches, has transformed how we process and interpret these data, enabling new insights into brain organization and function. Here, I draw on my expertise in PET and MRI research to explore key AI methodologies that are reshaping neuroimaging analysis.

Figure 1.3 depicts the paradigm shift from descriptive to predictive modeling in clinical neuroscience. The traditional descriptive approach (left) focuses on group-level statistical differences with limited clinical translation, while the modern

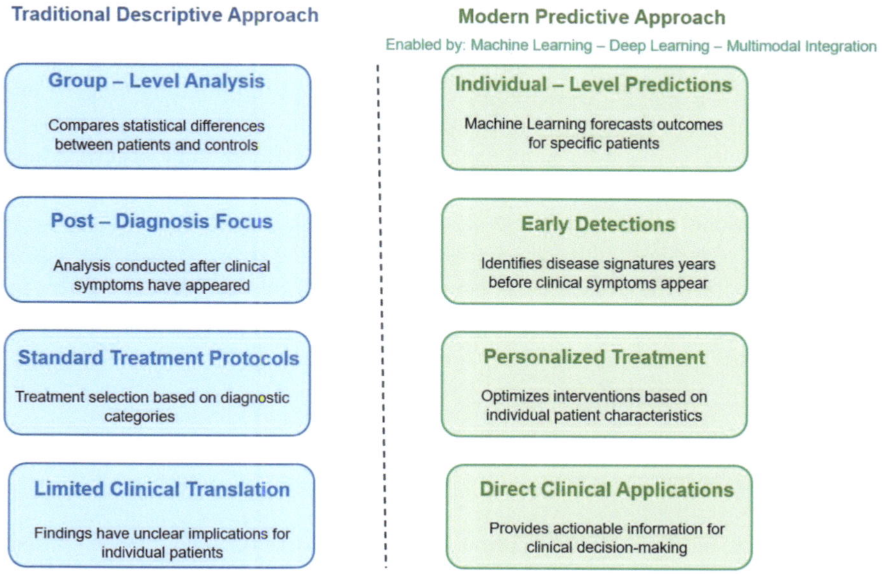

Fig. 1.3 From descriptive to predictive models in neurology

predictive approach (right) emphasizes individual-level forecasting and personalized interventions. This transformation represents a fundamental change in how we approach neurological disorders—moving from simply documenting differences between groups to forecasting specific outcomes for individual patients. As illustrated, this shift enables earlier detection of disease, more personalized treatment selection, and more actionable clinical information, ultimately paving the way for precision medicine in neurology.

1.4.1 Machine Learning for Neuroimaging Classification

Supervised machine learning approaches have become essential tools for classification tasks in neuroimaging, including disease diagnosis, the prediction of outcomes, and the identification of brain states. These methods learn patterns from labeled training data that can then be applied to new, unseen cases.

Support vector machines (SVMs) were among the first machine learning methods widely adopted in neuroimaging analysis. SVMs identify a hyperplane that maximally separates different classes in a high-dimensional feature space, making them well-suited to neuroimaging data where the number of features (voxels) typically exceeds the number of samples (subjects). For example, SVMs have been successfully applied to distinguish patients with Alzheimer's disease from healthy controls based on structural MRI features [48], achieving accuracy rates exceeding 90% in some studies.

Random forests, another popular machine learning approach, use an ensemble of decision trees to classify data. This method has proven effective for neuroimaging classification tasks due to its ability to handle nonlinear relationships and provide measures of feature importance. In my collaborative work on multiple sclerosis, we have used random forests to identify the most discriminative neuroimaging features for distinguishing between different disease subtypes [49], providing insights into the heterogeneous pathology of this condition.

More recently, gradient boosting methods such as XGBoost and LightGBM have gained popularity for neuroimaging classification due to their robust performance and computational efficiency. These algorithms sequentially build an ensemble of weak predictive models, with each model focusing on the errors of its predecessors. Using these approaches, researchers have developed highly accurate classifiers for various neurological conditions, including Parkinson's disease [50], autism spectrum disorder [51], and schizophrenia [52].

1.4.2 Deep Learning for Image Segmentation and Reconstruction

Deep learning, particularly convolutional neural networks (CNNs), has revolutionized image analysis in many domains, and neuroimaging is no exception. CNNs automatically learn hierarchical features from data, progressing from simple edge detectors in early layers to complex pattern recognizers in deeper layers. This architecture makes them ideally suited for neuroimaging analysis, where relevant features span multiple scales and levels of abstraction.

In structural neuroimaging, deep learning has transformed brain segmentation, the process of delineating different brain structures and tissue types. The U-Net architecture, originally developed for biomedical image segmentation [53], has been particularly influential. Adaptations of U-Net have achieved state-of-the-art performance in segmenting brain structures such as the hippocampus, a region critically involved in memory formation and often affected in neurodegenerative disorders [54]. These automated segmentation methods not only increase throughput but also improve consistency across datasets, an important consideration for large-scale studies.

Deep learning has also advanced the reconstruction of neuroimaging data, particularly in contexts where data acquisition is constrained by time or other factors. For example, in PET imaging, deep learning methods have been developed to enhance image quality from low-dose scans [55], potentially reducing radiation exposure for patients. Similarly, in MRI, deep learning approaches have enabled acceleration of acquisition times by reconstructing high-quality images from undersampled k-space data [56]. These advances make neuroimaging more accessible and comfortable for patients, an important consideration for clinical applications.

In my own research on PET neuroimaging in Long COVID, we have explored how deep learning methods can help extract more information from limited data, a significant advantage when studying conditions where large datasets may not yet be available [42]. By applying techniques such as transfer learning, where models pretrained on large datasets are fine-tuned for specific applications, we can develop effective analysis tools even with relatively small sample sizes.

1.4.3 Multimodal Integration and Dimensionality Reduction

Different neuroimaging modalities provide complementary information about the brain. Structural MRI reveals anatomical details, functional MRI captures blood-oxygen-level-dependent signals related to neural activity, and PET can measure metabolic activity or specific molecular targets depending on the radiotracer used. Integrating these diverse data types to develop a comprehensive understanding of brain function is a significant challenge, one where AI approaches have proven particularly valuable.

Traditional approaches to multimodal integration often rely on separate analyses of each modality followed by a heuristic combination of results. In contrast, machine learning methods can learn optimal ways to combine information across modalities. For example, multimodal deep learning architectures can process different data types in parallel pathways before integrating features at higher levels [57]. Using such approaches, researchers have developed improved diagnostic models for conditions such as Alzheimer's disease, where combining structural MRI, PET, and cerebrospinal fluid biomarkers yields higher accuracy than any single modality alone [58].

Dimensionality reduction techniques represent another important class of methods for neuroimaging analysis, particularly for functional data where temporal dynamics add additional complexity. These techniques identify lower-dimensional representations of high-dimensional data that preserve essential structure while discarding noise. Popular approaches include principal component analysis (PCA), independent component analysis (ICA), and, more recently, manifold learning methods such as t-distributed stochastic neighbor embedding (t-SNE) and uniform manifold approximation and projection (UMAP).

In recent work, my colleagues and I have explored the potential of dimension reduction techniques specifically for neuroimaging analysis of Long COVID fatigue [42]. By applying these methods to multimodal PET and MRI data, we can identify patterns of altered brain metabolism and structure that may serve as biomarkers for this condition. This approach has proven particularly valuable for identifying subtle, distributed changes that might not be apparent through conventional analysis methods.

1.4.4 Graph-Based Methods for Connectivity Analysis

The brain is fundamentally a network, with function emerging from patterns of connectivity between different regions. Graph theoretical approaches, which represent the brain as a set of nodes (regions) connected by edges (connections), have become increasingly important for analyzing brain connectivity. When combined with machine learning, these methods enable powerful analyses of how connectivity patterns relate to cognition, behavior, and disease.

In graph-based analyses, machine learning algorithms can identify subgraphs or connectivity patterns that discriminate between different groups or predict individual differences in behavior. For example, spectral graph convolutional networks have been used to analyze functional connectivity in autism spectrum disorder, identifying altered connectivity patterns that predict symptom severity [59]. Similar approaches have been applied to study connectivity changes in aging, depression, and other conditions.

Deep learning approaches specifically designed for graph data, such as graph neural networks (GNNs), represent a particularly promising direction for connectivity analysis. These models can learn directly from graph-structured data, automatically identifying relevant features at both local and global scales. For instance, GNNs have been used to predict brain age from structural connectivity patterns [60], providing a potential biomarker for neurodegenerative disorders.

In our research on multiple sclerosis and Long COVID, we have begun exploring how graph-based methods can help characterize the network-level impacts of these conditions [49]. By applying machine learning to connectivity data derived from multiple imaging modalities, we can develop more comprehensive models of how these conditions affect brain function at the system level.

1.4.5 Interpretable AI for Neuroimaging

As AI methods become increasingly sophisticated, ensuring their interpretability becomes more challenging yet also more important, particularly in clinical contexts. Researchers have therefore developed various approaches to make AI models for neuroimaging more transparent and explicable.

Attention mechanisms, which highlight regions of an input that most strongly influence a model's predictions, have proven valuable for interpreting neuroimaging models. For example, gradient-weighted class activation mapping (Grad-CAM) has been used to visualize which brain regions contribute most to a deep learning model's diagnostic decisions [61]. These approaches not only increase transparency but may also help identify novel biomarkers by drawing attention to regions not previously associated with a condition.

Another approach involves designing inherently interpretable models. For instance, sparse linear models with regularization can identify a minimal set of predictive features while maintaining good performance [62]. These models can be more readily interpreted by examining the weights assigned to different brain regions or connections.

In my own research, we have emphasized the importance of interpretable AI for clinical applications. When developing models to predict fatigue severity in Long COVID or multiple sclerosis, we prioritize approaches that not only provide accurate predictions but also offer insights into the underlying neurobiological mechanisms [63]. This emphasis on interpretability helps ensure that our AI tools contribute to mechanistic understanding rather than functioning as black boxes.

The AI methodologies described above have dramatically enhanced our ability to extract meaningful information from neuroimaging data. In the next section, we will explore how these advances are enabling a shift from descriptive to predictive models in the study of neurological disorders.

1.5 The Shift from Descriptive to Predictive Models in Neurological Disorders

Traditionally, neuroscience has focused on describing the structural and functional characteristics of the brain in health and disease. This descriptive approach has yielded valuable insights into the neural correlations of various conditions, identifying regions, networks, or systems that differ between patient populations and healthy controls. However, such group-level differences often have limited utility for individual patients and may not directly inform treatment decisions or prognosis.

In recent years, there has been a paradigm shift toward predictive modeling in clinical neuroscience—a shift largely enabled by advances in AI. Rather than simply describing differences between groups, predictive models aim to forecast outcomes for individual patients: Will this person develop a particular condition? How will their disease progress? How will they respond to a specific treatment? This transition from description to prediction represents a crucial step toward personalized medicine in neurology.

1.5.1 From Group Differences to Individual Predictions

Conventional neuroimaging studies typically compare group averages, identifying statistically significant differences between patients and controls or between different patient subgroups. While informative at the population level, these analyses often have limited translational value due to the substantial overlap between groups and the heterogeneity within diagnostic categories.

Predictive modeling takes a fundamentally different approach, using machine learning to identify patterns that can forecast outcomes for individual patients. Traditional approaches simply note that hippocampal volume is on average smaller in Alzheimer's disease patients compared to healthy controls. In contrast, a predictive model uses multiple data sources: neuroimaging features, genetic markers, and clinical variables. This comprehensive approach can estimate an individual's risk of developing the disease within a specific timeframe. This shift is evident in the evolution of research on multiple sclerosis (MS), a condition that has been a focus of my own work. Early neuroimaging studies in MS primarily documented differences in lesion load, brain volume, and connectivity between patients and controls. While valuable, these findings had limited prognostic utility due to the variable relationship between radiological findings and clinical outcomes—the so-called "clinico-radiological paradox" [64].

More recent research has employed machine learning to develop predictive models that can forecast individual disease trajectories in MS. For instance, Zhao et al. [40] developed a random forest model that predicts disability progression using a combination of clinical, MRI, and genetic markers. Similarly, Tacchino et al. [65] demonstrated that deep learning models trained on multimodal MRI data can predict cognitive impairment in MS patients with high accuracy. These predictive approaches offer more direct clinical value by helping to identify patients who might benefit from early aggressive treatment.

1.5.2 Predictive Biomarkers for Early Detection

Perhaps the most significant impact of predictive modeling has been in the early detection of neurological disorders, particularly neurodegenerative conditions. By the time clinical symptoms become apparent in diseases like Alzheimer's, Parkinson's, or amyotrophic lateral sclerosis (ALS), substantial neural damage has often already occurred. Predictive models offer the possibility of identifying individuals at risk before significant damage has accumulated, potentially enabling earlier intervention.

In Alzheimer's disease research, predictive modeling has been particularly fruitful. Machine learning algorithms trained on neuroimaging data can now identify patterns associated with preclinical Alzheimer's—changes that occur years or even decades before clinical symptoms emerge [41]. For example, Davatzikos et al. [66] developed a machine learning approach called SPARE-AD (Spatial Pattern of Abnormality for Recognition of Early AD) that uses structural MRI to predict conversion from mild cognitive impairment (MCI) to Alzheimer's disease. Subsequent work has extended this approach to incorporate multiple biomarkers, further improving predictive accuracy [67].

Similar progress has been made in Parkinson's disease, where machine learning models can now detect early signs of neurodegeneration based on subtle changes in brain structure and function. For instance, Adeli et al. [68] developed a multimodal deep learning model that integrates structural and functional MRI to detect prodromal Parkinson's disease with high sensitivity. These approaches could potentially identify patients who might benefit from neuroprotective therapies, if and when such therapies become available.

In my own research on Long COVID, we have applied predictive modeling to identify individuals at risk of developing persistent neurological symptoms following SARS-CoV-2 infection [63]. By analyzing patterns of metabolic activity in the brain using PET imaging, we can identify signatures that predict which patients are likely to experience long-term fatigue and cognitive dysfunction. This approach could help target interventions to those most likely to develop persistent symptoms, potentially improving outcomes.

1.5.3 Treatment Response Prediction

Another important application of predictive modeling is forecasting how individual patients will respond to specific treatments. This approach, often termed "precision medicine," aims to match patients with the therapies most likely to benefit them, minimizing ineffective treatments and potentially harmful side effects.

In depression, for example, machine learning models have been developed to predict response to different antidepressant medications based on EEG patterns [43]. Similarly, in epilepsy, predictive models can help identify patients likely to benefit from surgical intervention [69]. These approaches have the potential to substantially improve patient outcomes by avoiding the trial-and-error approach that often characterizes treatment selection in neurology and psychiatry.

In my collaborative research on transcranial direct current stimulation (tDCS) for Long COVID management, we have explored how AI can help optimize stimulation parameters based on individual patient characteristics [70]. By analyzing patterns of brain metabolism and connectivity, we can potentially identify the optimal stimulation targets and parameters for each patient, maximizing efficacy and minimizing side effects. This personalized approach represents a significant advance over one-size-fits-all treatment protocols.

1.5.4 Disease Trajectory Prediction

Neurological disorders often exhibit substantial heterogeneity in progression rates and patterns, posing challenges for prognosis and treatment planning. Predictive models trained on longitudinal data can help forecast disease trajectories, identifying patients at risk of rapid progression or specific complications.

In ALS, for example, machine learning models have been developed to predict survival time based on clinical, genetic, and neuroimaging features [71]. These models outperform traditional prognostic tools and can help guide decisions about interventions such as ventilatory support or feeding tube placement. Similar approaches have been applied in Huntington's disease [72], MS [40], and other progressive neurological conditions.

An important recent development in this area is the application of deep learning to model disease progression over time. Recurrent neural networks (RNNs) and their variants, such as long short-term memory (LSTM) networks, are particularly well suited to modeling temporal sequences. For instance, Fisher et al. [21] used an LSTM network to model the progression of Alzheimer's disease based on longitudinal neuroimaging data, enabling the prediction of future brain atrophy patterns. These approaches allow for more nuanced forecasting of disease trajectories, potentially accounting for nonlinear progression patterns and interactions between different disease mechanisms.

1.5.5 Challenges and Limitations

Despite the promise of predictive modeling in neurology, significant challenges remain. Perhaps the most fundamental is the need for large, well-characterized datasets for model training and validation. While initiatives like the Alzheimer's Disease Neuroimaging Initiative (ADNI) and the Human Connectome Project have made valuable data resources available, many neurological conditions remain understudied, particularly rare disorders and those affecting underrepresented populations.

Another challenge concerns model generalizability. Models trained on data from a specific population or acquired with particular protocols may not perform well when applied to different settings. This issue, often termed "dataset shift," can limit the clinical utility of predictive models and requires careful validation across diverse populations and acquisition protocols [73].

Interpretability represents a third significant challenge. As predictive models become more complex, understanding the basis for their predictions becomes more difficult. Yet in clinical contexts, interpretability is often crucial for building trust and guiding interventions. Various approaches to explainable AI, as discussed in the previous section, are being developed to address this challenge, but further work is needed to make complex models fully transparent to clinicians and patients.

Finally, there is the challenge of implementation. Translating predictive models from research settings to clinical practice requires addressing issues of workflow integration, regulatory approval, and clinician education. While progress has been made in developing AI systems approved for clinical use, broader implementation will require continued efforts to demonstrate clinical utility and cost-effectiveness.

Despite these challenges, the shift from descriptive to predictive models represents a fundamental transformation in how we approach neurological disorders. By enabling early detection, personalized treatment selection, and more accurate prognosis, predictive models have the potential to significantly improve patient outcomes and resource allocation in neurology.

1.6 Case Study: Early Detection of Neurodegenerative Patterns Using Machine Learning

To illustrate the practical impact of AI in neuroscience, particularly the transition from descriptive to predictive approaches, let us examine a specific case study: the application of machine learning for the early detection of Alzheimer's disease (AD). This case exemplifies how AI methodologies can identify subtle patterns that precede clinical symptoms, potentially enabling earlier intervention.

1.6.1 The Challenge of Early Alzheimer's Detection

Alzheimer's disease poses a particularly compelling case for early detection via AI. The pathological process begins years, even decades, before clinical symptoms become apparent [74]. By the time a patient presents with memory complaints, significant neuronal damage has already occurred. If treatments are to be effective, they likely need to be administered during this preclinical phase, before extensive irreversible damage has accumulated [75].

Traditional neuroimaging analyses have identified group-level differences between patients with AD and healthy controls, including hippocampal atrophy, cortical thinning in temporal and parietal regions, and altered patterns of brain connectivity [76]. However, these group-level differences have limited utility for individual patients due to substantial overlap between diagnostic categories and the heterogeneous nature of the disease.

1.6.2 Machine Learning Approaches for Alzheimer's Detection

Several machine learning frameworks have demonstrated promise for the early detection of AD, particularly for identifying individuals with mild cognitive impairment (MCI) who are likely to progress to AD.

The SPARE-AD (Spatial Pattern of Abnormality for Recognition of Early AD) framework, developed by Davatzikos and colleagues [66], represents an early and influential approach. This method uses a support vector machine to identify patterns of brain atrophy characteristic of AD. When applied to individuals with MCI, SPARE-AD can predict with approximately 75% accuracy which patients will progress to AD within 3 years. Subsequent refinements of this approach have incorporated multiple biomarkers, including cerebrospinal fluid measures of amyloid and tau, to further improve predictive accuracy [77].

Deep learning methods have further advanced the field. For example, Ding et al. [78] developed a deep convolutional neural network that predicts conversion from MCI to AD using structural MRI data. Their model achieved an impressive 82.4% accuracy in identifying MCI converters, outperforming conventional machine learning approaches. The model identified relevant brain regions, including the hippocampus, amygdala, and entorhinal cortex, consistent with known pathological patterns in AD.

Multimodal approaches that integrate data from different imaging modalities have proven particularly powerful. For instance, Liu et al. [58] developed a deep learning model that combines structural MRI, FDG-PET, and cerebrospinal fluid biomarkers to predict AD progression. This multimodal approach achieved higher accuracy than any single modality alone, highlighting the value of integrating complementary information sources.

1.6.3 Beyond Imaging: Integrating Multiple Data Types

Recent AI approaches for AD detection have expanded beyond neuroimaging to incorporate various data types, including genetic information, cognitive assessments, electronic health records, and even digital biomarkers derived from smartphone interactions or speech patterns.

Particularly notable is the Alzheimer's disease prediction of longitudinal evolution (TADPOLE) challenge, which invited researchers worldwide to develop algorithms for predicting AD progression using the ADNI dataset [79]. The winning approaches combined multiple data modalities and employed ensemble methods that integrated predictions from several models. These approaches achieved remarkable accuracy in predicting cognitive scores, diagnostic status, and patterns of brain atrophy up to 5 years in advance.

Genetic data represent another valuable source of information for AD prediction. The ε4 allele of the apolipoprotein E gene (APOE4) is a well-established risk factor for AD, but its predictive value is limited when considered in isolation. By integrating genetic data with neuroimaging and other biomarkers, machine learning models can achieve more nuanced risk stratification. For example, Khanna et al. [80] developed a random forest model that combines genetic, neuroimaging, and cognitive data to predict AD progression, achieving higher accuracy than models using any single data type alone.

Digital biomarkers derived from speech patterns and language use have emerged as particularly promising for early AD detection. Natural language processing (NLP) techniques can identify subtle linguistic changes that may precede clinically apparent cognitive decline. For instance, Fraser et al. [81] developed a machine learning approach that analyzes speech samples to differentiate between healthy controls, individuals with MCI, and AD patients. The model identified features such as grammatical complexity, vocabulary richness, and hesitation patterns that correlated with cognitive status. This approach is especially valuable as it is noninvasive, potentially scalable, and could be implemented in primary care settings.

1.6.4 Practical Impact and Future Directions

The practical impact of these AI approaches for early AD detection is beginning to be realized in both research and clinical settings. Several predictive models have been translated into research tools that are being used in clinical trials to identify suitable participants for early intervention studies. For example, the European Prevention of Alzheimer's Dementia (EPAD) initiative uses predictive modeling to identify individuals at high risk of developing AD for inclusion in prevention trials [82].

In clinical practice, commercial tools such as Neuroreader and NeuroQuant, which use machine learning to quantify brain volumes from MRI scans, have received regulatory approval and are being used to support AD diagnosis. While not yet fully predictive, these tools represent important steps toward implementing AI in routine clinical neurology.

Future directions in this field include the development of more interpretable models that can provide clinicians with clear rationales for predictions, the validation of models across diverse populations to ensure generalizability, and the integration of predictive models with emerging therapeutic approaches. As potentially disease-modifying treatments for AD become available, accurate early detection will become increasingly important for identifying patients who might benefit from intervention before significant neurodegeneration has occurred.

1.6.5 Lessons from the Case Study

This case study illustrates several key themes in the application of AI to neuroscience. First, it demonstrates the value of predictive modeling over purely descriptive approaches. Rather than simply documenting differences between groups, these models forecast outcomes for individual patients, providing actionable information for clinical decision-making.

Second, it highlights the importance of multimodal integration. By combining different types of data—structural and functional neuroimaging, genetic information, cognitive assessments, and digital biomarkers—AI models can achieve higher predictive accuracy than would be possible using any single data source.

Third, it underscores the potential for AI to identify subtle patterns that might not be apparent through conventional analysis. The early signs of neurodegenerative processes may be distributed across brain regions and manifest in ways that are difficult to detect without advanced computational methods.

Finally, it illustrates the practical challenges of implementing AI in clinical neuroscience. Despite impressive results in research settings, translation to routine clinical practice requires addressing issues of interpretability, generalizability, regulatory approval, and workflow integration.

This case study of early AD detection exemplifies both the promise and the challenges of AI applications in neuroscience. Similar approaches are being developed for other neurological conditions, from Parkinson's disease to multiple sclerosis to psychiatric disorders, suggesting a broader transformation in how we approach brain health and disease.

1.7 Conclusion

Artificial intelligence has fundamentally transformed our approach to neuroscience research and clinical practice. From the early computational models of the mid-twentieth century to today's sophisticated deep learning systems, AI methodologies have enhanced our ability to analyze complex neural data, identify patterns associated with various brain states and disorders, and develop predictive models with clinical utility.

In neuroimaging, AI has automated time-consuming analytical processes, extracted more information from existing data, and enabled integration across multiple modalities. These advances have not only increased research productivity but also opened new avenues for understanding brain structure and function. The shift from descriptive to predictive modeling represents a particularly significant development, promising more personalized and proactive approaches to neurological care.

Yet significant challenges remain. Ensuring the interpretability of increasingly complex models, validating their performance across diverse populations, and successfully implementing them in clinical workflows all require continued attention. Moreover, as we will explore in subsequent chapters, the application of AI in neuroscience raises important ethical questions about privacy, consent, and the potential for bias.

Looking ahead, we can anticipate even closer integration between AI and neuroscience. Advances in computing power, data sharing, and algorithmic innovation will likely enable more sophisticated analyses of neural data at multiple scales, from single neurons to whole-brain networks. The bidirectional relationship between these fields will continue, with neuroscience informing the development of more brain-like AI systems and AI enhancing our understanding of the brain.

As we navigate this exciting frontier, it will be essential to maintain a balanced perspective that acknowledges both the transformative potential of AI in neuroscience and the limitations and challenges of current approaches. By doing so, we can harness these powerful methodologies to advance our understanding of the brain and improve outcomes for patients with neurological disorders.

1.8 Summary

This chapter presents a comprehensive overview of how artificial intelligence has fundamentally transformed neuroscience research and clinical practice, opening new pathways to understanding the brain's intricate complexity. Beginning with Santiago Ramón y Cajal's metaphor of the brain as "unexplored continents," this chapter traces how AI has become an essential navigation tool for mapping these neural territories.

The historical trajectory reveals a fascinating evolution spanning eight decades, from Warren McCulloch and Walter Pitts' 1943 mathematical neural network models through the perceptron era of the 1950s, the cognitive revolution and backpropagation advances of the 1970s–1980s, the emergence of computational neuroscience in the 1990s–2000s, to today's deep learning revolution. Each phase brought increasingly sophisticated tools for understanding brain function, culminating in today's powerful pattern recognition and predictive modeling capabilities.

This chapter illuminates the current landscape where AI applications span from basic research to breakthrough clinical technologies. In neuroimaging, machine learning and deep learning have automated complex analyses, enhanced accuracy in brain segmentation, and enabled unprecedented decoding of mental states from brain activity patterns. The author draws on expertise in PET and MRI research to demonstrate how AI facilitates multimodal integration, combining structural imaging, functional data, genetic information, and clinical assessments to create comprehensive models of neurological conditions.

A particularly significant development highlighted is the paradigm shift from descriptive to predictive modeling. Rather than merely documenting group-level differences between patients and controls, modern AI approaches forecast individual outcomes—predicting disease progression, treatment responses, and risk factors. This transformation enables personalized medicine approaches that can identify patients likely to benefit from specific interventions while avoiding ineffective treatments.

This chapter's centerpiece case study on early Alzheimer's disease detection exemplifies these advances in action. Machine learning models can now identify preclinical patterns years before symptoms emerged, achieving impressive accuracy rates by integrating neuroimaging data with genetic markers, cognitive assessments, and even digital biomarkers derived from speech patterns. This early detection capability could prove crucial as potentially disease-modifying treatments become available.

Throughout, this chapter emphasizes the bidirectional relationship between AI and neuroscience—while AI tools enhance brain research, neuroscientific insights continue inspiring new AI architectures like spiking neural networks and neuromorphic computing systems. The author also acknowledges persistent challenges, including ensuring model interpretability, validating performance across diverse populations, and successfully implementing AI tools in clinical workflows.

This chapter concludes by positioning this AI-neuroscience convergence as a transformative frontier that promises more sophisticated analyses across multiple scales of brain organization, from individual neurons to whole-brain networks, ultimately advancing both our understanding of the mind and our ability to treat neurological disorders.

References

1. Herculano-Houzel S. The remarkable, yet not extraordinary, human brain as a scaled-up primate brain and its associated cost. Proc Natl Acad Sci USA. 2012;109(Suppl 1):10661–8.
2. McCulloch WS, Pitts W. A logical calculus of the ideas immanent in nervous activity. 1943. Bull Math Biol. 1990;52(1-2):99–115. discussion 73–97.
3. Rosenblatt F. The perceptron: a probabilistic model for information storage and organization in the brain. Psychol Rev. 1958;65(6):386–408.
4. Minsky M, Papert S. Perceptrons: an introduction to computational geometry. Cambridge, MA: MIT Press; 1969.
5. Hodgkin AL, Huxley AF. A quantitative description of membrane current and its application to conduction and excitation in nerve. J Physiol. 1952;117(4):500–44.
6. Gardner H. The mind's new science: a history of the cognitive revolution. New York: Basic Books; 1985.
7. Marr D. Vision: a computational investigation into the human representation and processing of visual information. Cambridge, MA: MIT Press; 1982.
8. Rumelhart DE, Hinton GE, Williams RJ. Learning representations by back-propagating errors. Nature. 1986;323(6088):533–6.
9. Hopfield JJ. Neural networks and physical systems with emergent collective computational abilities. Proc Natl Acad Sci USA. 1982;79(8):2554–8.
10. Sejnowski TJ, Koch C, Churchland PS. Computational neuroscience. Science. 1988;241(4871):1299–306.
11. Eliasmith C, Anderson CH. Neural engineering: computation, representation, and dynamics in neurobiological systems. Cambridge, MA: MIT Press; 2003.
12. Vapnik VN. The nature of statistical learning theory. New York: Springer; 1995.
13. Gabor AJ, Leach RR, Dowla FU. Automated seizure detection using a self-organizing neural network. Electroencephalogr Clin Neurophysiol. 1996;99(3):257–66.
14. Vinitski S, Gonzalez C, Mohamed F, Iwanaga T, Knobler RL, Khalili K, Mack J. Improved intracranial lesion characterization by tissue segmentation based on a 3D feature map. Magn Reson Med. 1997;37(3):457–69.
15. Krizhevsky A, Sutskever I, Hinton GE. ImageNet classification with deep convolutional neural networks. Adv Neural Inf Proces Syst. 2012;25:1097–105.
16. Litjens G, Kooi T, Bejnordi BE, Setio AAA, Ciompi F, Ghafoorian M, van der Laak JAWM, van Ginneken B, Sánchez CI. A survey on deep learning in medical image analysis. Med Image Anal. 2017;42:60–88.
17. Yamins DL, Hong H, Cadieu CF, Solomon EA, Seibert D, DiCarlo JJ. Performance-optimized hierarchical models predict neural responses in higher visual cortex. Proc Natl Acad Sci USA. 2014;111(23):8619–24.
18. Kell AJE, Yamins DLK, Shook EN, Norman-Haignere SV, McDermott JH. A task-optimized neural network replicates human auditory behavior, predicts brain responses, and reveals a cortical processing hierarchy. Neuron. 2018;98(3):630–644.e16.
19. Shin HC, Tenenholtz NA, Rogers JK, Schwarz CG, Senjem ML, Gunter JL, et al. Medical image synthesis for data augmentation and anonymization using generative adversarial networks. In: International workshop on simulation and synthesis in medical imaging. Cham: Springer; 2018. p. 1–11.

20. Palazzo S, Spampinato C, Kavasidis I, Giordano D, Shah M. Generative adversarial networks conditioned by brain signals. In: Proceedings of the IEEE international conference on computer vision. Venice: IEEE; 2020. p. 3410–8.
21. Fisher CK, Smith AM, Walsh JR, Coalition Against Major Diseases, Abbott, Alliance for Aging Research, Alzheimer's Association, Alzheimer's Foundation of America, AstraZeneca Pharmaceuticals LP, Bristol-Myers Squibb Company, Critical Path Institute, CHDI Foundation, Inc., Eli Lilly and Company, F. Hoffmann-La Roche Ltd, Forest Research Institute, Genentech, Inc., GlaxoSmithKline, Johnson & Johnson, National Health Council, Novartis Pharmaceuticals Corporation, Parkinson's Action Network, Parkinson's Disease Foundation, Pfizer, Inc., Sanofi-aventis, Collaborating Organizations: Clinical Data Interchange Standards Consortium (CDISC), Ephibian, Metrum Institute. Machine learning for comprehensive forecasting of Alzheimer's Disease progression. Sci Rep. 2019;9(1):13622.
22. Khayretdinova M, Shovkun A, Degtyarev V, Kiryasov A, Pshonkovskaya P, Zakharov I. Predicting age from resting-state scalp EEG signals with deep convolutional neural networks on TD-brain dataset. Front Aging Neurosci. 2022;14:1019869.
23. Zhang G, Zhou J, He G, Zhu H. Deep fusion of multi-modal features for brain tumor image segmentation. Heliyon. 2023;9(8):e19266.
24. Zelger P, Arnold M, Rossi S, Seebacher J, Muigg F, Graf S, Rodríguez-Sánchez A. Beyond averaging: a transformer approach to decoding event related brain potentials. NeuroImage. 2025;308:121049.
25. Akkus Z, Galimzianova A, Hoogi A, Rubin DL, Erickson BJ. Deep learning for brain MRI segmentation: state of the art and future directions. J Digit Imaging. 2017;30(4):449–59.
26. Guha Roy A, Conjeti S, Navab N, Wachinger C. Alzheimer's disease neuroimaging initiative. QuickNAT: a fully convolutional network for quick and accurate segmentation of neuroanatomy. NeuroImage. 2019;186:713–27.
27. Norman KA, Polyn SM, Detre GJ, Haxby JV. Beyond mind-reading: multi-voxel pattern analysis of fMRI data. Trends Cogn Sci. 2006;10(9):424–30.
28. Horikawa T, Kamitani Y. Generic decoding of seen and imagined objects using hierarchical visual features. Nat Commun. 2017;8:15037.
29. Bullmore E, Sporns O. Complex brain networks: graph theoretical analysis of structural and functional systems. Nat Rev Neurosci. 2009;10(3):186–98.
30. Khosla M, Jamison K, Kuceyeski A, Sabuncu MR. Ensemble learning with 3D convolutional neural networks for functional connectome-based prediction. NeuroImage. 2019;199:651–62.
31. Pandarinath C, O'Shea DJ, Collins J, Jozefowicz R, Stavisky SD, Kao JC, Trautmann EM, Kaufman MT, Ryu SI, Hochberg LR, Henderson JM, Shenoy KV, Abbott LF, Sussillo D. Inferring single-trial neural population dynamics using sequential auto-encoders. Nat Methods. 2018;15(10):805–15.
32. Cunningham JP, Yu BM. Dimensionality reduction for large-scale neural recordings. Nat Neurosci. 2014;17(11):1500–9.
33. Januszewski M, Kornfeld J, Li PH, Pope A, Blakely T, Lindsey L, Maitin-Shepard J, Tyka M, Denk W, Jain V. High-precision automated reconstruction of neurons with flood-filling networks. Nat Methods. 2018;15(8):605–10.
34. Van Essen DC, Smith SM, Barch DM, Behrens TE, Yacoub E, Ugurbil K, WU-Minn HCP Consortium. The WU-Minn Human Connectome Project: an overview. NeuroImage. 2013;80:62–79.
35. Hochberg LR, Bacher D, Jarosiewicz B, Masse NY, Simeral JD, Vogel J, Haddadin S, Liu J, Cash SS, van der Smagt P, Donoghue JP. Reach and grasp by people with tetraplegia using a neurally controlled robotic arm. Nature. 2012;485(7398):372–5.
36. Willett FR, Avansino DT, Hochberg LR, Henderson JM, Shenoy KV. High-performance brain-to-text communication via handwriting. Nature. 2021;593(7858):249–54.
37. Musk E, Neuralink. An integrated brain-machine interface platform with thousands of channels. J Med Internet Res. 2019;21(10):e16194.

38. Oxley TJ, Yoo PE, Rind GS, Ronayne SM, Lee CMS, Bird C, Hampshire V, Sharma RP, Morokoff A, Williams DL, MacIsaac C, Howard ME, Irving L, Vrljic I, Williams C, John SE, Weissenborn F, Dazenko M, Balabanski AH, Friedenberg D, Burkitt AN, Wong YT, Drummond KJ, Desmond P, Weber D, Denison T, Hochberg LR, Mathers S, O'Brien TJ, May CN, Mocco J, Grayden DB, Campbell BCV, Mitchell P, Opie NL. Motor neuroprosthesis implanted with neurointerventional surgery improves capacity for activities of daily living tasks in severe paralysis: first in-human experience. J Neurointerv Surg. 2021;13(2):102–8.
39. Lotte F, Bougrain L, Cichocki A, Clerc M, Congedo M, Rakotomamonjy A, Yger F. A review of classification algorithms for EEG-based brain-computer interfaces: a 10 year update. J Neural Eng. 2018;15(3):031005.
40. Zhao Y, Healy BC, Rotstein D, Guttmann CR, Bakshi R, Weiner HL, Brodley CE, Chitnis T. Exploration of machine learning techniques in predicting multiple sclerosis disease course. PLoS One. 2017;12(4):e0174866.
41. Giorgio J, Landau SM, Jagust WJ, Tino P, Kourtzi Z, Alzheimer's Disease Neuroimaging Initiative. Modelling prognostic trajectories of cognitive decline due to Alzheimer's disease. Neuroimage Clin. 2020;26:102199.
42. Rudroff T, Rainio O, Klén R, Tuulari JJ. The untapped potential of dimension reduction in neuroimaging: AI-driven multimodal analysis of long COVID fatigue. Brain Sci. 2024;14(12):1209.
43. Widge AS, Bilge MT, Montana R, Chang W, Rodriguez CI, Deckersbach T, Carpenter LL, Kalin NH, Nemeroff CB. Electroencephalographic biomarkers for treatment response prediction in major depressive illness: a meta-analysis. Am J Psychiatry. 2019;176(1):44–56.
44. Butson CR, Cooper SE, Henderson JM, McIntyre CC. Patient-specific analysis of the volume of tissue activated during deep brain stimulation. NeuroImage. 2007;34(2):661–70.
45. Tavanaei A, Ghodrati M, Kheradpisheh SR, Masquelier T, Maida A. Deep learning in spiking neural networks. Neural Netw. 2019;111:47–63.
46. Davies M, Srinivasa N, Lin TH, Chinya G, Cao Y, Choday SH, et al. Loihi: a neuromorphic manycore processor with on-chip learning. IEEE Micro. 2018;38(1):82–99.
47. Mozafari M, Ganjtabesh M, Nowzari-Dalini A, Masquelier T. SpykeTorch: efficient simulation of convolutional spiking neural networks with at most one spike per neuron. Front Neurosci. 2019;13:625.
48. Klöppel S, Stonnington CM, Chu C, Draganski B, Scahill RI, Rohrer JD, Fox NC, Jack CR Jr, Ashburner J, Frackowiak RS. Automatic classification of MR scans in Alzheimer's disease. Brain. 2008;131(Pt 3):681–9.
49. Thompson AJ, Banwell BL, Barkhof F, Carroll WM, Coetzee T, Comi G, Correale J, Fazekas F, Filippi M, Freedman MS, Fujihara K, Galetta SL, Hartung HP, Kappos L, Lublin FD, Marrie RA, Miller AE, Miller DH, Montalban X, Mowry EM, Sorensen PS, Tintoré M, Traboulsee AL, Trojano M, Uitdehaag BMJ, Vukusic S, Waubant E, Weinshenker BG, Reingold SC, Cohen JA. Diagnosis of multiple sclerosis: 2017 revisions of the McDonald criteria. Lancet Neurol. 2018;17(2):162–73.
50. Dadu A, Satone V, Kaur R, Hashemi SH, Leonard H, Iwaki H, Makarious MB, Billingsley KJ, Bandres-Ciga S, Sargent LJ, Noyce AJ, Daneshmand A, Blauwendraat C, Marek K, Scholz SW, Singleton AB, Nalls MA, Campbell RH, Faghri F. Identification and prediction of Parkinson's disease subtypes and progression using machine learning in two cohorts. npj Parkinsons Dis. 2022;8(1):172.
51. Heinsfeld AS, Franco AR, Craddock RC, Buchweitz A, Meneguzzi F. Identification of autism spectrum disorder using deep learning and the ABIDE dataset. Neuroimage Clin. 2017;17:16–23.
52. Rozycki M, Satterthwaite TD, Koutsouleris N, Erus G, Doshi J, Wolf DH, Fan Y, Gur RE, Gur RC, Meisenzahl EM, Zhuo C, Yin H, Yan H, Yue W, Zhang D, Davatzikos C. Multisite machine learning analysis provides a robust structural imaging signature of schizophrenia detectable across diverse patient populations and within individuals. Schizophr Bull. 2018;44(5):1035–44.

53. Ronneberger O, Fischer P, Brox T. U-net: convolutional networks for biomedical image segmentation. In: International conference on medical image computing and computer-assisted intervention. Cham: Springer; 2015. p. 234–41.
54. Thyreau B, Sato K, Fukuda H, Taki Y. Segmentation of the hippocampus by transferring algorithmic knowledge for large cohort processing. Med Image Anal. 2018;43:214–28.
55. Chen KT, Gong E, de Carvalho Macruz FB, Xu J, Boumis A, Khalighi M, Poston KL, Sha SJ, Greicius MD, Mormino E, Pauly JM, Srinivas S, Zaharchuk G. Ultra-low-dose ^{18}F-Florbetaben amyloid PET imaging using deep learning with multi-contrast MRI inputs. Radiology. 2019;290(3):649–56.
56. Schlemper J, Caballero J, Hajnal JV, Price AN, Rueckert D. A deep cascade of convolutional neural networks for dynamic MR image reconstruction. IEEE Trans Med Imaging. 2018;37(2):491–503.
57. Pereira CR, Pereira DR, Silva FA, Masieiro JP, Weber SA, Hook C, Papa JP. A new computer vision-based approach to aid the diagnosis of Parkinson's disease. Comput Methods Prog Biomed. 2016;136:79–88.
58. Liu S, Liu S, Cai W, Che H, Pujol S, Kikinis R, Feng D, Fulham MJ, ADNI. Multimodal neuroimaging feature learning for multiclass diagnosis of Alzheimer's disease. IEEE Trans Biomed Eng. 2015;62(4):1132–40.
59. Ktena SI, Parisot S, Ferrante E, Rajchl M, Lee M, Glocker B, Rueckert D. Metric learning with spectral graph convolutions on brain connectivity networks. NeuroImage. 2018;169:431–42.
60. Zhou J, Cui G, Zhang Z, Yang C, Liu Z, Wang L, et al. Graph neural networks: a review of methods and applications. AI Open. 2020;1:57–81.
61. Khosravi M, Zare Z, Mojtabaeian SM, Izadi R. Artificial intelligence and decision-making in healthcare: a thematic analysis of a systematic review of reviews. Health Serv Res Manag Epidemiol. 2024;11:23333928241234863.
62. Seo B, Li J. Explainable machine learning by SEE-Net: closing the gap between interpretable models and DNNs. Sci Rep. 2024;14(1):26302.
63. Rudroff T. Frontal-striatal glucose metabolism and fatigue in patients with multiple sclerosis, long COVID, and COVID-19 recovered controls. Exp Brain Res. 2024;242(9):2125–36.
64. Barkhof F. The clinico-radiological paradox in multiple sclerosis revisited. Curr Opin Neurol. 2002;15(3):239–45.
65. Storelli L, Azzimonti M, Gueye M, Vizzino C, Preziosa P, Tedeschi G, De Stefano N, Pantano P, Filippi M, Rocca MA. A deep learning approach to predicting disease progression in multiple sclerosis using magnetic resonance imaging. Investig Radiol. 2022;57(7):423–32.
66. Davatzikos C, Xu F, An Y, Fan Y, Resnick SM. Longitudinal progression of Alzheimer's-like patterns of atrophy in normal older adults: the SPARE-AD index. Brain. 2009;132(Pt 8):2026–35.
67. Westman E, Muehlboeck JS, Simmons A. Combining MRI and CSF measures for classification of Alzheimer's disease and prediction of mild cognitive impairment conversion. NeuroImage. 2012;62(1):229–38.
68. Adeli E, Shi F, An L, Wee CY, Wu G, Wang T, Shen D. Joint feature-sample selection and robust diagnosis of Parkinson's disease from MRI data. NeuroImage. 2016;141:206–19.
69. Gleichgerrcht E, Munsell B, Bhatia S, Vandergrift WA 3rd, Rorden C, McDonald C, Edwards J, Kuzniecky R, Bonilha L. Deep learning applied to whole-brain connectome to determine seizure control after epilepsy surgery. Epilepsia. 2018;59(9):1643–54.
70. Rudroff T, Rainio O, Klén R. Leveraging artificial intelligence to optimize transcranial direct current stimulation for long COVID management: a forward-looking perspective. Brain Sci. 2024;14(8):831.
71. Westeneng HJ, Debray TPA, Visser AE, van Eijk RPA, Rooney JPK, Calvo A, Martin S, McDermott CJ, Thompson AG, Pinto S, et al. Prognosis for patients with amyotrophic lateral sclerosis: development and validation of a personalized prediction model. Lancet Neurol. 2018;17(5):423–33.
72. Paulsen JS, Long JD, Ross CA, Harrington DL, Erwin CJ, Williams JK, Westervelt HJ, Johnson HJ, Aylward EH, Zhang Y, Bockholt HJ, Barker RA, PREDICT-HD Investigators

and Coordinators of the Huntington Study Group. Prediction of manifest Huntington's disease with clinical and imaging measures: a prospective observational study. Lancet Neurol. 2014;13(12):1193–201.
73. Guan MY, Gulshan V, Dai AM, Hinton GE. Who said what: modeling individual labelers improves classification. In: Proceedings of the AAAI conference on artificial intelligence, vol. 33; 2019. p. 3802–9.
74. Jack CR Jr, Knopman DS, Jagust WJ, Shaw LM, Aisen PS, Weiner MW, Petersen RC, Trojanowski JQ. Hypothetical model of dynamic biomarkers of the Alzheimer's pathological cascade. Lancet Neurol. 2010;9(1):119–28.
75. Pais M, Martinez L, Ribeiro O, Loureiro J, Fernandez R, Valiengo L, Canineu P, Stella F, Talib L, Radanovic M, Forlenza OV. Early diagnosis and treatment of Alzheimer's disease: new definitions and challenges. Braz J Psychiatry. 2020;42(4):431–41.
76. Frisoni GB, Fox NC, Jack CR Jr, Scheltens P, Thompson PM. The clinical use of structural MRI in Alzheimer disease. Nat Rev Neurol. 2010;6(2):67–77.
77. Toledo JB, Da X, Bhatt P, Wolk DA, Arnold SE, Shaw LM, Trojanowski JQ, Davatzikos C. Alzheimer's Disease Neuroimaging Initiative. Relationship between plasma analytes and SPARE-AD defined brain atrophy patterns in ADNI. PLoS One. 2013;8(2):e55531.
78. Ding Y, Sohn JH, Kawczynski MG, Trivedi H, Harnish R, Jenkins NW, Lituiev D, Copeland TP, Aboian MS, Mari Aparici C, Behr SC, Flavell RR, Huang SY, Zalocusky KA, Nardo L, Seo Y, Hawkins RA, Hernandez Pampaloni M, Hadley D, Franc BL. A deep learning model to predict a diagnosis of Alzheimer disease by using ^{18}F-FDG PET of the brain. Radiology. 2019;290(2):456–64.
79. Marinescu RV, Oxtoby NP, Young AL, Bron EE, Toga AW, Weiner MW, … Alexander DC. TADPOLE challenge: prediction of longitudinal evolution in Alzheimer's disease. arXiv:1805.03909 [Preprint]. 2018.
80. Khanna S, Domingo-Fernández D, Iyappan A, Emon MA, Hofmann-Apitius M, Fröhlich H. Using multi-scale genetic, neuroimaging and clinical data for predicting Alzheimer's disease and reconstruction of relevant biological mechanisms. Sci Rep. 2018;8(1):11173.
81. Fraser KC, Meltzer JA, Rudzicz F. Linguistic features identify Alzheimer's disease in narrative speech. J Alzheimer's Dis. 2016;49(2):407–22.
82. Ritchie CW, Molinuevo JL, Truyen L, Satlin A, Van der Geyten S, Lovestone S, European Prevention of Alzheimer's Dementia (EPAD) Consortium. Development of interventions for the secondary prevention of Alzheimer's dementia: the European Prevention of Alzheimer's Dementia (EPAD) project. Lancet Psychiatry. 2016;3(2):179–86.

Multimodal Analysis and Neurological Disorders

2

A chain is only as strong as its weakest link, but a network is as strong as its strongest connections.

—Attributed to George F. Kennan (modified)

2.1 Introduction

The human brain remains one of the most complex systems known to science. It consists of a dynamic network of approximately 86 billion neurons forming trillions of connections [1]. This complexity is reflected in the diversity of neurological disorders, which can affect multiple aspects of brain structure and function through various pathophysiological mechanisms. Understanding these conditions requires a comprehensive approach that captures different dimensions of neural dysfunction.

Traditional neuroimaging approaches have often relied on single modalities to investigate brain structure or function. These include structural magnetic resonance imaging (MRI) to examine anatomical features, functional MRI (fMRI) to assess BOLD signals as a proxy for neural activity, diffusion tensor imaging (DTI) to map white matter tracts, and positron emission tomography (PET) to measure metabolic activity or specific molecular targets. While each of these modalities provides valuable insights, they capture only partial aspects of the complex neural processes underlying health and disease.

Multimodal neuroimaging analysis, the integration of data from multiple imaging techniques, offers a more comprehensive approach. By combining complementary information across modalities, researchers can develop richer models of brain structure and function, potentially revealing patterns that might remain obscure when examining any single modality in isolation. As the philosopher Aristotle noted over two millennia ago, "The whole is greater than the sum of its parts." This

T. Rudroff, *AI for the Healthy Brain and Neurological Disorders*, Neurocultural Health and Wellbeing, https://doi.org/10.1007/978-3-032-11596-6_2

principle aptly describes the promise of multimodal neuroimaging, where the integration of diverse data types can yield insights beyond what could be gleaned from analyzing each modality separately.

In recent years, artificial intelligence (AI), particularly machine learning and deep learning approaches, has transformed multimodal neuroimaging analysis. These computational methods excel at identifying complex patterns across high-dimensional datasets, making them well suited to the challenge of integrating information across different imaging modalities. Moreover, AI approaches can potentially uncover subtle, distributed patterns of abnormality that might elude conventional analytical methods, offering new opportunities for understanding neurological disorders.

This chapter explores the intersection of multimodal neuroimaging, AI, and neurological disorders. We begin by examining the technical foundations of multimodal analysis, considering both traditional approaches and emerging AI-driven methods. We then discuss how AI can overcome limitations in conventional neuroimaging interpretation, enabling more sensitive and specific detection of neural abnormalities. Next, we explore dimension reduction techniques, which play a crucial role in managing the high-dimensional data generated by multimodal imaging. We then present a case study on Long COVID fatigue, drawing on recent research using multimodal PET and MRI analysis to understand this emerging neurological condition. Finally, we consider future directions for multimodal analysis in clinical applications, examining both the potential and challenges of translating these approaches to routine clinical practice.

Throughout this chapter, we will see how multimodal analysis, enhanced by AI, is transforming our understanding of neurological disorders—moving from isolated snapshots of brain structure or function to integrated models that capture the multifaceted nature of neural health and disease.

2.2 Technical Foundations of Multimodal Neuroimaging Analysis

Multimodal neuroimaging analysis faces the fundamental challenge of integrating data from different imaging techniques, each with distinct spatial and temporal resolutions, contrast mechanisms, and sources of noise. This section explores the technical foundations of multimodal integration, examining both traditional approaches and emerging computational methods.

2.2.1 Neuroimaging Modalities: Complementary Windows into Brain Structure and Function

Before discussing integration approaches, we must understand the complementary nature of different neuroimaging modalities. Each technique provides unique information about brain structure or function:

Structural MRI (*sMRI*) uses strong magnetic fields and radio waves to generate high-resolution images of brain anatomy. T1-weighted images provide excellent gray-white matter contrast, facilitating analysis of cortical thickness, subcortical volumes, and overall brain morphology. T2-weighted and FLAIR sequences are sensitive to tissue water content, making them valuable for detecting lesions and edema [2]. Structural MRI typically offers high spatial resolution (approximately 1 mm) but provides static anatomical information rather than dynamic functional data.

Functional MRI (*fMRI*) measures changes in blood-oxygen-level-dependent (BOLD) signals, which serve as an indirect marker of neural activity. fMRI detects local increases in oxygenated blood flow resulting from neural activation. This allows fMRI to map functional brain activity during tasks or in the resting state [3]. While fMRI offers good spatial resolution (2–3 mm), its temporal resolution is limited by the hemodynamic response (typically several seconds), and the BOLD signal represents an indirect measure of neural activity.

Diffusion MRI (*dMRI*), including diffusion tensor imaging (DTI), measures the diffusion of water molecules in tissue, providing information about white matter microstructure and connectivity. dMRI tracks the preferential diffusion of water along axonal bundles. This enables the reconstruction of white matter tracts and quantification of measures such as fractional anisotropy (FA) and mean diffusivity (MD). These measures reflect fiber integrity [4]. Diffusion MRI thus complements structural and functional techniques by mapping the brain's structural connectivity.

Positron Emission Tomography (*PET*) uses radioactive tracers to measure various aspects of brain function and molecular composition. PET can assess various aspects of brain function depending on the tracer employed. These include glucose metabolism (using 18F-fluorodeoxyglucose, or FDG), amyloid or tau deposition (with tracers such as 11C-PIB or 18F-flortaucipir), neurotransmitter systems (e.g., 11C-raclopride for dopamine receptors), and neuroinflammation (e.g., 11C-PK11195 for microglial activation) [5]. While PET offers unique molecular specificity, its spatial resolution (4–5 mm) is lower than MRI, and it involves radiation exposure.

Electroencephalography (*EEG*) and *Magnetoencephalography* (*MEG*) measure electrical activity and associated magnetic fields generated by neuronal firing, providing direct indicators of neural activity with millisecond temporal resolution. These techniques excel at capturing rapid temporal dynamics of brain activity but offer lower spatial resolution compared to MRI, particularly for deep brain structures [6].

Functional Near-Infrared Spectroscopy (*fNIRS*) uses near-infrared light to measure changes in hemoglobin concentration and oxygenation in the cortex, providing an alternative means of assessing functional brain activity. While fNIRS has lower spatial resolution and depth penetration than fMRI, it offers advantages in terms of portability, cost, and tolerance of movement [7].

These diverse modalities capture complementary aspects of brain structure and function across different spatial and temporal scales. The challenge of multimodal integration lies in combining these diverse data types in a manner that leverages their respective strengths while managing their limitations and inherent differences. To effectively harness the complementary information provided by these diverse neuroimaging modalities, researchers must follow a systematic pipeline that addresses the technical challenges at each stage of analysis. Figure 2.1 presents a comprehensive overview of the multimodal neuroimaging analysis pipeline employed in our Long COVID fatigue research, illustrating the progression from raw data acquisition to clinical insights.

As depicted in Fig. 2.1, the pipeline begins with the simultaneous or sequential acquisition of multiple imaging modalities, each capturing distinct aspects of brain structure and function. The subsequent preprocessing stage ensures data quality and compatibility across modalities, while dimension reduction techniques manage the curse of dimensionality inherent in high-dimensional neuroimaging data. The critical multimodal integration step employs various fusion strategies to combine information across modalities, ultimately yielding insights into the frontal-striatal-thalamic network dysfunction characteristic of Long COVID fatigue. This systematic approach achieves a 65% variance explanation in fatigue symptoms—a 25% improvement over the best single modality—demonstrating the power of integrated multimodal analysis.

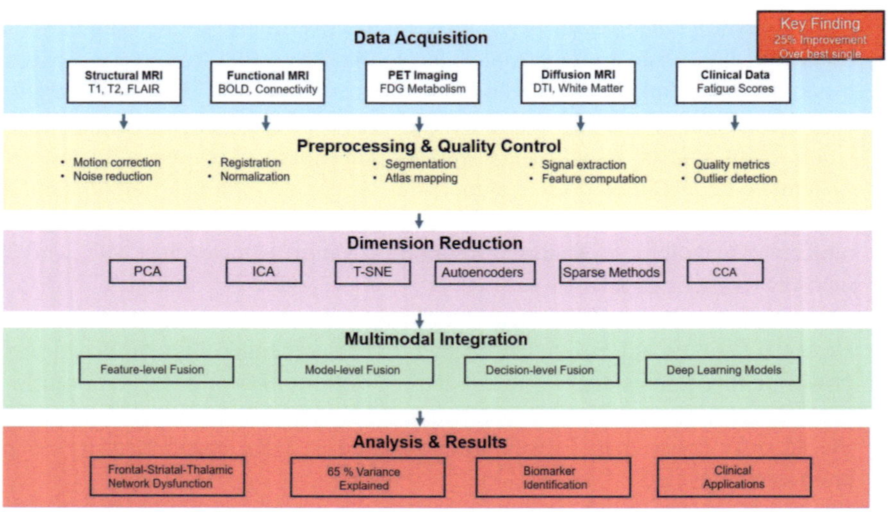

Fig. 2.1 Multimodal neuroimaging analysis pipeline for Long COVID fatigue

2.2.2 Traditional Approaches to Multimodal Integration

Traditional approaches to multimodal neuroimaging integration can be broadly categorized into three strategies: data fusion at the acquisition level, analysis of each modality separately followed by interpretation-level integration, and model-driven integration using computational frameworks.

Acquisition-level fusion involves the simultaneous or near-simultaneous collection of data from multiple modalities, such as with combined PET/MRI scanners or concurrent EEG-fMRI recordings. This approach minimizes registration issues and temporal discrepancies between modalities, although it requires specialized equipment and may involve trade-offs in the quality of individual modalities compared to dedicated systems [8]. For example, PET/MRI scanners enable the simultaneous acquisition of structural, functional, and molecular imaging data, facilitating direct comparison between metabolic activity (PET) and anatomical or functional information (MRI) [9]. However, MRI-compatible PET detectors may have lower sensitivity than standalone PET systems, and MR sequences may require adaptation to accommodate the PET hardware.

Separate analysis with interpretation-level integration represents a more common traditional approach, where each modality is processed and analyzed independently, and the results are subsequently integrated through expert interpretation. For example, a clinician might examine structural MRI for anatomical abnormalities, FDG-PET for metabolic changes, and DTI for white matter integrity, synthesizing these findings to form a comprehensive assessment. While straightforward to implement, this approach relies heavily on expert judgment and may miss subtle intermodality associations that are not immediately apparent upon visual inspection or independent statistical analysis [10].

Model-driven integration employs computational frameworks to combine information across modalities based on biological or methodological models. Examples include biophysical models that combine EEG and fMRI data based on neurovascular coupling principles [11], multimodal extensions of independent component analysis (ICA) that identify common sources across modalities [12], and Bayesian frameworks that integrate prior knowledge with multimodal observations [13]. These approaches offer more principled integration than interpretation-level methods but often require strong assumptions about the associations between modalities and may be computationally intensive.

While these traditional approaches have yielded valuable insights, they face significant limitations when dealing with the high-dimensional, complex data generated by modern neuroimaging techniques. Traditional methods often struggle to capture nonlinear associations between modalities, may be sensitive to noise or artifacts in individual modalities, and typically require extensive preprocessing and feature engineering. Moreover, they may miss subtle, distributed patterns of abnormalities that span multiple modalities and brain regions. These limitations have motivated the development of advanced computational approaches, particularly those based on machine learning and deep learning.

2.2.3 Machine Learning Approaches to Multimodal Integration

Machine learning approaches offer several advantages for multimodal neuroimaging analysis, including the ability to automatically discover complex patterns across modalities, handle high-dimensional data without extensive feature engineering, and potentially identify subtle, distributed abnormalities. These methods can be broadly categorized as feature-level fusion, model-level fusion, and decision-level fusion strategies.

Feature-level fusion involves combining features extracted from different modalities into a single feature set, which is then used for subsequent analysis or classification. This approach allows the learning algorithm to discover relationships between features across modalities. Traditional feature-level fusion often employs feature selection or dimension reduction techniques to manage the high dimensionality resulting from combining multiple modalities. For example, Zhang et al. [14] combined structural MRI and FDG-PET features for Alzheimer's disease classification, using multi-kernel learning to weight features from different modalities. More recently, deep learning approaches have enabled end-to-end feature learning and fusion, automatically extracting relevant features from raw or minimally processed imaging data across modalities [15].

Model-level fusion maintains separate processing streams for each modality before integrating them at an intermediate stage of the analysis pipeline. This approach allows for modality-specific preprocessing and feature extraction while still enabling cross-modal integration. For instance, multimodal deep learning architectures often process different modalities through parallel pathways before merging features at higher layers [16]. This strategy can accommodate the distinct characteristics of each modality while capturing intermodality relationships. A specific example is the multimodal convolutional neural network developed by Nie et al. [17], which uses separate convolutional pathways for T1, T2, and FLAIR MRI before fusion through fully connected layers for brain tumor segmentation.

Decision-level fusion involves making separate predictions for each modality and then combining these predictions to form a final decision. This approach is particularly useful when modalities differ substantially in their content of information or quality, as it allows for adaptive weighting of different sources. Ensemble methods, such as stacking or voting, are common implementations of decision-level fusion. For example, Kohannim et al. [18] used a decision fusion approach for Alzheimer's disease classification, training separate classifiers on MRI, PET, genetic, and cognitive measures before combining their outputs through a weighted voting scheme.

Recent advances in deep learning have expanded the possibilities for multimodal integration, enabling more sophisticated fusion strategies. *Cross-modal encoding* approaches learn representations in one modality that predict or generate data in another modality, potentially capturing fundamental relationships between different aspects of brain structure and function [19]. *Multimodal autoencoders* learn latent representations that capture shared information across modalities, potentially revealing common factors underlying different measurements [20]. *Graph neural networks* can model both within-modality and cross-modality relationships as edges in a complex network, offering a flexible framework for representing multimodal brain data [21].

2.2.4 Preprocessing and Registration Challenges

A critical prerequisite for effective multimodal integration is accurate spatial and temporal alignment of data across modalities. Each neuroimaging technique has distinct spatial and temporal characteristics, necessitating careful preprocessing and registration to enable meaningful integration.

Spatial registration involves aligning images from different modalities to a common coordinate system, ensuring that the same anatomical location corresponds to the same coordinate across all modalities. This process typically involves estimating a spatial transformation that maps one image to another, which may be rigid (preserving distances between points) or nonrigid (allowing local deformations). Registration between structural MRI and other modalities, such as PET or fMRI, is particularly challenging due to differences in contrast mechanisms, spatial resolution, and geometric distortions. Advanced registration algorithms, such as symmetric diffeomorphic normalization (SyN) implemented in ANTs [22], have improved the accuracy of cross-modal registration, but challenges remain, particularly for patients with significant structural abnormalities or when registering between fundamentally different modalities like EEG and MRI.

Temporal alignment is crucial when combining modalities with different temporal resolutions, such as EEG (milliseconds) and fMRI (seconds). This process may involve downsampling the higher temporal resolution data, modeling the temporal associations between modalities (e.g., the hemodynamic response function relating neural activity to BOLD signals), or identifying temporal landmarks for alignment [23]. Simultaneous acquisition, as with concurrent EEG-fMRI, can mitigate some temporal alignment issues but introduces technical challenges such as MRI-induced artifacts in EEG recordings.

Harmonization across different scanners, protocols, or sites represents another significant challenge for multimodal integration, particularly for multicenter studies or when combining legacy data with new acquisitions. Variations in hardware, acquisition parameters, and reconstruction algorithms can introduce systematic differences that confound biological effects of interest. Statistical approaches for harmonization, such as ComBat [24, 25], and more recently, deep learning-based methods for image-to-image translation [26], have been developed to address these issues, although perfect harmonization remains an open challenge.

2.2.5 Quantitative Measures and Derived Features

Beyond raw imaging data, multimodal analysis often incorporates quantitative measures and derived features that capture specific aspects of brain structure or function. These features provide a more interpretable and potentially more robust basis for integration than raw voxel intensities.

Structural features derived from sMRI include cortical thickness, surface area, subcortical volumes, and morphometric measures such as gyrification and sulcal depth [27]. These measures capture different aspects of brain anatomy and may be affected differently by various pathological processes. For example, cortical thickness and surface area have distinct genetic influences and developmental trajectories and may show different patterns of change in conditions like Alzheimer's disease or schizophrenia [28].

Connectivity measures derived from fMRI or dMRI represent the brain as a network of regions connected by functional or structural links. Functional connectivity, typically estimated as the correlation between BOLD time series from different regions, captures statistical dependencies in neural activity [29]. Structural connectivity, reconstructed from diffusion MRI through tractography, estimates the physical connections between brain regions [30]. Graph theoretical measures derived from these networks, such as modularity, efficiency, and centrality, provide quantitative characterizations of brain organization that can be compared across individuals or groups [31].

Metabolic and molecular features derived from PET include standardized uptake values (SUVs), binding potentials, or kinetic parameters for specific tracers. These measures quantify processes such as glucose metabolism, neurotransmitter receptor density, or protein deposition, providing insights into the molecular basis of brain function and dysfunction [32]. For example, in research on fatigue in multiple sclerosis and Long COVID, researchers have used FDG-PET to quantify glucose metabolism in specific brain regions and networks, revealing distinct patterns of metabolic alteration associated with different conditions [33].

Electrophysiological features derived from EEG or MEG include spectral power in different frequency bands, coherence measures, and event-related potentials or fields. These features capture different aspects of neural dynamics, from ongoing oscillatory activity to specific responses to stimuli or events [34]. They provide a direct measure of neural activity with excellent temporal resolution, complementing the indirect, slower hemodynamic measures provided by fMRI.

Integrating these diverse quantitative measures presents both opportunities and challenges. On the one hand, derived features may be more interpretable and less noisy than raw imaging data, potentially simplifying integration. On the other hand, each measure captures only specific aspects of brain structure or function, and relationships between different measures may be complex and nonlinear. Moreover, derived features depend on the specific processing pipelines and parameters used for their extraction, introducing additional sources of variability that must be considered in multimodal analyses.

The technical foundations described above provide the framework for multimodal neuroimaging analysis. In the next section, we will explore how AI approaches can overcome traditional limitations in neuroimaging interpretation, enabling more sensitive and specific detection of neural abnormalities.

2.3 How AI Overcomes Traditional Limitations in Neuroimaging Interpretation

Conventional approaches to neuroimaging interpretation face several significant limitations. These challenges are particularly pronounced when dealing with complex, high-dimensional data generated by multimodal imaging. Artificial intelligence, especially machine learning and deep learning methods, offers powerful tools to address these limitations, enabling more sensitive, specific, and comprehensive analysis of neuroimaging data. This section explores how AI approaches overcome key challenges in neuroimaging interpretation, with a focus on multimodal integration for neurological disorders.

Artificial intelligence offers a paradigm shift in how we approach multimodal neuroimaging analysis, providing sophisticated solutions to the fundamental limitations of traditional methods. Figure 2.2 presents a comprehensive overview of how AI approaches address each major challenge in neuroimaging interpretation, from the curse of dimensionality to the need for personalized predictions.

As illustrated in Fig. 2.2, AI methods systematically overcome traditional limitations through six key innovation areas. Feature learning and dimensionality reduction techniques, particularly convolutional neural networks and self-supervised learning, automatically extract relevant patterns from high-dimensional data without requiring manual feature engineering. These approaches enable individual patient prediction through personalized models that move beyond group-level comparisons to provide clinically actionable insights.

The framework demonstrates how complex pattern recognition methods, including deep learning and graph neural networks, capture distributed abnormalities across multiple brain regions and modalities—patterns that might remain invisible

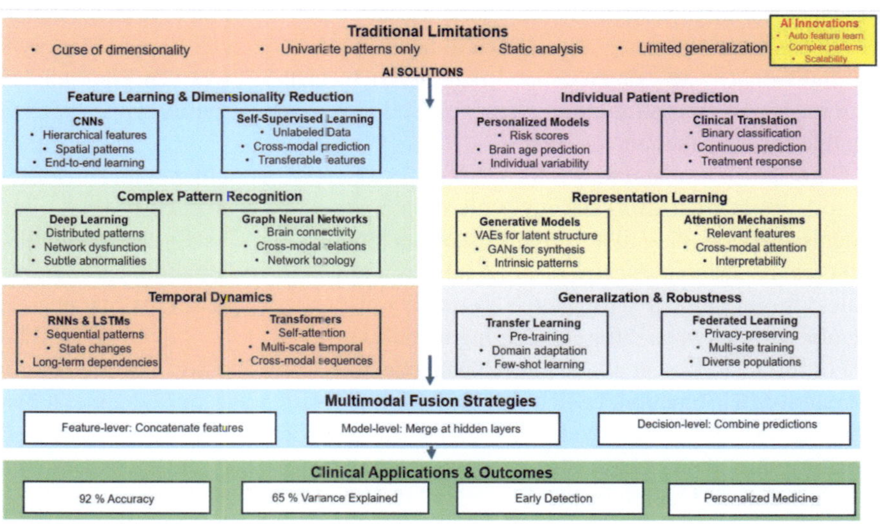

Fig. 2.2 AI approaches to multimodal integration in neuroimaging

to conventional univariate analyses. Representation learning through generative models and attention mechanisms extracts meaningful features while maintaining interpretability, while temporal modeling approaches like transformers and RNNs capture the dynamic nature of brain function.

These AI innovations culminate in flexible fusion strategies that integrate information at feature, model, or decision levels, ultimately achieving the remarkable clinical outcomes observed in our Long COVID research: 92% classification accuracy and 65% variance explained in fatigue symptoms. The following sections detail how each of these AI approaches specifically addresses the limitations of traditional neuroimaging analysis.

2.3.1 Limitation 1: The Curse of Dimensionality

Neuroimaging data is inherently high-dimensional. A typical structural MRI might include millions of voxels, while multimodal datasets combining several imaging techniques multiply this complexity. Traditional statistical approaches, such as mass-univariate analyses, face the "curse of dimensionality" when dealing with such data. As dimensionality increases, the amount of data needed to maintain statistical power grows exponentially [35]. This challenge is compounded in neurological disorders, which often involve subtle, distributed patterns of abnormality rather than focal lesions.

2.3.2 AI Solution: Feature Learning and Dimensionality Reduction

Machine learning, particularly deep learning, addresses the curse of dimensionality through automatic feature learning and dimensionality reduction. Instead of analyzing individual voxels or predefined regions, these methods can identify complex patterns across the entire brain, effectively reducing dimensionality while preserving relevant information.

Convolutional neural networks (CNNs), for example, learn hierarchical features from neuroimaging data, starting with simple edge detectors in early layers and progressively building more complex representations [36]. This approach naturally captures the spatial structure of brain images and can identify patterns at multiple scales. In multimodal settings, CNNs can learn cross-modal features that encode relationships between different imaging modalities.

For instance, Wen et al. [37] developed a deep learning approach for integrating structural MRI, functional MRI, and diffusion tensor imaging in Alzheimer's disease classification. Their model automatically learned relevant features from each modality and their relationships, achieving higher accuracy than conventional approaches based on predefined features. Similarly, Li et al. [38] employed a deep neural network to integrate multimodal MRI for autism spectrum disorder classification, demonstrating superior performance compared to traditional machine learning with handcrafted features.

Self-supervised learning represents a particularly promising approach for neuroimaging, allowing models to learn meaningful representations from unlabeled data before fine-tuning on specific tasks [39]. This approach is especially valuable in neuroimaging, where labeled data may be scarce but unlabeled images are often abundant. By learning to predict one modality from another or to reconstruct masked portions of images, self-supervised models can capture fundamental properties of brain structure and function that generalize across tasks and conditions.

2.3.3 Limitation 2: Group-Level Versus Individual Analysis

Traditional neuroimaging analyses often focus on group-level comparisons, such as contrasting patients with controls to identify statistically significant differences. While valuable for research, this approach has limited clinical utility, as it does not directly address the needs of individual patients. Clinicians need to know whether a specific patient's brain shows abnormalities and what these might mean for diagnosis, prognosis, or treatment.

2.3.4 AI Solution: Predictive Modeling for Individual Patients

Machine learning enables a shift from descriptive, group-level analyses to predictive modeling for individual patients. Rather than simply identifying average differences between groups, these approaches can predict individual diagnoses, prognoses, or treatment responses based on patterns in neuroimaging data.

Arbabshirani et al. [40] reviewed machine learning applications in brain imaging, finding that predictive models could accurately classify individual patients across various neurological and psychiatric disorders. In schizophrenia, machine learning models based on structural and functional MRI can predict diagnosis with accuracies exceeding 80% [41]. Similarly, in Alzheimer's disease, multimodal predictive models combining MRI, PET, and cerebrospinal fluid biomarkers can identify individuals at high risk of progression from mild cognitive impairment to dementia [42].

Beyond binary classification, AI approaches can generate continuous predictions or risk scores, potentially capturing the spectrum of disease severity or progression risk. For instance, Cole et al. [43] developed a CNN-based approach to predict "brain age" from structural MRI, with the difference between predicted and chronological age serving as a biomarker of accelerated aging associated with various neurological conditions.

Importantly, recent advances in AI have enabled the development of personalized predictive models that account for individual variability. Drysdale et al. [44] proposed a personalized depression prediction model that adapts to individual characteristics by incorporating demographic and clinical information alongside neuroimaging data. This approach acknowledges the heterogeneity within diagnostic categories and moves toward truly personalized medicine.

2.3.5 Limitation 3: Univariate Versus Multivariate Patterns

Traditional statistical approaches to neuroimaging often employ mass-univariate analysis, examining each voxel or region independently. This approach may miss complex, multivariate patterns of abnormality that span multiple brain areas or involve interactions between regions. This limitation is particularly relevant for neurological disorders, which often affect brain networks rather than isolated regions.

2.3.6 AI Solution: Capturing Complex, Distributed Patterns

Machine learning methods excel at identifying multivariate patterns in neuroimaging data. Rather than examining each voxel or feature in isolation, these approaches can capture complex associations across the entire brain, potentially revealing subtle, distributed abnormalities that might be missed by conventional analyses.

Multivariate pattern analysis (MVPA) techniques, including support vector machines, random forests, and deep learning models, have been widely applied to neuroimaging data, demonstrating superior sensitivity compared to univariate approaches [45]. For example, in multiple sclerosis, machine learning models can identify distributed patterns of structural and functional alteration associated with cognitive impairment, even in patients with minimal conventional MRI abnormalities [46].

Deep learning approaches, particularly CNNs, can automatically learn complex spatial patterns from raw imaging data without requiring predefined features or regions of interest. This capability is especially valuable for disorders with subtle, diffuse abnormalities. For instance, Payan and Montana [47] applied deep learning to detect early signs of Alzheimer's disease in structural MRI, identifying patterns of atrophy that traditional methods might miss.

In multimodal settings, AI approaches can capture not only within-modality patterns but also relationships between modalities. Graph neural networks (GNNs) represent a particularly promising approach for modeling these complex associations, representing both brain regions and modalities as nodes in a graph with learnable connections [48]. This framework naturally accommodates the complex, network-like structure of the brain and the relationships between different imaging techniques.

2.3.7 Limitation 4: Extracting Meaning from Complex Signals

Neuroimaging generates complex signals that can be challenging to interpret, particularly when integrating multiple modalities with different temporal and spatial characteristics. Traditional approaches often rely on simplified models or predefined features that may not capture the full complexity of the underlying neural processes.

2.3.8 AI Solution: Learning Representations from Data

AI methods can learn meaningful representations directly from complex neuroimaging signals, potentially capturing subtle features and associations that might be missed by conventional analyses. This capability is particularly valuable for integrating complementary information across modalities.

Deep generative models, such as variational autoencoders (VAEs) and generative adversarial networks (GANs), can learn latent representations that capture the underlying structure of neuroimaging data [49]. These unsupervised approaches do not require labeled data and can potentially reveal intrinsic patterns in brain structure and function. For example, Biffi et al. [50] used a VAE to learn a latent representation of brain morphology from structural MRI, identifying patterns associated with Alzheimer's disease and other neurodegenerative conditions.

Attention mechanisms in deep learning models can identify which aspects of the imaging data are most relevant for a particular task, potentially highlighting key regions or features for diagnosis or prognosis. For instance, in a multimodal setting, attention mechanisms can learn to focus on specific modalities or brain regions depending on the task and input data, providing both improved performance and interpretability.

Self-supervised contrastive learning represents another promising approach for extracting meaningful representations from neuroimaging data. By learning to distinguish between different views or transformations of the same image, these methods can capture invariant properties of brain structure and function without requiring labeled data [51]. For example, Taleb et al. [39] demonstrated that contrastive learning could identify meaningful representations from brain MRI that transferred effectively to various downstream tasks, including disease classification.

2.3.9 Limitation 5: Static Versus Dynamic Analysis

Traditional neuroimaging analyses often treat the brain as a static entity, focusing on fixed anatomical structures or time-averaged functional patterns. However, the brain is inherently dynamic, with complex temporal patterns that may be crucial for understanding both normal function and disease-related abnormalities.

2.3.10 AI Solution: Modeling Temporal Dynamics

AI approaches, particularly recurrent neural networks (RNNs) and transformer models, excel at capturing temporal dynamics in neuroimaging data. These methods can model complex sequences and long-range dependencies, potentially revealing temporal patterns associated with neurological disorders.

Long short-term memory (LSTM) networks and other RNN variants have been applied to functional MRI data to model brain dynamics during rest and task performance [52]. These approaches can capture temporal dependencies at multiple scales, from short-term fluctuations to longer-term state changes. For example, researchers have used LSTM networks to model dynamic functional connectivity in schizophrenia, identifying temporal patterns that distinguished patients from controls with high accuracy.

More recently, transformer models, which have revolutionized natural language processing, have been adapted for neuroimaging analysis. These models use self-attention mechanisms to capture long-range dependencies in sequential data, making them well suited for modeling complex temporal patterns in brain activity. Researchers have employed transformer models to analyze dynamic functional connectivity in autism spectrum disorder, demonstrating superior performance compared to conventional static analyses.

In multimodal settings, these temporal modeling approaches can integrate information across modalities with different temporal resolutions. For example, researchers have developed multimodal recurrent neural networks that combine EEG and fMRI data for seizure prediction, leveraging the complementary temporal and spatial information provided by these modalities.

2.3.11 Limitation 6: Interpretability and Clinical Translation

Traditional machine learning models, particularly deep learning approaches, are often criticized as "black boxes" that provide little insight into their decision-making processes. This lack of interpretability poses challenges for clinical translation, as clinicians typically require explanations for diagnostic or prognostic assessments.

2.3.12 AI Solution: Explainable AI for Neuroimaging

Recent advances in explainable AI (XAI) have led to the development of methods that can provide insights into model decisions while maintaining high performance. These approaches are essential for the clinical translation of AI-based neuroimaging analysis.

Saliency maps and attention visualization techniques can highlight which regions or features of a brain image most strongly influence a model's prediction. For example, researchers have used gradient-weighted class activation mapping (Grad-CAM) to visualize which brain regions contributed most to Alzheimer's disease classification in deep learning models. Similarly, in multimodal settings, attention mechanisms can reveal which modalities and features are most informative for a particular task or patient.

More sophisticated XAI approaches include concept-based explanations, which relate model decisions to high-level clinical concepts. For instance, a model might explain its prediction of Alzheimer's disease in terms of hippocampal atrophy, white matter integrity, and glucose hypometabolism—concepts that are directly interpretable to clinicians. Such explanations bridge the gap between AI models and clinical reasoning, potentially facilitating the integration of AI into clinical practice.

Another approach involves developing inherently interpretable models that balance performance with explainability. For example, researchers have proposed interpretable convolutional neural networks for neuroimaging, which incorporate domain knowledge and provide transparent decision processes while maintaining competitive performance. Similarly, interpretable deep learning frameworks for multimodal neuroimaging analysis have been developed, where each layer and feature has a clear neurobiological interpretation.

2.3.13 Limitation 7: Generalization Across Populations and Settings

Traditional neuroimaging analyses often suffer from limited generalizability, with findings from one cohort or study site failing to replicate in different populations or settings. This challenge is particularly acute for AI approaches, which may learn scanner-specific or cohort-specific patterns rather than true biological signals.

2.3.14 AI Solution: Transfer Learning and Domain Adaptation

Transfer learning and domain adaptation techniques can improve the generalizability of AI models across different populations, scanners, and protocols. These approaches leverage knowledge gained from one dataset or domain to improve performance on another, potentially overcoming the limitations of small or site-specific training data.

Pretraining on large, diverse datasets followed by fine-tuning on specific tasks has proven effective in neuroimaging applications. For example, models pretrained on general brain imaging datasets can be fine-tuned for specific disorders or populations, requiring less task-specific data while maintaining good performance. This approach is particularly valuable for rare conditions or understudied populations where labeled data may be scarce.

Domain adaptation techniques explicitly address differences between source and target domains, such as different scanners or protocols. For instance, researchers have proposed adversarial domain adaptation approaches for cross-site neuroimaging analysis, which learn scanner-invariant features that generalize across different acquisition settings. Similarly, domain adaptation methods for harmonizing diffusion MRI data across sites have been developed, enabling more robust multicenter studies.

Federated learning represents another promising approach for improving generalizability while addressing privacy concerns. Rather than centralizing data from multiple sites, federated learning trains models across decentralized datasets, with only model parameters shared between sites. This approach enables the development of models that leverage data from diverse populations and settings without requiring direct data sharing, potentially facilitating larger and more representative studies.

By addressing these fundamental limitations of traditional neuroimaging analysis, AI approaches enable more sensitive, specific, and comprehensive characterization of neurological disorders. The next section will focus specifically on dimension reduction techniques, which play a crucial role in managing the high-dimensional data generated by multimodal imaging.

2.4 Dimension Reduction Techniques in Neuroimaging

Neuroimaging data, particularly when collected across multiple modalities, present formidable analytical challenges due to their high dimensionality, complex structure, and heterogeneous nature. Dimension reduction techniques—methods that extract lower-dimensional representations while preserving essential information—play a crucial role in addressing these challenges. This section outlines key dimension reduction approaches for neuroimaging analysis, with a particular focus on applications in multiple sclerosis and Long COVID.

2.4.1 The Necessity of Dimension Reduction in Neuroimaging

Before delving into specific techniques, it is worth considering why dimension reduction is essential for neuroimaging analysis, particularly in multimodal settings. Several factors motivate the use of dimension reduction:

The curse of dimensionality, as discussed earlier, poses fundamental statistical challenges when analyzing high-dimensional data with limited samples. A typical neuroimaging dataset might include thousands or millions of voxels across multiple modalities but only dozens or hundreds of subjects. Without dimension reduction, models are prone to overfitting, learning noise patterns that do not generalize to new data [35].

Computational efficiency represents another practical consideration. Many machine learning algorithms scale poorly with dimensionality, becoming computationally prohibitive for raw neuroimaging data. Dimension reduction can dramatically reduce computation time and memory requirements, enabling more complex analyses or larger studies.

Noise reduction is an important benefit of many dimension reduction techniques. By projecting data onto a lower-dimensional subspace that captures most of the variance or relevant structure, these methods can filter out noise components, potentially improving the signal-to-noise ratio and model performance [45].

Visualization and interpretation become more feasible with reduced dimensionality. While the original high-dimensional space is impossible to visualize directly, lower-dimensional projections can reveal patterns and relationships in the data that might otherwise remain obscure. This capability is particularly valuable for exploring complex multimodal datasets.

Feature extraction and biomarker discovery often rely on dimension reduction to identify meaningful, compact representations of neuroimaging data. These lower-dimensional features can serve as biomarkers for disease detection, progression monitoring, or treatment response prediction.

To illustrate how these principles are applied in practice, Fig. 2.3 presents the complete dimension reduction pipeline employed in our multimodal neuroimaging analysis of Long COVID fatigue. This systematic approach demonstrates how we transform the overwhelming complexity of raw neuroimaging data into clinically meaningful, interpretable features.

As depicted in Fig. 2.3, our dimension reduction strategy addresses the formidable challenge of analyzing millions of voxels across multiple modalities. Beginning with high-dimensional inputs—structural MRI's approximately two million voxels, functional MRI's 90 regions across 500 timepoints, PET's 500,000 voxels, and diffusion MRI's complex directional data—we employ a multistage approach that preserves critical information while dramatically reducing computational complexity.

The pipeline utilizes both linear methods (PCA, ICA, and CCA) and nonlinear approaches (t-SNE, autoencoders, and UMAP), each suited to capturing different aspects of the data structure. Our staged strategy—first reducing dimensionality within each modality, then aligning features across modalities, and finally performing joint reduction—proved essential for managing the curse of dimensionality while maintaining the ability to explain 65% of the variance in fatigue symptoms.

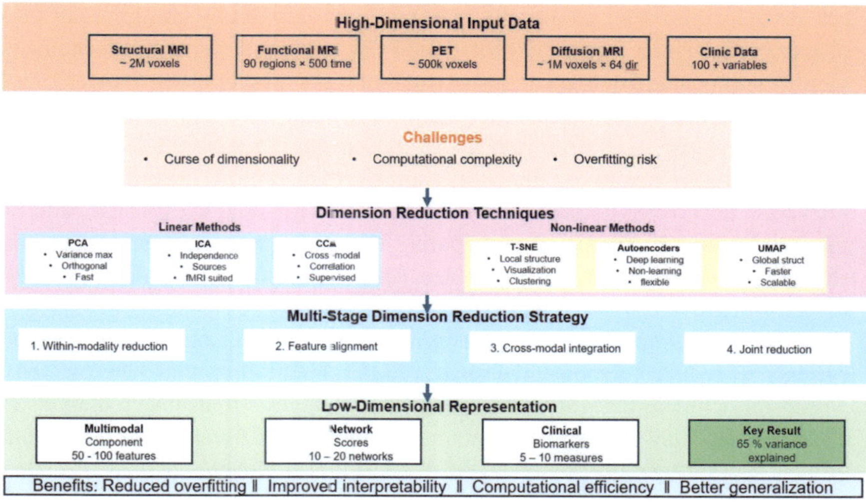

Fig. 2.3 Dimension reduction process in multimodal neuroimaging

This systematic dimension reduction not only made the analysis computationally feasible but also enhanced interpretability, transforming raw imaging data into 50–100 multimodal components, 10–20 network scores, and 5–10 clinical biomarkers that directly relate to Long COVID pathophysiology. The following sections detail specific techniques employed at each stage of this pipeline.

2.4.2 Methodological Insights and Lessons Learned

Research on Long COVID fatigue has yielded several methodological insights regarding multimodal neuroimaging analysis and dimension reduction:

1. *Value of multimodal integration*: The multimodal approach consistently outperformed any single modality in explaining fatigue symptoms. The multimodal component derived from linked ICA explained 65% of the variance in fatigue severity, compared to 52% for FDG-PET alone, 38% for functional connectivity, 32% for structural measures, and 30% for diffusion metrics. This finding underscores the complementary nature of different imaging modalities and the value of integrated analysis.
2. *Staged dimension reduction strategy*: The multi-stage approach—first reducing dimensionality within each modality before integrating across modalities—proved more effectively than either applying dimension reduction to concatenated raw data or analyzing each modality in isolation. This strategy managed the "curse of dimensionality" while preserving modality-specific information and enabling meaningful cross-modal integration.
3. *Importance of spatial regularization*: For neuroimaging data, incorporating spatial structure into dimension reduction improved both interpretability and predictive performance. Spatially regularized PCA and structured sparse techniques yielded components with clearer neuroanatomical interpretations than standard approaches, facilitating translation to clinical applications.
4. *Balance between supervised and unsupervised approaches*: While supervised methods directly linked to clinical measures provided more immediate clinical relevance, unsupervised dimension reduction often revealed unexpected patterns that might have been missed with purely supervised approaches. A balanced strategy incorporating both supervised and unsupervised techniques proved most informative.
5. *Heterogeneity within diagnostic categories*: The spread of Long COVID fatigue patients in reduced dimensional spaces highlighted substantial heterogeneity within this clinical category. This observation suggests that binary diagnostic classifications may obscure important neurobiological differences within patient groups, supporting more personalized, dimensional approaches to characterizing neurological disorders.
6. *Nonlinear associations across modalities*: While linear dimension reduction techniques (e.g., PCA and CCA) captured substantial cross-modal relationships, nonlinear approaches (e.g., t-SNE, UMAP, and VAEs) revealed additional pat-

terns, particularly in patient subgroups. This finding suggests that the associations between different aspects of brain structure and function may be complex and nonlinear, particularly in pathological states.

7. *Challenges of longitudinal analysis*: In a subset of patients followed longitudinally, we observed that dimension reduction techniques optimized for cross-sectional data sometimes failed to capture subtle changes over time. Specialized approaches for longitudinal dimension reduction, such as temporal ICA or recurrent neural networks, may be necessary to fully characterize disease progression and recovery.

These methodological insights have informed our subsequent research and may guide future applications of multimodal neuroimaging and dimension reduction in clinical neuroscience. The balance between dimensionality, interpretability, and predictive power remains a central challenge in this field, requiring careful consideration of the specific research questions and data characteristics in each application.

2.5 Future Directions in Multimodal Analysis for Clinical Applications

The application of multimodal neuroimaging and dimension reduction techniques to neurological disorders is a rapidly evolving field with substantial potential for clinical impact. Based on our research and the broader literature, several promising directions for future development emerge:

2.5.1 Integration of Neuroimaging with Other Data Types

While our focus has been on multimodal neuroimaging, the integration of neuroimaging with other data types holds tremendous promise for understanding neurological disorders more comprehensively:

Genomic and transcriptomic integration: Combining neuroimaging with genetic and gene expression data could illuminate the molecular mechanisms underlying observed brain abnormalities. For example, relating the Long COVID fatigue pattern to genetic variants affecting immune response or neurotransmitter systems might reveal subtypes with different underlying mechanisms and treatment responses. Dimension reduction approaches specifically designed for integrating imaging and genomic data, such as sparse canonical correlation analysis (sCCA) or multiview matrix factorization, could facilitate these analyses.

Wearable and mobile monitoring data: Continuous monitoring through wearables and smartphones can provide real-world measures of activity, sleep, autonomic function, and symptom fluctuations. Integrating these dynamic, ecologically valid measures with neuroimaging through time-aware dimension reduction could reveal how brain abnormalities manifest in daily life and respond to real-world stressors. In our Long COVID research, we have begun collecting wearable data to relate brain glucose metabolism to daily activity patterns and fatigue fluctuations.

Electronic health records and clinical data: Leveraging structured and unstructured data from electronic health records could provide rich contextual information and longitudinal clinical trajectories. Dimension reduction approaches that can handle mixed data types, missing values, and temporal dependencies would be particularly valuable for integrating these diverse data sources with neuroimaging.

Digital biomarkers: Computerized assessments of cognition, motor function, speech, and other domains can provide objective, sensitive measures of neurological function. These digital biomarkers could complement neuroimaging data, potentially serving as proxies for brain function in settings where imaging is impractical or unavailable. For example, in our Long COVID research, we found that smartphone-based assessments of processing speed correlated with frontal-striatal metabolism and could potentially serve as a remote monitoring tool.

2.5.2 Advanced AI Approaches for Multimodal Integration

Artificial intelligence, particularly deep learning, offers increasingly sophisticated approaches for multimodal integration and dimension reduction:

Graph neural networks (GNNs) provide a natural framework for modeling brain networks and their relationship to other data modalities [48]. By representing brain regions, connections, and various measures as nodes and edges in a graph, GNNs can learn complex relationships across modalities while respecting the inherent structure of brain data. For example, a heterogeneous graph could include nodes representing brain regions, with edges representing structural and functional connectivity, and additional node features representing metabolic activity, genetic information, or clinical variables.

Multimodal transformers adapt the transformer architecture, which has revolutionized natural language processing, to neuroimaging and other biomedical data. These models use attention mechanisms to capture relationships between different data types and across different spatial and temporal scales. The flexibility of transformer architectures makes them well suited for integrating heterogeneous data in clinical neuroscience.

Self-supervised learning approaches enable models to learn meaningful representations from unlabeled or partially labeled data, addressing the challenge of limited labeled samples in clinical studies [39]. Contrastive learning, in particular, has shown promise for neuroimaging applications, learning representations that capture clinically relevant features without requiring large, labeled datasets. These approaches could be especially valuable for rare conditions or understudied populations.

Federated learning permits model training across decentralized datasets without sharing raw data, addressing privacy concerns and enabling larger, more diverse studies. This approach could facilitate multisite neuroimaging studies of neurological disorders, pooling data across institutions while respecting privacy regulations and data ownership. For conditions like Long COVID, where collaborative research is essential due to the novelty and global impact, federated learning could accelerate scientific progress while protecting patient privacy.

2.5.3 Clinical Translation and Implementation

Translating the insights gained from multimodal neuroimaging and dimension reduction to clinical practice requires addressing several challenges:

Standardization and harmonization of imaging protocols, preprocessing pipelines, and analytical approaches are essential for reproducible, generalizable findings [25]. Initiatives like the Brain Imaging Data Structure (BIDS) and various harmonization techniques aim to address these challenges, but further work is needed to establish standards for multimodal integration and dimension reduction specifically.

Interpretable AI approaches that provide clear explanations for their predictions are crucial for clinical acceptance and regulatory approval. Methods that balance predictive performance with interpretability, such as attention visualization, concept-based explanations, or inherently interpretable models, will be particularly valuable for clinical applications of multimodal neuroimaging.

Clinical decision support systems that integrate multimodal neuroimaging findings into existing clinical workflows could facilitate adoption by healthcare providers. These systems could range from simple visualization tools that highlight relevant brain patterns to sophisticated AI-driven platforms that provide diagnostic and treatment recommendations based on integrated multimodal data.

Accessible and scalable approaches that do not require extensive technical expertise or computational resources would broaden the impact of multimodal neuroimaging. Cloud-based platforms, automated preprocessing and analysis pipelines, and simplified interfaces could make these advanced techniques accessible to a wider range of clinicians and researchers, particularly in resource-limited settings.

2.5.4 Emerging Applications in Neurological Disorders

Beyond Long COVID and multiple sclerosis, multimodal neuroimaging and dimension reduction approaches hold promise for various neurological disorders:

Neurodegenerative diseases like Alzheimer's and Parkinson's involve complex, multifaceted pathological processes that affect multiple aspects of brain structure and function [42]. Multimodal approaches could provide more comprehensive characterization of disease subtypes, progression trajectories, and treatment responses, potentially enabling earlier diagnosis and more personalized interventions.

Psychiatric disorders such as depression, schizophrenia, and bipolar disorder have historically been defined by symptoms rather than underlying neurobiology. Multimodal neuroimaging combined with dimension reduction could help identify biologically defined subtypes within these heterogeneous categories, potentially leading to more targeted treatments and improved outcomes.

Traumatic brain injury (TBI) produces complex, heterogeneous patterns of damage that can be difficult to characterize with any single imaging modality. Multimodal approaches integrating structural, functional, and molecular imaging could provide a more comprehensive assessment of injury severity, recovery potential, and treatment needs, informing personalized rehabilitation strategies.

Developmental disorders like autism spectrum disorder involve atypical brain development that affects multiple neural systems and processes. Multimodal approaches could help characterize developmental trajectories, identify predictive biomarkers for early intervention, and monitor response to therapy, potentially improving long-term outcomes for affected individuals.

2.5.5 Personal Research Agenda

Building on our work in Long COVID and multiple sclerosis, future research agenda for advancing multimodal analysis in clinical neuroscience includes several key directions:

Longitudinal multimodal imaging of Long COVID recovery: By following patients over time with repeated multimodal imaging, we aim to characterize neural recovery trajectories, identify predictors of persistent symptoms, and monitor response to various interventions. This longitudinal approach requires specialized dimension reduction techniques that can capture temporal dynamics while integrating across modalities.

Transdiagnostic approaches to fatigue and cognitive dysfunction: Comparing multimodal neuroimaging patterns across conditions characterized by fatigue and cognitive symptoms—including Long COVID, multiple sclerosis, chronic fatigue syndrome, and depression—could reveal common and distinct mechanisms, potentially leading to more targeted interventions. Dimension reduction approaches that can identify shared and condition-specific components across patient groups will be particularly valuable for this transdiagnostic perspective.

Neuromodulation targeting based on multimodal signatures: Using multimodal neuroimaging to guide neuromodulation interventions such as transcranial direct current stimulation (tDCS) or transcranial magnetic stimulation (TMS) could improve treatment efficacy for symptoms like fatigue and cognitive dysfunction. By targeting stimulation based on individual multimodal patterns rather than standard anatomical locations, we may achieve more consistent and substantial therapeutic effects.

Integration of neuroimaging with digital biomarkers: Developing models that relate multimodal neuroimaging patterns to digital biomarkers derived from smartphones, wearables, and computerized assessments could enable remote monitoring and personalized intervention without requiring repeated imaging. This approach could make precision neurology more accessible and scalable, particularly for conditions requiring long-term management.

Hippocampus-inspired AI for multimodal integration: Drawing inspiration from how the hippocampus integrates multisensory information and forms coherent representations, we are developing neuromorphic computing approaches specifically designed for multimodal neuroimaging data. These biologically informed models may offer advantages for capturing complex, cross-modal relationships in brain data, potentially leading to more interpretable and neurobiologically plausible representations.

These research directions aim to advance both the methodological foundations of multimodal neuroimaging analysis and its clinical applications, ultimately contributing to more precise diagnosis, prognosis, and treatment of neurological disorders.

2.6 Conclusion

Multimodal analysis of neuroimaging data represents a powerful approach for characterizing the complex neural underpinnings of neurological disorders. By integrating complementary information across different imaging modalities and employing sophisticated dimension reduction techniques to manage high-dimensional data, we can obtain more comprehensive and nuanced understandings of conditions like Long COVID fatigue, multiple sclerosis, and other neurological disorders.

Our research has demonstrated the value of this approach, revealing a multimodal neural signature of Long COVID fatigue that spans metabolism, structure, connectivity, and white matter integrity. This signature, centered on frontal-striatal-thalamic circuits, provides insights into the mechanisms underlying fatigue symptoms and suggests potential targets for therapeutic intervention. The dimension reduction techniques employed in this work—from modality-specific approaches like spatially regularized PCA to joint multimodal methods like linked ICA and deep learning—enabled the extraction of meaningful patterns from complex, high-dimensional data, highlighting relationships that might have remained obscure through conventional analyses.

Looking ahead, the integration of neuroimaging with other data types, the development of advanced AI approaches for multimodal analysis, and the translation of these methods to clinical practice hold tremendous promise for advancing precision neurology. By characterizing individual patients in terms of their specific neural signatures rather than broad diagnostic categories, we may eventually achieve more personalized, effective approaches to diagnosing and treating neurological disorders.

As we continue to develop and refine these approaches, maintaining a balance between methodological innovation and clinical relevance will be essential. The ultimate goal is not merely to extract complex patterns from neuroimaging data but also to translate these patterns into meaningful insights that improve patient care and outcomes. By bridging the gap between advanced computational methods and clinical neuroscience, multimodal neuroimaging analysis and dimension reduction techniques can contribute to a more precise, personalized understanding of the brain in health and disease.

2.7 Summary

This chapter examines how multimodal neuroimaging analysis, enhanced by artificial intelligence, advances our understanding and diagnosis of neurological disorders. This chapter covers the technical foundations of integrating diverse imaging modalities (MRI, fMRI, PET, DTI, and EEG/MEG), demonstrating how each provides complementary information about brain structure and function.

This chapter explains how AI approaches overcome key limitations of traditional neuroimaging analysis, including the curse of dimensionality, group-level versus individual analysis, and the challenge of capturing complex, distributed brain patterns. Machine learning methods enable automatic feature learning, personalized predictions, and the modeling of temporal brain dynamics while improving interpretability through explainable AI techniques.

A significant focus is placed on dimension reduction techniques essential for managing high-dimensional neuroimaging data. This chapter presents a systematic pipeline that transforms millions of voxels into clinically meaningful biomarkers, illustrated through a Long COVID fatigue case study that achieved 65% variance explanation in symptoms—a 25% improvement over single-modality approaches.

Key findings include the identification of frontal-striatal-thalamic network dysfunction in Long COVID fatigue, with 92% classification accuracy demonstrating the clinical potential of multimodal integration. This chapter concludes by exploring future directions, including integration with genomic data, wearable monitoring, and advanced AI architectures like graph neural networks and multimodal transformers.

The central message emphasizes that multimodal neuroimaging analysis represents a paradigm shift from isolated brain measurements toward integrated models that capture the multifaceted nature of neurological disorders, ultimately enabling more personalized and effective diagnostic and treatment approaches.

References

1. Herculano-Houzel S. The human brain in numbers: a linearly scaled-up primate brain. Front Hum Neurosci. 2009;3:31.
2. Wardlaw JM, Smith EE, Biessels GJ, Cordonnier C, Fazekas F, Frayne R, et al. Neuroimaging standards for research into small vessel disease and its contribution to ageing and neurodegeneration. Lancet Neurol. 2013;12(8):822–38.
3. Ogawa S, Lee TM, Kay AR, Tank DW. Brain magnetic resonance imaging with contrast dependent on blood oxygenation. Proc Natl Acad Sci. 1990;87(24):9868–72.
4. Basser PJ, Mattiello J, LeBihan D. MR diffusion tensor spectroscopy and imaging. Biophys J. 1994;66(1):259–67.
5. Phelps ME. Positron emission tomography provides molecular imaging of biological processes. Proc Natl Acad Sci. 2000;97(16):9226–33.
6. Hari R, Puce A. MEG-EEG primer. Oxford University Press; 2017.
7. Ferrari M, Quaresima V. A brief review on the history of human functional near-infrared spectroscopy (fNIRS) development and fields of application. NeuroImage. 2012;63(2):921–35.
8. Catana C. Principles of simultaneous PET/MR imaging. Magn Reson Imaging Clin. 2015;23(2):187–200.
9. Delso G, Fürst S, Jakoby B, Ladebeck R, Ganter C, Nekolla SG, Ziegler SI. Performance measurements of the Siemens mMR integrated whole-body PET/MR scanner. J Nucl Med. 2011;52(12):1914–22.
10. Deco G, Jirsa VK, McIntosh AR. Emerging concepts for the dynamical organization of resting-state activity in the brain. Nat Rev Neurosci. 2011;12(1):43–56.
11. Friston KJ, Harrison L, Penny W. Dynamic causal modelling. NeuroImage. 2003;19(4):1273–302.
12. Calhoun VD, Liu J, Adalı T. A review of group ICA for fMRI data and ICA for joint inference of imaging, genetic, and ERP data. NeuroImage. 2009;45(1):S163–72.
13. Penny WD, Friston KJ, Ashburner JT, Kiebel SJ, Nichols TE, editors. Statistical parametric mapping: the analysis of functional brain images. Elsevier; 2011.
14. Zhang D, Wang Y, Zhou L, Yuan H, Shen D. Multimodal classification of Alzheimer's disease and mild cognitive impairment. NeuroImage. 2011;55(3):856–67.
15. Plis SM, Hjelm DR, Salakhutdinov R, Allen EA, Bockholt HJ, Long JD, Calhoun VD. Deep learning for neuroimaging: a validation study. Front Neurosci. 2014;8:229.
16. Suk HI, Lee SW, Shen D, Alzheimer's Disease Neuroimaging Initiative. Hierarchical feature representation and multimodal fusion with deep learning for AD/MCI diagnosis. Neuroimage. 2014;101:569–82.
17. Nie D, Zhang H, Adeli E, Liu L, Shen D. 3D deep learning for multi-modal imaging-guided survival time prediction of brain tumor patients. In: International conference on medical image computing and computer-assisted intervention. Springer; 2016. p. 212–20.
18. Kohannim O, Hua X, Hibar DP, Lee S, Yanovsky I, Leporé N, Thompson PM. Boosting power for clinical trials using classifiers based on multiple biomarkers. Neurobiol Aging. 2010;31(8):1429–42.
19. Ngiam J, Khosla A, Kim M, Nam J, Lee H, Ng AY. Multimodal deep learning. In: Proceedings of the 28th international conference on machine learning. Madison: Omnipress; 2011. p. 689–96.

20. Wang W, Arora R, Livescu K, Bilmes J. On deep multi-view representation learning. In: International conference on machine learning. PMLR; 2015. p. 1083–92.
21. Ktena SI, Parisot S, Ferrante E, Rajchl M, Lee M, Glocker B, Rueckert D. Metric learning with spectral graph convolutions on brain connectivity networks. NeuroImage. 2018;169:431–42.
22. Avants BB, Epstein CL, Grossman M, Gee JC. Symmetric diffeomorphic image registration with cross-correlation: evaluating automated labeling of elderly and neurodegenerative brain. Med Image Anal. 2008;12(1):26–41.
23. Jorge J, van der Zwaag W, Figueiredo P. EEG–fMRI integration for the study of human brain function. NeuroImage. 2014;102:24–34.
24. Johnson WE, Li C, Rabinovic A. Adjusting batch effects in microarray expression data using empirical Bayes methods. Biostatistics. 2007;8(1):118–27.
25. Fortin JP, Parker D, Tunç B, Watanabe T, Elliott MA, Ruparel K, Shinohara RT. Harmonization of multi-site diffusion tensor imaging data. NeuroImage. 2017;161:149–70.
26. Zhu JY, Park T, Isola P, Efros AA. Unpaired image-to-image translation using cycle-consistent adversarial networks. In: Proceedings of the IEEE international conference on computer vision. IEEE; 2017. p. 2223–32.
27. Fischl B. FreeSurfer. Neuroimage. 2012;62(2):774–81.
28. Panizzon MS, Fennema-Notestine C, Eyler LT, Jernigan TL, Prom-Wormley E, Neale M, et al. Distinct genetic influences on cortical surface area and cortical thickness. Cereb Cortex. 2009;19(11):2728–35.
29. Biswal B, Zerrin Yetkin F, Haughton VM, Hyde JS. Functional connectivity in the motor cortex of resting human brain using echo-planar MRI. Magn Reson Med. 1995;34(4):537–41.
30. Tournier JD, Mori S, Leemans A. Diffusion tensor imaging and beyond. Magn Reson Med. 2011;65(6):1532–56.
31. Rubinov M, Sporns O. Complex network measures of brain connectivity: uses and interpretations. NeuroImage. 2010;52(3):1059–69.
32. Raichle ME, Mintun MA. Brain work and brain imaging. Annu Rev Neurosci. 2006;29:449–76.
33. Chalah MA, Riachi N, Ahdab R, Créange A, Lefaucheur JP, Ayache SS. Fatigue in multiple sclerosis: neural correlates and the role of non-invasive brain stimulation. Front Cell Neurosci. 2015;9:460.
34. Cohen MX. Analyzing neural time series data: theory and practice. MIT Press; 2014.
35. Bellman R. Dynamic programming. Princeton University Press; 1957.
36. LeCun Y, Bengio Y, Hinton G. Deep learning. Nature. 2015;521(7553):436–44.
37. Wen J, Thibeau-Sutre E, Diaz-Melo M, Samper-González J, Routier A, Bottani S, Alzheimer's Disease Neuroimaging Initiative. Convolutional neural networks for classification of Alzheimer's disease: overview and reproducible evaluation. Med Image Anal. 2020;63:101694.
38. Li X, Morgan PS, Ashburner J, Smith J, Rorden C. The first step for neuroimaging data analysis: DICOM to NIfTI conversion. J Neurosci Methods. 2016;264:47–56.
39. Taleb A, Loetzsch W, Danz N, Severin J, Gaertner T, Bergner B, Lippert C. 3D self-supervised methods for medical imaging. Adv Neural Inf Process Syst. 2020;33:18158–72.
40. Arbabshirani MR, Plis S, Sui J, Calhoun VD. Single subject prediction of brain disorders in neuroimaging: promises and pitfalls. NeuroImage. 2017;145:137–65.
41. Schnack HG, Kahn RS. Detecting neuroimaging biomarkers for psychiatric disorders: sample size matters. Front Psych. 2016;7:50.
42. Jack CR Jr, Bennett DA, Blennow K, Carrillo MC, Dunn B, Haeberlein SB, et al. NIA-AA research framework: toward a biological definition of Alzheimer's disease. Alzheimers Dement. 2018;14(4):535–62.
43. Cole JH, Poudel RP, Tsagkrasoulis D, Caan MW, Steves C, Spector TD, Montana G. Predicting brain age with deep learning from raw imaging data results in a reliable and heritable biomarker. NeuroImage. 2017;163:115–24.

44. Drysdale AT, Grosenick L, Downar J, Dunlop K, Mansouri F, Meng Y, et al. Resting-state connectivity biomarkers define neurophysiological subtypes of depression. Nat Med. 2017;23(1):28–38.
45. Pereira F, Mitchell T, Botvinick M. Machine learning classifiers and fMRI: a tutorial overview. NeuroImage. 2009;45(1):S199–209.
46. Wottschel V, Alexander DC, Kwok PP, Chard DT, Stromillo ML, De Stefano N, Ciccarelli O. Predicting outcome in clinically isolated syndrome using machine learning. NeuroImage Clin. 2015;7:281–7.
47. Payan A, Montana G. Predicting Alzheimer's disease: a neuroimaging study with 3D convolutional neural networks. arXiv:1502.02506 [Preprint]. 2015.
48. Zhou J, Cui G, Hu S, Zhang Z, Yang C, Liu Z, et al. Graph neural networks: a review of methods and applications. AI Open. 2020;1:57–81.
49. Kingma DP, Welling M. Auto-encoding variational bayes. arXiv:1312.6114 [Preprint]. 2013.
50. Biffi C, Oktay O, Tarroni G, Bai W, De Marvao A, Doumou G, et al. Learning interpretable anatomical features through deep generative models: application to cardiac remodeling. In: International conference on medical image computing and computer-assisted intervention. Springer; 2018. p. 464–71.
51. Chen T, Kornblith S, Norouzi M, Hinton G. A simple framework for contrastive learning of visual representations. In: International conference on machine learning. PMLR; 2020. p. 1597–607.
52. Hochreiter S, Schmidhuber J. Long short-term memory. Neural Comput. 1997;9(8):1735–80.

Biomimetic AI and the Brain

3

The brain is wider than the sky,For, put them side by side,The one the other will includeWith ease, and you beside.—Emily Dickinson

3.1 Introduction

The human brain remains nature's most sophisticated information processing system. This remarkably efficient organ consumes merely 20 watts of power while performing complex computations that still surpass artificial systems in many domains [1]. Despite tremendous advances in artificial intelligence, our most sophisticated AI models face significant efficiency challenges. They require orders of magnitude more energy and training examples to achieve what humans can accomplish with minimal exposure and energy consumption. This disparity has inspired researchers to look more closely at the brain's architecture and operational principles to inform the development of more efficient, adaptive, and robust artificial systems—an approach often termed "biomimetic AI."

Biomimetic AI refers to artificial intelligence systems that draw inspiration from biological neural systems, incorporating principles and mechanisms observed in the brain to enhance their functionality and efficiency. This approach stands in contrast to purely engineering-driven methods that may achieve impressive results but often through brute-force computation rather than the elegant efficiency of biological systems. The field represents a bidirectional relationship between neuroscience and AI: neuroscientific discoveries inform AI development, while advances in AI provide new tools and frameworks for understanding the brain.

Figure 3.1 illustrates the core concept of this chapter—the bidirectional flow of insights between biological brains and artificial intelligence systems. The left panel represents key features of biological neural systems, including neuroplasticity, learning mechanisms, and memory systems. The right panel shows corresponding

T. Rudroff, *AI for the Healthy Brain and Neurological Disorders*, Neurocultural Health and Wellbeing, https://doi.org/10.1007/978-3-032-11596-6_3

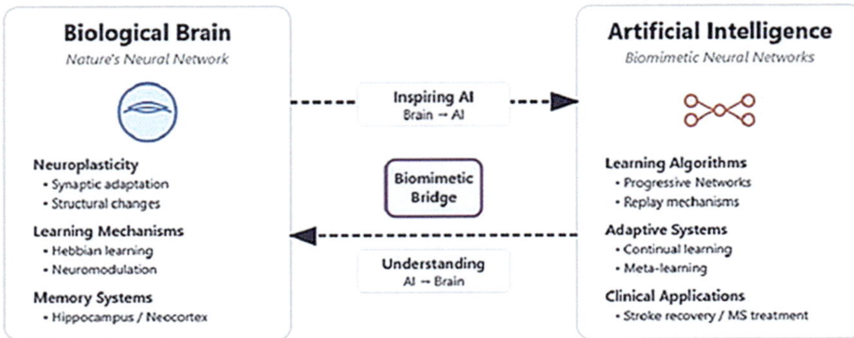

Fig. 3.1 Biomimetic bridge between neuroscience and artificial intelligence

AI implementations, including learning algorithms, adaptive systems, and clinical applications. The central "biomimetic bridge" emphasizes how these two domains inform and enhance each other, with arrows indicating the flow of inspiration from brain to AI and the flow of understanding from computational models back to neuroscience.

One of the most fundamental challenges in both biological and artificial learning systems is the stability-plasticity dilemma: how to remain adaptable enough to acquire new information while maintaining the stability needed to preserve existing knowledge [2]. The human brain has evolved sophisticated mechanisms to balance these competing demands, employing various forms of neuroplasticity that operate across different time scales and neural structures. Understanding and implementing these mechanisms in artificial systems represents a significant opportunity for advancing AI capabilities.

This chapter explores the convergence of neuroscience and artificial intelligence through the lens of biomimetic approaches. We begin by examining the stability-plasticity dilemma that faces both biological and artificial neural networks, discussing how the brain solves this challenge and how similar solutions might be implemented in AI systems. We then delve into specific principles of neuroplasticity that have inspired machine learning algorithms, including Hebbian learning, homeostatic plasticity, and neuromodulation. Next, we consider how broader insights from brain organization and function are influencing the next generation of AI architectures and approaches. Finally, we present a case study of how biomimetic AI approaches are being applied to understand and treat disorders characterized by neuroplasticity impairments, such as multiple sclerosis and poststroke conditions.

Throughout this chapter, we highlight how biomimetic approaches not only advance AI capabilities but also deepen our understanding of the brain itself, creating a virtuous cycle of discovery and innovation that benefits both fields. As we continue to unravel the brain's mysteries and refine our artificial systems, the gap between biological and artificial intelligence may narrow, potentially leading to more efficient, adaptable, and human-like AI technologies that can address complex real-world problems.

3.2 The Stability-Plasticity Dilemma in Both Biological and Artificial Neural Networks

3.2.1 Understanding the Dilemma

The stability-plasticity dilemma represents a fundamental challenge for any learning system, whether biological or artificial: how to remain sufficiently plastic to incorporate new information while maintaining enough stability to preserve previously acquired knowledge [2, 3]. This challenge manifests across multiple levels of organization and time scales in the brain and presents similar difficulties in artificial neural networks.

In technical terms, excessive plasticity leads to catastrophic forgetting or interference, where new learning erases or disrupts previously acquired information [4]. Conversely, excessive stability results in a reduced capacity to learn from new experiences or adapt to changing environments. This dilemma is not merely a technical challenge but reflects a fundamental trade-off between adaptability and reliability that any sophisticated learning system must navigate.

The stability-plasticity dilemma becomes particularly acute in continual or lifelong learning scenarios, where systems must acquire knowledge sequentially rather than simultaneously. The human brain excels at this type of learning, seamlessly integrating new information without the wholesale disruption of existing knowledge. In contrast, many artificial neural networks struggle when trained sequentially on different tasks, often exhibiting catastrophic forgetting as new learning overwrites previously learned parameters [5].

Figure 3.2 illustrates this fundamental trade-off between stability and plasticity in learning systems (adapted from [6]). As shown in the figure, the relationship between these two properties follows a Pareto frontier, where improvements in one dimension necessarily lead to compromises in the other. The inflection point on this curve represents a critical balance, the point where the trade-off dynamics shift from gradual to more pronounced changes. This visualization underscores why achieving optimal learning requires carefully navigating between excessive stability (which prevents new learning) and excessive plasticity (which leads to forgetting). The challenge for both biological and artificial systems lies in finding the optimal position along this frontier that enables the rapid acquisition of new information while preserving existing knowledge.

3.2.2 The Brain's Solutions to the Stability-Plasticity Dilemma

The brain employs multiple complementary mechanisms to address the stability-plasticity dilemma, operating across different spatial and temporal scales [7]. These mechanisms collectively enable the remarkable balance between adaptability and stability that characterizes human cognition.

Complementary Learning Systems represent one of the brain's primary solutions to the stability-plasticity dilemma. According to this theory, the brain employs two

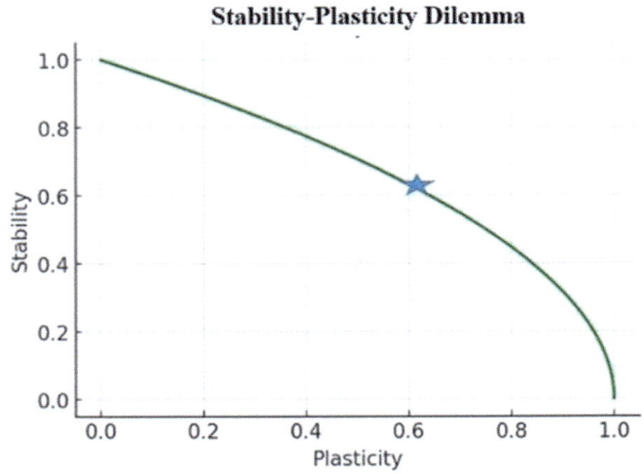

Fig. 3.2 Stability-plasticity trade-off in learning systems. The graph illustrates the fundamental trade-off between stability (*y*-axis) and plasticity (*x*-axis) in learning systems. The green curve represents the Pareto frontier, showing the optimal trade-off where increasing one capability necessarily reduces the other. The blue dot marks the inflection point, where the trade-off dynamics shift from gradual to more pronounced changes, representing a critical balance between stability and plasticity. (Adapted from Rudroff et al. [6])

distinct but complementary memory systems. The hippocampal system handles the rapid learning of specific experiences. The neocortical system gradually integrates these experiences into generalized knowledge structures [8]. The hippocampus can quickly form connections based on new experiences. This occurs without disrupting existing cortical representations. Memory consolidation then occurs through a gradual process. Hippocampal representations are integrated into neocortical networks during periods of offline processing, such as sleep [9].

Figure 3.3 illustrates how the complementary learning systems in the brain have inspired similar architectures in artificial intelligence. The biological system (top) shows the hippocampus acting as a fast learning system that rapidly encodes new experiences, which are then gradually consolidated into the neocortex through replay events during sleep or rest. This process allows the brain to quickly capture specific experiences while slowly integrating them into stable, long-term knowledge structures. The AI system (bottom) mirrors this architecture with an experience buffer that stores recent experiences for fast access and a neural network that slowly updates its parameters through experience replay during offline training. The timeline demonstrates how both systems progress from initial encoding through consolidation to stable knowledge integration.

Evidence for this dual-system approach comes from multiple sources, including patients with hippocampal damage who can no longer form new episodic memories but retain access to previously consolidated semantic knowledge [10]. Neuroimaging studies have further confirmed the differential engagement of hippocampal and neocortical structures during different phases of learning and memory.

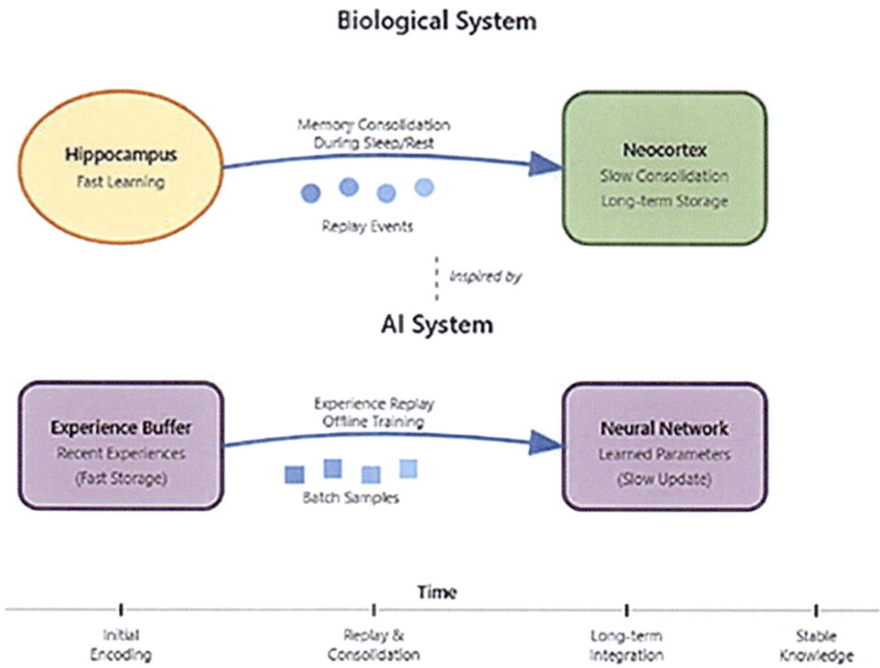

Fig. 3.3 Complementary learning systems

Synaptic Consolidation provides another mechanism for balancing stability and plasticity at the cellular level. After initial learning, synapses undergo a process of consolidation involving protein synthesis that stabilizes the changes. This leads to a temporary reduction in plasticity at specific synapses, protecting the newly formed memories from interference.

Metaplasticity refers to the "plasticity of synaptic plasticity," where the threshold for inducing subsequent plasticity is adjusted based on prior activity [11]. This mechanism ensures that neurons do not become too excitable or too resistant to change, maintaining a balanced capacity for learning while preventing runaway plasticity.

Structural Plasticity through dendritic spine formation and elimination offers another solution. Rather than modifying existing connections, the brain can form entirely new connections or eliminate unnecessary ones, providing an additional degree of freedom for learning without interference [12]. Long-term imaging studies have shown that while some dendritic spines are highly stable and persist for months or even years, others are more dynamic, turning over within days or weeks. This combination of stable and dynamic elements provides a physical substrate for balancing stability and plasticity.

3.2.3 Stability-Plasticity Challenges in Artificial Neural Networks

Artificial neural networks face stability-plasticity challenges that parallel those observed in biological systems although often with different manifestations due to their distinct architectures and learning mechanisms.

Catastrophic Forgetting represents the most direct manifestation of the stability-plasticity dilemma in artificial neural networks. When trained sequentially on multiple tasks, many neural networks exhibit dramatic forgetting of previously learned tasks as the parameters are adjusted to optimize performance on new ones [4]. This occurs because most artificial neural networks use shared parameters for all tasks, and updates to these parameters to improve performance on one task can inadvertently disrupt performance on previously learned tasks.

3.2.4 Biomimetic Approaches to the Stability-Plasticity Dilemma

Drawing inspiration from biological solutions, researchers have developed various approaches to address the stability-plasticity dilemma in artificial neural networks.

Complementary Learning Systems Implementations have been developed that mimic the hippocampal-neocortical distinction observed in the brain. During learning, experiences are first encoded in the hippocampal network and then gradually transferred to the cortical network through a process resembling system consolidation, often implemented as replay of previous experiences interleaved with new learning.

The *Elastic Weight Consolidation (EWC)* algorithm represents another biologically inspired approach, drawing on the concept of synaptic consolidation [5]. EWC identifies the parameters that are most important for previously learned tasks and selectively reduces their plasticity when learning new tasks. This is achieved by modifying the loss function to include a penalty term for changes to important parameters, weighted by their importance for previous tasks. This approach has shown promising results in reducing catastrophic forgetting while allowing for the acquisition of new skills.

Progressive Neural Networks add new columns of units for each new task while maintaining lateral connections to previously trained columns, allowing new tasks to leverage representations from earlier tasks without modifying them [13]. Similarly, other approaches selectively expand network capacity when necessary to accommodate new tasks while maintaining performance on previous ones.

By drawing inspiration from the brain's solutions to the stability-plasticity dilemma, these biomimetic approaches have significantly advanced the field of continual learning in artificial neural networks. However, substantial challenges remain, and the gap between biological and artificial systems in terms of lifelong learning capabilities remains significant. Further integration of neuroscientific insights with machine learning innovations promises to yield even more effective solutions to this fundamental dilemma.

3.3 Neuroplasticity Principles Applied to Machine Learning Algorithm Design

Neuroplasticity represents a cornerstone of learning and adaptation in biological systems. This refers to the brain's ability to reorganize itself by forming new neural connections throughout life. Various forms of neuroplasticity operate at different temporal and spatial scales. These range from rapid changes in synaptic efficacy to long-term structural remodeling of neural circuits. These mechanisms have inspired numerous approaches in machine learning algorithm design, seeking to capture the efficiency and adaptability of biological learning systems.

Figure 3.4 provides an overview of how key neuroplasticity mechanisms have been translated into machine learning algorithms. This mapping illustrates the direct relationships between biological learning principles and their artificial intelligence implementations. As shown in the figure, fundamental neuroplasticity mechanisms—Hebbian learning, spike-timing-dependent plasticity (STDP), homeostatic plasticity, and neuromodulation—have each inspired specific computational approaches in modern AI systems. These biological-to-artificial correspondences form the foundation for the detailed discussions that follow in this section.

3.3.1 Hebbian Learning and Its Variants

Hebbian learning, encapsulated in the phrase "neurons that fire together, wire together," represents one of the earliest and most influential neuroplasticity principles applied to machine learning [14]. This principle suggests that synaptic

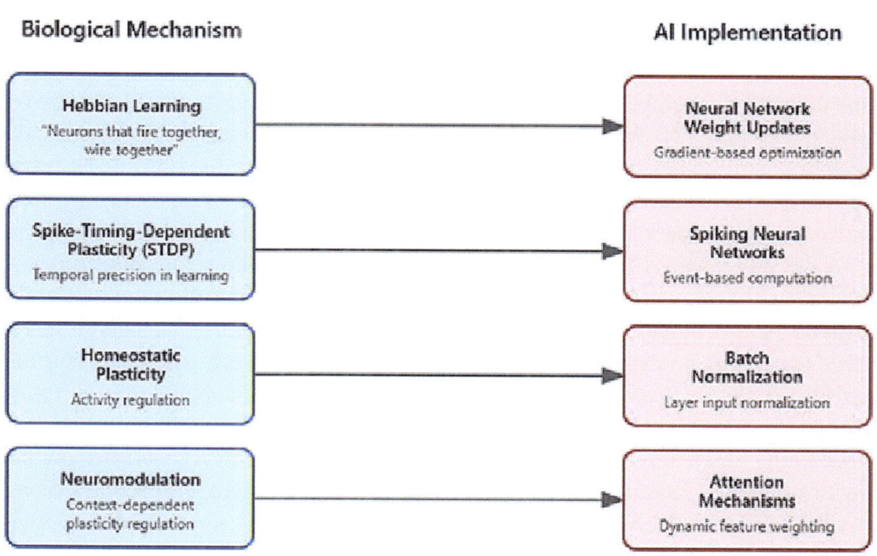

Fig. 3.4 How neuroplasticity mechanisms inspire AI

connections are strengthened when pre- and postsynaptic neurons are simultaneously active, providing a local, unsupervised learning rule that can extract statistical regularities from input patterns.

In artificial neural networks, Hebbian learning has been implemented through various algorithms, including basic Hebbian learning rules, self-organizing maps (SOMs), and competitive learning approaches [15]. These methods share the fundamental principle of strengthening connections between units that are simultaneously active, allowing networks to develop representations that capture the statistical structure of their inputs without explicit supervision.

However, pure Hebbian learning suffers from inherent instability, as it creates a positive feedback loop where stronger connections lead to increased correlation in activation, which further strengthens connections. This can result in runaway synaptic strengths and pathological network activity. Biological systems address this issue through various regulatory mechanisms, including synaptic scaling and metaplasticity.

Machine learning implementations have incorporated similar regulatory mechanisms to address the limitations of pure Hebbian learning. The Oja rule, for example, introduces a normalization term that prevents the unbounded growth of synaptic weights while enabling the extraction of principal components from input data [16].

More recent approaches have extended these principles to deep learning architectures. Local learning rules inspired by Hebbian plasticity have been developed for training deep neural networks, offering alternatives to backpropagation that better align with the local nature of biological learning.

3.3.2 Spike-Timing-Dependent Plasticity (STDP)

Spike-Timing-Dependent Plasticity (STDP) represents a temporally precise form of Hebbian learning observed in biological neurons, where the direction and magnitude of synaptic weight changes depend on the precise timing of pre- and postsynaptic spikes [17, 18]. When a presynaptic spike precedes a postsynaptic spike, the synapse is strengthened (long-term potentiation, LTP); when the order is reversed, the synapse is weakened (long-term depression, LTD). The magnitude of these changes typically decreases with increasing temporal separation between spikes.

STDP has inspired the development of learning algorithms for spiking neural networks (SNNs), which model neural dynamics more realistically than traditional artificial neural networks by explicitly representing the timing of neural spikes rather than using continuous-valued activations. These networks can process information efficiently using sparse, temporally coded signals, potentially offering advantages in terms of energy efficiency and the processing of temporal data.

Learning algorithms based on STDP have been applied to various tasks, including unsupervised feature learning, pattern recognition, and temporal sequence processing. For example, STDP-based learning has been used to train SNNs for image classification tasks, achieving competitive performance while maintaining biological plausibility and potential energy efficiency advantages.

Recent work has explored the integration of STDP-based learning with deep learning architectures, developing hybrid approaches that combine the temporal precision and biological plausibility of STDP with the powerful representational capabilities of deep networks. These approaches hold promise for neuromorphic computing applications, where energy efficiency and real-time processing of dynamic data are crucial.

3.3.3 Homeostatic Plasticity

Homeostatic plasticity encompasses various mechanisms that maintain neural activity within functional ranges, preventing either excessive excitation or quiescence [19]. These include synaptic scaling, where the strength of all synapses onto a neuron is globally adjusted to maintain a target activity level, and intrinsic plasticity, which modifies the inherent excitability of neurons based on their activity history.

In artificial neural networks, homeostatic mechanisms have been implemented to address issues of stability and convergence. Batch normalization, for example, normalizes the activations of neural network layers to have zero mean and unit variance, improving training stability and convergence [20]. This parallels aspects of synaptic scaling by preventing runaway activations and maintaining activity within a functional range.

More explicitly, biomimetic approaches have directly implemented homeostatic regulatory mechanisms in artificial neural networks. For example, researchers have developed learning algorithms that combine Hebbian plasticity with synaptic scaling and intrinsic plasticity to maintain network stability during continual learning [7]. This approach showed improved resistance to catastrophic forgetting compared to networks without homeostatic regulation.

3.3.4 Neuromodulation

Neuromodulatory systems in the brain, including dopaminergic, cholinergic, and noradrenergic pathways, play crucial roles in regulating plasticity based on factors such as reward, novelty, and behavioral state [21]. These systems effectively gate plasticity, allowing learning to be selectively enhanced or suppressed based on behavioral relevance and environmental context.

Reinforcement learning algorithms draw inspiration from dopaminergic signaling in the brain, which encodes reward prediction errors that drive learning [22]. Temporal difference (TD) learning algorithms, in particular, closely parallel the activity patterns observed in dopaminergic neurons, computing the difference between expected and received rewards to update value representations.

More recent approaches have implemented explicit neuromodulatory signals in artificial neural networks, allowing for dynamic regulation of learning based on task demands or environmental factors. Meta-learning systems use neuromodulatory networks to regulate plasticity in target networks based on task performance, enabling rapid adaptation to new tasks or changing environments.

Attention mechanisms in modern neural networks also parallel aspects of cholinergic neuromodulation, which enhances the processing of behaviorally relevant stimuli while suppressing irrelevant information [23]. Self-attention mechanisms, in particular, allow networks to dynamically focus processing on relevant features or tokens, improving performance on tasks requiring integration of information across spatial or temporal contexts.

3.3.5 Structural Plasticity

Structural plasticity in the brain involves the formation, elimination, and reorganization of physical connections between neurons, providing a mechanism for long-term memory storage and network optimization [12]. This form of plasticity operates over longer time scales than synaptic weight changes and offers additional degrees of freedom for learning and adaptation.

In artificial neural networks, structural plasticity has inspired approaches that dynamically modify network architecture during learning. Pruning methods, for example, systematically remove unnecessary connections or units from trained networks, improving efficiency while maintaining performance. This parallels the synaptic pruning observed during development and learning in biological systems, where redundant or unused connections are eliminated to optimize network function.

Conversely, network growth algorithms add new units or connections during training to expand representational capacity as needed. The Progressive Neural Networks approach, for example, adds new columns of units for each new task while maintaining lateral connections to previously trained columns, allowing for forward transfer without catastrophic forgetting [13].

Neural architecture search (NAS) represents a more general approach to structural plasticity, using evolutionary algorithms or reinforcement learning to discover optimal network architectures for specific tasks. While not explicitly biomimetic, these approaches embody the principle of structural adaptation to task demands that characterize biological neural systems.

By incorporating these various neuroplasticity principles into machine learning algorithm design, researchers have developed artificial systems that better capture the efficiency, adaptability, and robustness of biological learning. While significant gaps remain between biological and artificial learning systems, these biomimetic approaches represent important steps toward more brain-like artificial intelligence that can learn continuously, adapt to new environments, and maintain performance across diverse tasks.

3.4 How Studying the Brain Is Influencing the Next Generation of AI

The association between neuroscience and artificial intelligence continues to evolve, with new discoveries about brain function inspiring novel AI architectures, algorithms, and approaches. This section explores how various aspects of brain organization and function are influencing the development of next-generation AI systems, moving beyond individual plasticity mechanisms to broader principles of neural computation and cognition.

3.4.1 Brain-Inspired Architectural Principles

The brain's hierarchical and modular organization has inspired architectural innovations in artificial neural networks. The cerebral cortex is organized into a hierarchy of processing stages. Lower levels extract simple features while higher levels combine these into increasingly complex and abstract representations [24]. This hierarchical processing has directly inspired the architecture of convolutional neural networks (CNNs), which employ a series of feature extraction layers of increasing abstraction [25].

More recent work has explored the brain's recurrent connectivity patterns. In these patterns, information flows not only forward through the hierarchy but also laterally within levels and backward from higher to lower areas. These recurrent connections enable contextual modulation of processing, integration of top-down expectations with bottom-up sensory data, and maintenance of information over time. In artificial systems, recurrent neural networks (RNNs) and their variants like Long Short-Term Memory (LSTM) networks implement recurrent connectivity to process sequential data and maintain context over time [26].

The modular organization of the brain, with specialized regions for different functions, has inspired modular neural network architectures that employ distinct subnetworks for different subtasks or data modalities. These approaches can improve learning efficiency, reduce interference between tasks, and enable more interpretable representations.

More fundamentally, the brain's sparse connectivity, where each neuron connects to only a small fraction of possible targets, has inspired sparse artificial architectures that reduce computational requirements while maintaining performance. Similarly, the distributed nature of neural representations, where information is encoded across populations of neurons rather than in individual cells, has informed approaches like vector symbolic architectures that use distributed representations to enable symbolic-like reasoning with neural systems.

3.4.2 Predictive Processing and Active Inference

The brain appears to operate as a prediction machine, continuously generating expectations about sensory inputs and updating its internal models based on prediction errors [27, 28]. This predictive processing framework provides a unifying account of perception, action, and learning, suggesting that the brain minimizes surprise by either updating its models to better predict sensory data or acting to make sensory data better match its predictions.

This framework has inspired developments in generative models and unsupervised learning. Variational autoencoders (VAEs), for example, learn to generate sensory data from latent variables while also learning an approximate posterior distribution over those latent variables given sensory data, mirroring the bidirectional inferential processes posited by predictive coding theories [29].

Active inference, which extends predictive processing to the domain of action, suggests that organisms act to confirm their predictions and minimize surprise. This framework has inspired new approaches to reinforcement learning that emphasize exploration driven by prediction errors and information gain rather than solely by reward maximization.

More generally, the predictive processing framework has contributed to the growing emphasis on world models in artificial intelligence—internal models that capture the dynamics of the environment and enable planning and counterfactual reasoning. These approaches align with the brain's apparent use of internal models for prediction, planning, and imagination, potentially enabling more human-like reasoning and decision-making in artificial systems.

3.4.3 Attention and Working Memory

The brain's attentional mechanisms, which selectively enhance the processing of relevant information while suppressing irrelevant distractions, have inspired attention mechanisms in artificial neural networks. Self-attention mechanisms, in particular, have revolutionized natural language processing and are increasingly applied across domains from computer vision to reinforcement learning [23]. These mechanisms enable networks to dynamically weight the relevance of different input elements, effectively focusing computational resources on the most informative parts of the input.

Working memory, the brain's system for temporarily maintaining and manipulating information, has inspired architectures that include explicit memory components. Memory Networks and Neural Turing Machines, for example, augment neural networks with external memory resources that can be written to and read from, enabling more complex reasoning and longer-term dependence than traditional recurrent architectures.

The gating mechanisms in LSTM networks also draw inspiration from working memory, implementing controlled read, write, and forget operations that parallel aspects of executive control over working memory contents [26]. More recent transformer architectures can be viewed as implementing a form of content-addressable memory through their attention mechanisms, where each position can attend to any other position based on content similarity [23].

3.4.4 Social Learning and Cultural Transmission

Humans learn not only from individual experience but also from observing and imitating others, enabling the cultural transmission of knowledge across generations [30]. This social learning capability has inspired approaches like imitation learning and learning from demonstration in artificial intelligence, where agents learn policies from expert demonstrations rather than through trial and error.

More broadly, the social dimensions of human learning have influenced developments in multiagent learning systems, where agents learn from and with each other in collaborative or competitive environments. These approaches can accelerate learning and enable the emergence of complex behaviors and social norms that would be difficult to develop through individual learning alone.

The cumulative nature of cultural knowledge, where each generation builds on the achievements of previous ones, has parallels in approaches like knowledge distillation and progressive neural networks, where knowledge from previously trained models is transferred to new models or tasks [13]. These approaches enable a form of artificial cultural evolution, where knowledge accumulates and is refined across successive generations of models.

3.4.5 Embodied and Situated Cognition

The brain evolved not as an abstract reasoning device but as a control system for embodied agents navigating complex environments [31]. This embodied perspective has influenced the development of robotics and reinforcement learning systems that learn through physical interaction with their environments rather than from disembodied datasets.

Embodied approaches emphasize the role of sensorimotor contingencies in shaping cognition, suggesting that intelligent behavior emerges from the dynamic interaction between brain, body, and environment rather than from abstract computation alone. This perspective has inspired the development of robots with compliant, biomimetic bodies that can exploit the physical dynamics of their environments rather than fighting against them.

The situated nature of cognition, where thinking is embedded in specific environmental and social contexts, has influenced context-aware AI systems that adapt their behavior based on environmental factors, user characteristics, or social settings. These approaches recognize that intelligence is not a context-free capability but is always expressed in particular situations and toward particular goals.

More broadly, the emphasis on embodiment has contributed to a shift from narrowly defined tasks toward more open-ended learning objectives that emphasize adaptation, exploration, and environmental mastery [32]. Curiosity-driven learning algorithms, for example, generate intrinsic rewards based on novelty or learning progress, encouraging agents to explore and learn about their environments beyond specific task objectives.

3.4.6 Emotional and Motivational Systems

The brain's emotional and motivational systems play crucial roles in learning, decision-making, and behavioral regulation [33]. These systems assign value to experiences, guide attention and memory formation, and energize or inhibit behavior based on anticipated outcomes and internal states.

In artificial intelligence, these insights have inspired the development of intrinsic motivation systems that generate rewards based on factors like novelty, surprise, or mastery rather than solely on external rewards [32]. These approaches enable more autonomous and open-ended learning, potentially addressing limitations of traditional reinforcement learning in sparse-reward environments.

Emotional appraisal mechanisms have been implemented in artificial systems to modulate decision-making based on contextual factors and anticipated outcomes. These approaches can enhance adaptivity by enabling more flexible responses to environmental challenges, paralleling the role of emotions in human decision-making.

3.4.7 Neurodevelopmental Principles

The brain develops through a carefully orchestrated process that combines genetic guidance with activity-dependent refinement, resulting in neural circuits exquisitely adapted to their environmental niches. This developmental perspective has inspired approaches in artificial intelligence that emphasize the importance of developmental trajectories and learning curricula.

Curriculum learning, where agents are exposed to progressively more complex tasks or environments, draws inspiration from the staged nature of cognitive development [34]. By starting with simpler problems and gradually increasing complexity, these approaches can improve learning efficiency and generalization, similar to how human cognitive development progresses through distinct stages of increasing sophistication.

3.4.8 Integrative Cognitive Architectures

While many AI approaches draw inspiration from specific aspects of brain function, integrative cognitive architectures attempt to capture broader principles of cognitive organization and function. These architectures aim to unify multiple cognitive capabilities—perception, memory, reasoning, planning, and learning—within cohesive computational frameworks.

ACT-R (adaptive control of thought-rational) represents one influential cognitive architecture that combines symbolic rule-based processing with subsymbolic activation dynamics inspired by neural networks [35]. This hybrid approach enables both structured knowledge representation and graded, context-sensitive processing, addressing limitations of purely symbolic or purely connectionist approaches.

More recent approaches have developed comprehensive brain-inspired cognitive models [36]. By mapping computational requirements onto spiking neural networks with realistic constraints, these frameworks have enabled the development of some of the most comprehensive brain-inspired cognitive models to date.

3.4.9 Looking Forward: Challenges and Opportunities

Despite significant advances, substantial gaps remain between biological and artificial intelligence. Biological systems still outperform their artificial counterparts in terms of energy efficiency, generalization from limited data, robustness to perturbations, and integration across cognitive domains [37]. Addressing these gaps presents both challenges and opportunities for future research at the intersection of neuroscience and AI.

Energy efficiency represents one critical challenge. The human brain operates on approximately 20 watts of power while performing complex cognitive tasks, whereas current AI systems require orders of magnitude more energy for comparable performance. Neuromorphic computing approaches that implement brain-inspired processing on specialized hardware offer one promising direction for addressing this efficiency gap.

Generalization from limited data presents another challenge. Humans can learn new concepts from just a few examples, whereas current AI systems typically require thousands or millions of training examples to achieve comparable performance [37]. Meta-learning approaches inspired by human learning-to-learn capabilities offer one potential solution, enabling systems to rapidly adapt to new tasks based on prior experience.

Integration across cognitive domains—perception, memory, reasoning, language, and action—remains a significant challenge for artificial systems, which typically excel in narrow domains but struggle to transfer knowledge across tasks or modalities. Integrative cognitive architectures and multimodal learning approaches represent promising directions for addressing this challenge, potentially enabling more unified and flexible artificial cognition.

As neuroscience continues to advance our understanding of brain function and artificial intelligence develops increasingly sophisticated computational capabilities, the cross-fertilization between these fields holds tremendous promise for the next generation of intelligent systems. By drawing inspiration from the brain's architectural principles, learning mechanisms, and cognitive capabilities, researchers are developing artificial systems that better capture the efficiency, adaptability, and integrative nature of biological intelligence.

3.5 Case Study: Applications in Disorders with Neuroplasticity Impairments

Various neurological conditions are characterized by neuroplasticity impairments. These conditions include multiple sclerosis (MS), stroke, traumatic brain injury (TBI), and neurodegenerative diseases. These impairments can be manifested in several ways. They include reduced capacity for synaptic plasticity, aberrant rewiring, or insufficient compensatory mechanisms. These problems limit recovery and adaptation. Biomimetic AI approaches offer promising avenues for two key goals: understanding these impairments and developing targeted interventions. This section presents case studies of applications in multiple sclerosis and poststroke rehabilitation, illustrating how biomimetic AI can advance both our understanding of impaired neuroplasticity and our ability to address it therapeutically.

Figure 3.5 illustrates a closed-loop rehabilitation system that exemplifies the application of biomimetic AI principles in clinical practice. The system integrates the continuous monitoring of neural signals from patients with neurological impairments, real-time AI analysis using adaptive algorithms inspired by neuroplasticity mechanisms, and personalized therapeutic interventions. The closed-loop design enables continuous adaptation based on patient responses, mirroring the brain's own feedback mechanisms for learning and adaptation. This architecture demonstrates how theoretical concepts from earlier sections, including complementary learning systems, dynamic plasticity regulation, and homeostatic mechanisms, can be translated into practical clinical applications that enhance recovery outcomes.

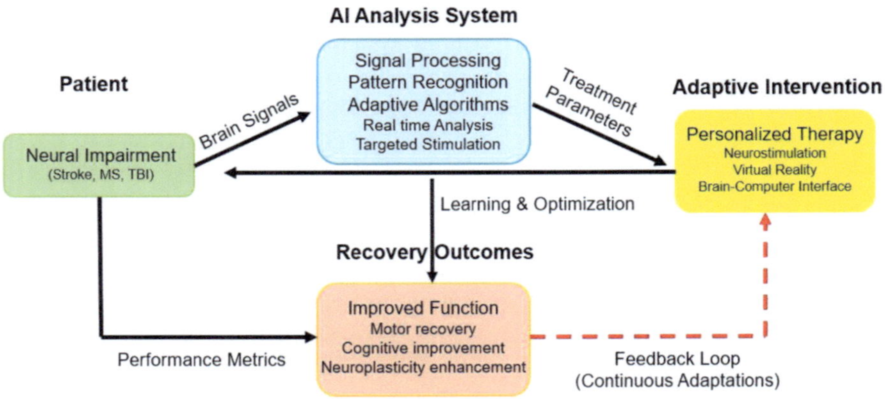

Fig. 3.5 Clinical application: closed-loop rehabilitation system

3.5.1 Multiple Sclerosis: Modeling Impaired Plasticity and Adaptation

Multiple sclerosis is a chronic autoimmune disease characterized by demyelination, axonal damage, and neurodegeneration, leading to progressive disability in many patients [38]. While MS has traditionally been viewed primarily as a demyelinating disease, growing evidence suggests that impaired neuroplasticity plays a significant role in disease progression and symptom manifestation [39].

Studies using transcranial magnetic stimulation (TMS) have demonstrated altered plasticity in MS patients, with impaired long-term potentiation (LTP)-like and long-term depression (LTD)-like effects following various stimulation protocols [40]. Functional MRI studies have further revealed altered patterns of brain activation and connectivity during cognitive and motor tasks, suggesting both impaired adaptive plasticity and potentially maladaptive reorganization [41].

Biomimetic computational models have been developed to understand these plasticity impairments and their relationships to MS symptoms. Predictive processing frameworks, which emphasize the brain's role in generating predictions and minimizing prediction errors, have been particularly fruitful in modeling MS-related cognitive and motor deficits [28]. These models conceptualize MS as disrupting the precision weighting of prediction errors, leading to the inefficient updating of internal models and consequent perceptual, cognitive, and motor deficits.

These computational approaches have informed the development of novel therapeutic strategies targeted at enhancing plasticity in MS. Transcranial electrical stimulation protocols based on computational models of cortical excitability have shown promise in enhancing motor learning and reducing fatigue in MS patients. Similarly, cognitive training programs designed based on predictive processing principles have demonstrated efficacy in improving attention and working memory in MS.

More directly, biomimetic AI approaches have been applied to develop personalized neurorehabilitation strategies for MS patients. For example, researchers have developed reinforcement learning-based systems that adaptively modify the difficulty of cognitive training tasks based on individual performance, mirroring aspects of intrinsic motivation and optimal challenge in biological learning [42]. This approach resulted in greater cognitive improvements compared to standard, non-adaptive training, suggesting that biomimetic learning principles can enhance rehabilitation outcomes.

Virtual reality (VR) systems incorporating biomimetic learning algorithms have been developed to address motor impairments in MS. These systems use computational models of motor learning and adaptation to create personalized training environments that progressively challenge patients while maintaining engagement and motivation. By incorporating principles of error-based learning, reward feedback, and multimodal sensory integration, these approaches aim to maximize plasticity induction and skill transfer to real-world activities.

Looking forward, the integration of multimodal neuroimaging, computational modeling, and AI-driven intervention design holds tremendous promise for personalized neurorehabilitation in MS. By characterizing individual patients' specific patterns of plasticity impairment and developing targeted interventions based on biomimetic learning principles, these approaches may enable more effective management of MS symptoms and potentially modification of disease progression.

3.5.2 Poststroke Rehabilitation: Enhancing Recovery Through Biomimetic Learning

Stroke represents a leading cause of long-term disability worldwide, with motor impairments being particularly common and debilitating sequelae [43]. Recovery from stroke depends critically on neuroplasticity—both spontaneous reorganization following brain injury and experience-dependent plasticity induced by rehabilitation [44]. However, the capacity for adaptive plasticity may be impaired after stroke due to factors including reduced neurotrophic support, altered excitatory-inhibitory balance, and structural damage to key network nodes [45].

Computational models of poststroke plasticity have provided important insights into recovery mechanisms and potential targets for intervention. These models span multiple levels of organization, from cellular and molecular processes of postinjury plasticity to network reorganization and behavioral adaptation [46].

At the cellular level, models of homeostatic plasticity have illuminated how changes in excitatory-inhibitory balance after stroke affect the potential for use-dependent plasticity during rehabilitation [47]. These models suggest that the critical period of heightened plasticity after stroke represents a window of both opportunity and vulnerability, where appropriate rehabilitation can enhance recovery while inappropriate experiences might induce maladaptive plasticity.

At the network level, models of structural and functional reorganization have captured how lesions affect global network properties and how compensatory rewiring may support recovery [48]. For example, connectome-based models have demonstrated how lesions to hub regions produce more severe and widespread network disruption than equivalent damage to peripheral nodes, potentially explaining the greater functional impact of strokes affecting certain brain regions.

Reinforcement learning models have been particularly useful for understanding poststroke motor learning and developing optimized rehabilitation approaches [49]. These models conceptualize stroke rehabilitation as a process of relearning optimal control policies after damage to sensorimotor circuits, with impairments potentially arising from disrupted reward processing, altered state representation, or impaired action selection.

Based on these computational insights, biomimetic AI approaches have been developed to enhance poststroke rehabilitation. Adaptive robotic therapy systems that incorporate principles of reinforcement learning, error-based learning, and challenge optimization have shown promise in improving motor recovery compared to conventional therapy [50]. These systems dynamically adjust assistance levels and task parameters based on individual performance and learning progress, mirroring aspects of optimal coaching in human motor learning.

Virtual reality systems incorporating neuroplasticity principles have been developed for upper limb rehabilitation after stroke. For example, researchers have created VR systems that apply principles of action observation, error-based learning, and reward feedback to enhance motor recovery [51]. These systems use motion capture to translate the patient's impaired limb movements into successful actions in the virtual environment, gradually reducing the assistance while maintaining reward feedback, thereby promoting use-dependent plasticity while minimizing frustration.

Brain-computer interfaces (BCIs) based on principles of operant conditioning and Hebbian plasticity have been developed to enhance motor recovery after stroke [52]. These systems detect motor imagery-related brain activity through EEG and provide feedback that reinforces patterns associated with successful movement, potentially strengthening the neural circuits involved in motor planning and execution. By incorporating principles of timing-dependent plasticity, where the temporal relationship between neural activity and feedback is carefully controlled, these approaches aim to maximize the Hebbian-like strengthening of motor circuits.

Pharmacological approaches to enhancing poststroke plasticity have also been informed by computational modeling [53]. Similarly, computational models of dopaminergic and cholinergic neuromodulation have informed approaches that combine pharmacological enhancement of plasticity with appropriately timed rehabilitation exercises.

Longitudinal models of stroke recovery have been developed to optimize the timing, intensity, and progression of rehabilitation interventions across the recovery trajectory [54]. These models incorporate principles of critical periods, homeostatic plasticity, and learning curve dynamics to predict how the efficacy of different interventions may change over time poststroke, potentially enabling more precise targeting of specific approaches to different recovery phases.

The integration of multimodal neuroimaging, computational modeling, and adaptive intervention design represents a promising frontier in poststroke rehabilitation. By characterizing individual patients' specific patterns of network disruption, residual functionality, and plasticity capacity, these approaches may enable truly personalized neurorehabilitation that maximizes recovery potential for each patient.

3.5.3 Shared Principles and Future Directions

The applications in MS and poststroke rehabilitation share several key principles that highlight the value of biomimetic AI approaches for addressing neuroplasticity impairments more generally:

1. *Multilevel modeling*: Both cases demonstrate the importance of modeling plasticity across multiple levels of organization, from cellular mechanisms to network dynamics to behavioral adaptation. This multilevel approach enables a more comprehensive understanding of plasticity impairments and a more targeted design of interventions.

2. *Closed-loop adaptation*: Adaptive systems that continuously monitor performance and adjust intervention parameters based on individual responses have shown promise in both MS and stroke rehabilitation. These closed-loop approaches mirror aspects of homeostatic regulation and meta-learning in biological systems.
3. *Personalization through computational modeling*: Both cases illustrate how computational models can characterize individual patterns of impairment and guide personalized intervention design, moving beyond one-size-fits-all approaches to rehabilitation.
4. *Integration of multiple plasticity principles*: The most effective approaches appear to incorporate multiple plasticity mechanisms simultaneously—Hebbian learning, reinforcement learning, homeostatic plasticity, and error-based learning—reflecting the multifaceted nature of plasticity in biological systems.
5. *Temporal specificity*: Both cases highlight the importance of timing in plasticity induction, whether in terms of critical periods postinjury or the precise timing of stimulation relative to endogenous activity or behavior.

Looking forward, several promising directions emerge for the application of biomimetic AI to neurological disorders characterized by plasticity impairments:

1. *Predictive biomarkers of plasticity capacity*: Computational models may enable the development of biomarkers that predict individual capacity for plasticity and recovery, potentially guiding treatment selection and prognostication [55].
2. *Multimodal closed-loop systems*: The integration of multiple sensing modalities with multiple intervention modalities may enable more precise targeting of plasticity mechanisms [56].
3. *AI-driven discovery of plasticity-enhancing interventions*: Machine learning approaches may accelerate the discovery of novel compounds or stimulation protocols that enhance plasticity in the injured or diseased brain, potentially by identifying unexpected patterns in high-dimensional datasets from preclinical and clinical studies [57].
4. *Bidirectional translation*: Insights from computational models of impaired plasticity may not only inform clinical applications but also advance our fundamental understanding of plasticity mechanisms in the healthy brain, creating a virtuous cycle of discovery between basic and applied research [58].

The application of biomimetic AI to neurological disorders characterized by plasticity impairments exemplifies the broader potential of brain-inspired approaches in both advancing our understanding of neural function and developing more effective interventions for neural dysfunction. By drawing on principles of neuroplasticity while leveraging the computational power and adaptability of AI systems, these approaches offer a promising path toward more personalized and effective treatments for a wide range of neurological conditions.

3.6 Conclusion

The convergence of neuroscience and artificial intelligence through biomimetic approaches represents a frontier with transformative potential for both fields. By drawing inspiration from the brain's architecture, learning mechanisms, and cognitive capabilities, researchers are developing artificial systems that better capture the efficiency, adaptability, and robustness of biological intelligence. Conversely, these biomimetic approaches provide new computational frameworks for understanding brain function and addressing neurological disorders.

The stability-plasticity dilemma exemplifies the shared challenges faced by biological and artificial learning systems, highlighting how insights from one domain can inform solutions in the other. The brain's sophisticated mechanisms for balancing adaptability with stability—complementary learning systems, synaptic consolidation, metaplasticity, structural plasticity, and network modularity—have inspired analogous approaches in artificial neural networks, including replay-based consolidation, elastic weight consolidation, attention-based consolidation, and dynamic architectural growth.

Various forms of neuroplasticity, from Hebbian learning and STDP to homeostatic regulation and neuromodulation, have similarly informed machine learning algorithm design, enabling more efficient, adaptive, and biologically plausible artificial learning. These biomimetic approaches not only advance AI capabilities but also provide computational frameworks for understanding how plasticity mechanisms interact in biological systems.

Broader principles of brain organization and function—hierarchical processing, predictive coding, attention mechanisms, embodied cognition, social learning, emotional regulation, and developmental trajectories—are influencing the next generation of AI systems, potentially enabling more human-like intelligence that combines the efficiency and adaptability of biological cognition with the precision and scalability of computational approaches.

The applications of biomimetic AI in neurological disorders characterized by plasticity impairments, such as multiple sclerosis and stroke, illustrate the practical impact of these approaches. By modeling impaired plasticity mechanisms and developing interventions based on biomimetic principles, researchers are advancing both our understanding of these conditions and our ability to address them therapeutically. The multilevel modeling, closed-loop adaptation, personalization, and temporal specificity emphasized in these applications reflect broader principles that may guide the development of biomimetic approaches to various neurological and psychiatric disorders.

Looking forward, the continued cross-fertilization between neuroscience and AI holds tremendous promise for addressing fundamental challenges in both fields. For neuroscience, computational models provide frameworks for integrating findings across levels of analysis and testing hypotheses about neural function that may be difficult to examine directly in biological systems. For AI, brain-inspired approaches offer potential solutions to longstanding challenges in energy efficiency, continual learning, generalization from limited data, and integration across cognitive domains.

As we continue to unravel the brain's complexity and refine our artificial systems, the gap between biological and artificial intelligence may narrow, potentially leading to AI systems that better complement human capabilities while advancing our understanding of our own minds. This virtuous cycle of discovery and innovation at the intersection of neuroscience and AI represents one of the most exciting frontiers in contemporary science and technology.

3.6.1 Summary

This chapter explores the fascinating convergence between neuroscience and artificial intelligence through the lens of biomimetic approaches—AI systems that draw inspiration from biological neural networks. This chapter opens with Emily Dickinson's poetic reflection on the brain's vastness, setting the stage for an examination of how nature's most sophisticated information processing system continues to inspire and inform artificial intelligence development.

This chapter systematically unpacks the bidirectional relationship between brain science and AI, where neuroscientific discoveries inform AI development while advances in AI provide new frameworks for understanding the brain itself. Central to this exploration is the stability-plasticity dilemma—a fundamental challenge facing both biological and artificial learning systems in balancing the need to acquire new information while preserving existing knowledge.

The author delves deep into how the brain elegantly solves this dilemma through mechanisms like complementary learning systems (where the hippocampus rapidly learns while the neocortex slowly consolidates), synaptic consolidation, metaplasticity, and structural reorganization. These biological solutions have inspired innovative AI approaches including experience replay, elastic weight consolidation, and dynamic architectural growth.

A substantial portion examines various forms of neuroplasticity and their translation into machine learning algorithms. From Hebbian learning's "neurons that fire together, wire together" principle to spike-timing-dependent plasticity, homeostatic regulation, and neuromodulation, each biological mechanism has spawned corresponding computational approaches that enhance AI capabilities while maintaining biological plausibility.

This chapter also surveys how broader principles of brain organization are shaping next-generation AI systems, including hierarchical processing architectures, predictive coding frameworks, attention mechanisms inspired by neural focus, embodied cognition approaches, and integrative cognitive architectures that attempt to unify multiple cognitive functions.

Particularly compelling are the clinical case studies demonstrating how biomimetic AI approaches are being applied to understand and treat neurological conditions characterized by impaired plasticity, such as multiple sclerosis and stroke. These applications showcase closed-loop rehabilitation systems that adaptively respond to patient needs, personalized neurorehabilitation strategies, and computational models that guide therapeutic interventions.

This chapter concludes by highlighting the transformative potential of this convergence, suggesting that as the gap between biological and artificial intelligence narrows, we may achieve AI systems that better complement human capabilities while simultaneously advancing our understanding of our own minds. This creates what the authors describe as a "virtuous cycle of discovery and innovation" that benefits both neuroscience and artificial intelligence fields.

References

1. Raichle ME, Gusnard DA. Appraising the brain's energy budget. Proc Natl Acad Sci. 2002;99(16):10237–9.
2. Grossberg S. How does a brain build a cognitive code? Psychol Rev. 1980;87(1):1–51.
3. Abraham WC, Robins A. Memory retention–the synaptic stability versus plasticity dilemma. Trends Neurosci. 2005;28(2):73–8.
4. McCloskey M, Cohen NJ. Catastrophic interference in connectionist networks: the sequential learning problem. Psychol Learn Motiv. 1989;24:109–65.
5. Kirkpatrick J, Pascanu R, Rabinowitz N, Veness J, Desjardins G, Rusu AA, et al. Overcoming catastrophic forgetting in neural networks. Proc Natl Acad Sci. 2017;114(13):3521–6.
6. Rudroff T, Rainio O, Klén R. Neuroplasticity meets artificial intelligence: a hippocampus-inspired approach to the stability-plasticity dilemma. Brain Sci. 2024;14(11):1111.
7. Zenke F, Poole B, Ganguli S. Continual learning through synaptic intelligence. In: International conference on machine learning. PMLR; 2017. p. 3987–95.
8. McClelland JL, McNaughton BL, O'Reilly RC. Why there are complementary learning systems in the hippocampus and neocortex: insights from the successes and failures of connectionist models of learning and memory. Psychol Rev. 1995;102(3):419–57.
9. Born J, Wilhelm I. System consolidation of memory during sleep. Psychol Res. 2012;76(2):192–203.
10. Squire LR, Zola SM. Structure and function of declarative and nondeclarative memory systems. Proc Natl Acad Sci. 1996;93(24):13515–22.
11. Abraham WC, Bear MF. Metaplasticity: the plasticity of synaptic plasticity. Trends Neurosci. 1996;19(4):126–30.
12. Holtmaat A, Svoboda K. Experience-dependent structural synaptic plasticity in the mammalian brain. Nat Rev Neurosci. 2009;10(9):647–58.
13. Rusu AA, Rabinowitz NC, Desjardins G, Soyer H, Kirkpatrick J, Kavukcuoglu K, et al. Progressive neural networks. arXiv preprint arXiv. 2016:1606.04671.
14. Hebb DO. The organization of behavior: a neuropsychological theory. Wiley; 1949.
15. Kohonen T. The self-organizing map. Proc IEEE. 1990;78(9):1464–80.
16. Oja E. Simplified neuron model as a principal component analyzer. J Math Biol. 1982;15(3):267–73.
17. Bi GQ, Poo MM. Synaptic modifications in cultured hippocampal neurons: dependence on spike timing, synaptic strength, and postsynaptic cell type. J Neurosci. 1998;18(24):10464–72.
18. Markram H, Gerstner W, Sjöström PJ. Spike-timing-dependent plasticity: a comprehensive overview. Front Synaptic Neurosci. 2012;4:2.
19. Turrigiano GG. The self-tuning neuron: synaptic scaling of excitatory synapses. Cell. 2008;135(3):422–35.
20. Ioffe S, Szegedy C. Batch normalization: accelerating deep network training by reducing internal covariate shift. In: International conference on machine learning. PMLR; 2015. p. 448–56.
21. Marder E. Neuromodulation of neuronal circuits: back to the future. Neuron. 2012;76(1):1–11.
22. Schultz W, Dayan P, Montague PR. A neural substrate of prediction and reward. Science. 1997;275(5306):1593–9.

23. Vaswani A, Shazeer N, Parmar N, Uszkoreit J, Jones L, Gomez AN, Kaiser Ł, Polosukhin I. Attention is all you need. Adv Neural Inf Process Syst. 2017;30. arXiv:1706.03762v7 [cs.CL].
24. Felleman DJ, Van Essen DC. Distributed hierarchical processing in the primate cerebral cortex. Cereb Cortex. 1991;1(1):1–47.
25. LeCun Y, Bengio Y, Hinton G. Deep learning. Nature. 2015;521(7553):436–44.
26. Hochreiter S, Schmidhuber J. Long short-term memory. Neural Comput. 1997;9(8):1735–80.
27. Clark A. Whatever next? Predictive brains, situated agents, and the future of cognitive science. Behav Brain Sci. 2013;36(3):181–204.
28. Friston K, FitzGerald T, Rigoli F, Schwartenbeck P, Pezzulo G. Active inference: a process theory. Neural Comput. 2017;29(1):1–49.
29. Kingma DP, Welling M. Auto-encoding variational Bayes. 2013. arXiv preprint arXiv:1312.6114.
30. Tomasello M, Kruger AC, Ratner HH. Cultural learning. Behav Brain Sci. 1993;16(3):495–511.
31. Pfeifer R, Bongard J. How the body shapes the way we think: a new view of intelligence. MIT Press; 2006.
32. Oudeyer PY, Kaplan F, Hafner VV. Intrinsic motivation systems for autonomous mental development. IEEE Trans Evol Comput. 2007;11(2):265–86.
33. LeDoux J. Rethinking the emotional brain. Neuron. 2012;73(4):653–76.
34. Bengio Y, Louradour J, Collobert R, Weston J. Curriculum learning. Am Podiatr Med Assoc. 2009;60(6). https://doi.org/10.1145/1553374.1553380.
35. Anderson JR, Bothell D, Byrne MD, Douglass S, Lebiere C, Qin Y. An integrated theory of the mind. Psychol Rev. 2004;111(4):1036–60.
36. Eliasmith C, Stewart TC, Choo X, Bekolay T, DeWolf T, Tang Y, Rasmussen D. A large-scale model of the functioning brain. Science. 2012;338(6111):1202–5.
37. Lake BM, Ullman TD, Tenenbaum JB, Gershman SJ. Building machines that learn and think like people. Behav Brain Sci. 2017;40:e253.
38. Compston A, Coles A. Multiple sclerosis. Lancet. 2008;372(9648):1502–17.
39. Ksiazek-Winiarek DJ, Szpakowski P, Glabinski A. Neural plasticity in multiple sclerosis: the functional and molecular background. Neural Plast. 2015;2015:307175.
40. Mori F, Kusayanagi H, Nicoletti CG, Weiss S, Marciani MG, Centonze D. Cortical plasticity predicts recovery from relapse in multiple sclerosis. Mult Scler J. 2014;20(4):451–7.
41. Rocca MA, Valsasina P, Martinelli V, Misci P, Falini A, Comi G, Filippi M. Large-scale neuronal network dysfunction in relapsing-remitting multiple sclerosis. Neurology. 2012;79(14):1449–57.
42. Tacchino A, Pedullà L, Bonzano L, Vassallo C, Battaglia MA, Mancardi G, et al. A new app for at-home cognitive training: description and pilot testing on patients with multiple sclerosis. JMIR Mhealth Uhealth. 2020;8(4):e14328.
43. Benjamin EJ, Blaha MJ, Chiuve SE, Cushman M, Das SR, Deo R, et al. Heart disease and stroke statistics—2017 update: a report from the American Heart Association. Circulation. 2017;135(10):e146–603.
44. Murphy TH, Corbett D. Plasticity during stroke recovery: from synapse to behaviour. Nat Rev Neurosci. 2009;10(12):861–72.
45. Carmichael ST. The 3 Rs of stroke biology: radial, relayed, and regenerative. Neurotherapeutics. 2016;13(2):348–59.
46. Reinkensmeyer DJ, Burdet E, Casadio M, Krakauer JW, Kwakkel G, Lang CE, et al. Computational neurorehabilitation: modeling plasticity and learning to predict recovery. J Neuroeng Rehabil. 2016;13(1):1–25.
47. Nahmani M, Turrigiano GG. Adult cortical plasticity following injury: recapitulation of critical period mechanisms? Neuroscience. 2014;283:4–16.

48. Falcon MI, Riley JD, Jirsa V, McIntosh AR, Shereen AD, Chen EE, Solodkin A. The virtual brain: modeling biological correlates of recovery after chronic stroke. Front Neurol. 2016;7:10.
49. Reinkensmeyer DJ, Guigon E, Maier MA. A computational model of use-dependent motor recovery following a stroke: optimizing corticospinal activations via reinforcement learning can explain residual capacity and other strength recovery dynamics. Neural Netw. 2012;29:60–9.
50. Krebs HI, Mernoff S, Fasoli SE, Hughes R, Stein J, Hogan N. A comparison of functional and impairment-based robotic training in severe to moderate chronic stroke: a pilot study. NeuroRehabilitation. 2015;23(1):81–7.
51. Maier M, Ballester BR, Duff A, Oller ED, Verschure PF. Effect of specific over nonspecific VR-based rehabilitation on poststroke motor recovery: a systematic meta-analysis. Neurorehabil Neural Repair. 2019;33(2):112–29.
52. Pichiorri F, Morone G, Petti M, Toppi J, Pisotta I, Molinari M, et al. Brain–computer interface boosts motor imagery practice during stroke recovery. Ann Neurol. 2015;77(5):851–65.
53. Cramer SC. Drugs to enhance motor recovery after stroke. Stroke. 2018;49(4):1192–8.
54. Hope TM, Friston K, Price CJ, Leff AP, Rotshtein P, Bowman H. Recovery after stroke: not so proportional after all? Brain. 2019;142(1):15–22.
55. Stinear CM, Barber PA, Petoe M, Anwar S, Byblow WD. The PREP algorithm predicts potential for upper limb recovery after stroke. Brain. 2012;135(8):2527–35.
56. Coscia M, Wessel MJ, Chaudary U, Millán JDR, Micera S, Guggisberg A, et al. Neurotechnology-aided interventions for upper limb motor rehabilitation in severe chronic stroke. Brain. 2019;142(8):2182–97.
57. Vamathevan J, Clark D, Czodrowski P, Dunham I, Ferran E, Lee G, et al. Applications of machine learning in drug discovery and development. Nat Rev Drug Discov. 2019;18(6):463–77.
58. Bassett DS, Zurn P, Gold JI. On the nature and use of models in network neuroscience. Nat Rev Neurosci. 2018;19(9):566–78.

AI in Clinical Neuroscience Practice

4

The future of medicine is inseparable from the future of technology.

—Eric Topol

4.1 Introduction

The integration of artificial intelligence into clinical neuroscience represents one of the most promising frontiers in modern medicine. The human brain generates vast amounts of data across multiple scales and modalities. This intricate architecture produces information ranging from molecular interactions to whole-brain dynamics. This complexity challenges traditional analytical approaches. However, it provides fertile ground for AI applications that excel at extracting patterns from high-dimensional, heterogeneous data. AI technologies are increasingly transforming how neurological disorders are diagnosed, monitored, and treated. Applications range from automated image analysis to predictive modeling of disease trajectories and optimization of treatment parameters.

The past decade has witnessed remarkable progress in the application of AI to various aspects of clinical neuroscience. Deep learning algorithms can now detect subtle patterns in neuroimaging data that may escape the human eye, potentially enabling earlier diagnosis of conditions like Alzheimer's disease or multiple sclerosis [1]. Machine learning models can integrate multimodal clinical data to predict individual disease trajectories, helping clinicians make more informed decisions about treatment timing and intensity [2]. Reinforcement learning approaches can optimize stimulation parameters for neuromodulation therapies, potentially enhancing efficacy while reducing side effects [3]. These advances suggest a future of more precise, personalized, and effective neurological care guided by AI systems.

T. Rudroff, *AI for the Healthy Brain and Neurological Disorders*, Neurocultural Health and Wellbeing, https://doi.org/10.1007/978-3-032-11596-6_4

However, the path from promising research to routine clinical implementation remains challenging. Neurological disorders are notoriously heterogeneous, with symptoms, progression trajectories, and treatment responses varying widely across individuals. Clinical data are often incomplete, inconsistent, or difficult to standardize across institutions. Regulatory frameworks for AI-based medical applications continue to evolve, creating uncertainty around validation requirements and approval pathways. Moreover, the "black box" nature of many advanced AI methods raises concerns about interpretability and trustworthiness, particularly in high-stakes medical decisions.

This chapter explores the current state and future potential of AI in clinical neuroscience practice. We begin by examining real-world applications across the care continuum, from diagnosis and prognosis to treatment planning and monitoring. We then focus on recent research in AI-optimized transcranial direct current stimulation (tDCS), illustrating how machine learning approaches can enhance the efficacy of noninvasive brain stimulation for conditions like depression, chronic pain, and cognitive impairment. Next, we explore broader approaches to personalized treatment in neurological disorders, considering how AI can help match patients with optimal interventions based on their individual characteristics. We then examine early detection systems for Alzheimer's disease, highlighting how AI analysis of neuroimaging biomarkers may enable diagnosis years before symptom onset. Finally, we address the key challenges in clinical implementation and validation, discussing potential solutions to accelerate the responsible integration of AI into neurological care.

Throughout this chapter, we emphasize the complementary relationship between AI systems and clinical expertise. Rather than replacing human judgment, AI tools are most powerful when they augment clinicians' capabilities, handling tasks that benefit from computational power while leaving the integration of technical outputs with broader patient contexts to experienced healthcare providers. This human-AI partnership promises to enhance both the precision and the humanity of neurological care, enabling more accurate and consistent analyses while freeing clinicians to focus on the interpersonal aspects of medicine that remain uniquely human.

The integration of artificial intelligence into clinical neuroscience is transforming how neurological disorders are diagnosed, monitored, and treated. Figure 4.1 provides an overview of AI applications across the neurology care continuum, illustrating how these technologies support clinicians at each stage from initial diagnosis to long-term monitoring. As we explore each of these applications in greater detail throughout this chapter, this framework helps conceptualize how various AI approaches address different clinical needs while collectively enhancing neurological care. The following sections examine specific implementations of AI across this continuum, beginning with real-world applications in diagnosis, prognosis, and treatment planning.

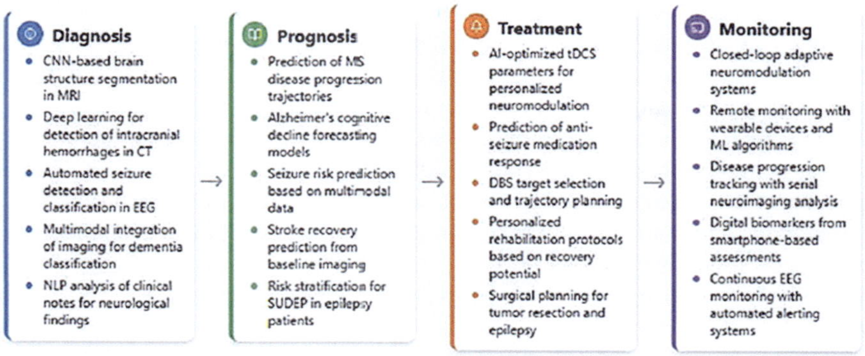

Fig. 4.1 AI applications across the neurology care continuum

4.2 Real-World Applications in Diagnosis, Prognosis, and Treatment Planning

Artificial intelligence is making significant inroads into clinical neuroscience practice across the entire care continuum. From initial diagnosis to long-term monitoring and treatment optimization, AI approaches are beginning to demonstrate real-world utility in various neurological conditions. This section explores key applications that are already impacting clinical practice or are on the cusp of clinical translation.

4.2.1 Diagnostic Applications

4.2.1.1 Neuroimaging Analysis and Interpretation

Neuroimaging plays a central role in the diagnosis of many neurological conditions. However, interpreting complex brain scans requires significant expertise and can be subject to interrater variability. AI approaches, particularly deep learning, have shown impressive performance in automating aspects of neuroimaging analysis across multiple modalities.

Convolutional neural networks (CNNs) can now automatically segment brain structures in magnetic resonance imaging (MRI). These systems achieve accuracy approaching or exceeding that of human experts [4]. These segmentation algorithms enable more precise volumetric measurements of structures like the hippocampus. Such measurements are critical for diagnosing conditions such as Alzheimer's disease and temporal lobe epilepsy. Commercial applications like Quantib™ ND and Icometrix have received regulatory approval and are being implemented in clinical settings, allowing for more consistent and efficient structural brain analyses [5].

For detecting acute abnormalities, AI systems have demonstrated particular utility in emergency settings. Deep learning algorithms can now identify intracranial hemorrhages from computed tomography (CT) scans with high sensitivity and specificity, potentially reducing diagnostic delays in time-critical conditions [6]. Similar approaches have been developed for detecting large vessel occlusions in stroke, helping to triage patients who might benefit from mechanical thrombectomy [7]. The FDA-cleared Viz. AI system, for example, automatically analyzes CT angiograms and notifies stroke teams of suspected large vessel occlusions, potentially reducing time to treatment [8].

In positron emission tomography (PET), AI algorithms can enhance the diagnostic accuracy of FDG-PET scans for neurodegenerative disorders. Deep learning approaches have demonstrated the ability to differentiate between Alzheimer's disease, frontotemporal dementia, and Lewy body dementia based on patterns of glucose metabolism, even in cases where visual interpretation is challenging [9].

4.2.1.2 Electroencephalography (EEG) Analysis

EEG remains a critical tool for diagnosing epilepsy and other neurological disorders, but its interpretation requires specialized expertise and can be time-consuming. AI approaches have demonstrated significant potential for automating various aspects of EEG analysis, from artifact removal to seizure detection and classification.

Deep learning algorithms can now detect epileptiform discharges and seizure activity in EEG recordings with sensitivity comparable to expert neurologists [10]. These algorithms have been implemented in commercial systems like Persyst, which received FDA clearance for automated seizure detection and has been integrated into clinical workflows at many epilepsy centers [11]. Beyond detection, AI approaches can also classify different seizure types based on EEG patterns, potentially informing more targeted treatment approaches.

For long-term monitoring, AI systems can analyze continuous EEG data in intensive care settings, helping to detect subclinical seizures and assess brain function in comatose patients [12]. Machine learning approaches can identify patterns associated with increased risk of seizures or neurological deterioration, potentially enabling earlier intervention.

4.2.1.3 Clinical Decision Support Systems

Integrating diverse clinical data beyond imaging and electrophysiology, AI-based clinical decision support systems (CDSS) are beginning to assist neurologists in diagnostic reasoning. These systems can combine patient history, examination findings, laboratory results, and imaging data to suggest potential diagnoses or recommend additional tests.

In the evaluation of cognitive disorders, AI systems have been developed to analyze patterns of cognitive test performance and identify signatures associated with different types of dementia [13]. These approaches can help distinguish Alzheimer's disease from other causes of cognitive impairment, potentially reducing diagnostic uncertainty and inappropriate treatment.

For movement disorders, machine learning approaches can analyze data from wearable sensors or smartphone-based assessments to quantify motor symptoms and support diagnosis [14]. These systems can detect subtle motor abnormalities that might not be apparent during brief clinical examinations, potentially enabling earlier diagnosis of conditions like Parkinson's disease or differentiating between similar movement disorders.

4.2.2 Prognostic Applications

4.2.2.1 Predicting Disease Progression

Predicting the course of neurological disorders represents a significant challenge, given their heterogeneous nature and the complex interplay of genetic, environmental, and physiological factors. AI approaches have shown promise for developing more accurate prognostic models that can inform treatment planning and patient counseling.

In multiple sclerosis (MS), machine learning models have been developed to predict disease progression based on clinical, imaging, and laboratory data [15]. These models can identify patients at higher risk for rapid disability progression, potentially enabling more aggressive early treatment for those who might benefit most. For example, Eshaghi et al. [16] developed a deep learning approach that predicts future brain atrophy patterns in MS patients from baseline MRI scans, providing a potential imaging biomarker for disease activity and treatment response.

For neurodegenerative disorders, AI approaches can model disease trajectories and predict the rate of cognitive or functional decline. In Alzheimer's disease, machine learning models incorporating baseline cognitive assessments, neuroimaging features, genetic data, and cerebrospinal fluid biomarkers can predict cognitive trajectories with greater accuracy than traditional statistical methods [17].

In epilepsy, AI models can predict seizure risk based on various factors, including EEG patterns, sleep quality, stress levels, and medication adherence [18]. These approaches may eventually enable more personalized seizure forecasting, allowing patients to adjust activities or take preventive measures during periods of increased risk.

4.2.3 Treatment Planning Applications

4.2.3.1 Therapy Selection and Optimization

Choosing the optimal treatment for individual patients represents a complex decision-making process that must consider multiple factors, including disease subtype, comorbidities, potential side effects, and patient preferences. AI approaches can support this process by identifying patterns associated with treatment response or failure across large patient populations.

In epilepsy, machine learning models have been developed to predict response to specific anti-seizure medications based on clinical features, EEG patterns, and genetic factors [19]. These models may eventually help clinicians select optimal first-line therapy, potentially reducing the trial-and-error approach that currently characterizes medication selection.

For multiple sclerosis, AI models can predict response to disease-modifying therapies based on baseline clinical and MRI characteristics [20]. These approaches may help optimize treatment selection, balancing efficacy against safety concerns like the risk of progressive multifocal leukoencephalopathy with certain medications.

4.2.3.2 Surgical Planning and Guidance

Neurosurgical procedures require precise planning and execution, given the complex and eloquent nature of neural structures. AI approaches are increasingly being integrated into surgical workflows, from preoperative planning to intraoperative guidance and postoperative monitoring.

For tumor resection, AI systems can automatically segment brain tumors and surrounding critical structures from MRI scans, facilitating surgical planning and risk assessment [21]. Advanced approaches can also predict the location of functional areas based on structural and functional imaging data, potentially reducing the need for invasive mapping procedures.

In epilepsy surgery, AI approaches can help localize seizure onset zones based on intracranial EEG recordings, potentially improving surgical targeting [22]. Machine learning models can also predict surgical outcomes based on preoperative factors, helping to identify the most promising candidates for intervention.

For deep brain stimulation (DBS), AI systems can assist with target selection and trajectory planning, helping to optimize electrode placement while avoiding critical structures [23]. Machine learning approaches can also predict clinical outcomes based on electrode location and stimulation parameters, potentially guiding postoperative programming.

These diverse applications across the care continuum illustrate the growing role of AI in clinical neuroscience practice. From supporting initial diagnosis to optimizing long-term management, AI approaches are increasingly demonstrating practical utility in real-world clinical settings. However, the implementation of these technologies varies widely across applications and institutions, with some becoming standard of care while others remain primarily research tools.

4.3 AI-Optimized Transcranial Direct Current Stimulation

Figure 4.2 illustrates the AI-driven process for optimizing transcranial direct current stimulation (tDCS), highlighting how artificial intelligence transforms this neuromodulation technique from a standardized approach to a personalized therapeutic intervention. This process integrates patient-specific data, computational modeling, machine learning predictions, and adaptive systems to enhance targeting precision and therapeutic outcomes. As we examine each stage of this process in detail, this visualization provides a framework for understanding how AI methods address the

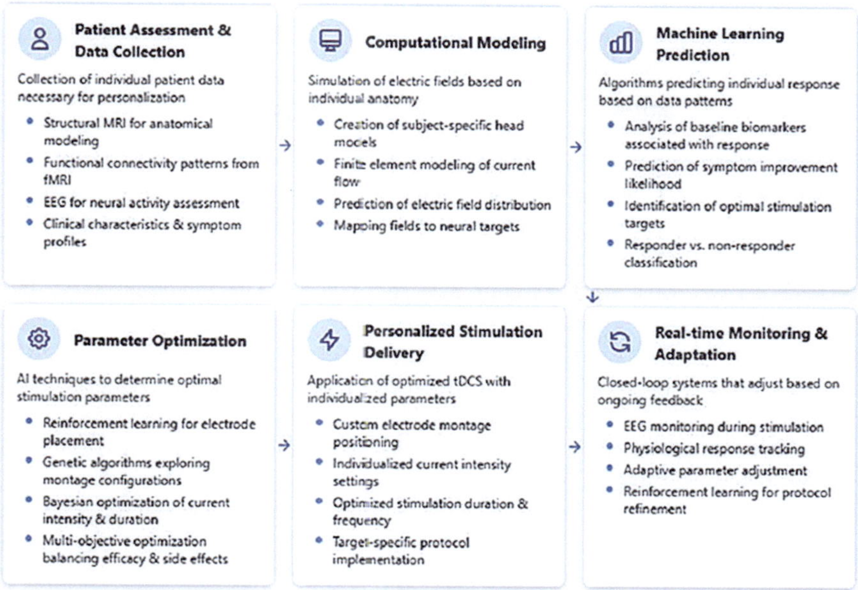

Fig. 4.2 AI-optimized transcranial direct current stimulation (tDCS) process

significant interindividual variability and targeting challenges that have historically limited tDCS efficacy.

Transcranial direct current stimulation (tDCS) represents a promising noninvasive neuromodulation technique. It shows potential for treating various neurological and psychiatric conditions. tDCS applies weak electrical currents to the scalp to modulate cortical excitability and plasticity. This approach can potentially enhance recovery from brain injury or improve symptoms in conditions ranging from depression to chronic pain [24]. However, traditional tDCS approaches face significant challenges in optimizing stimulation parameters for individual patients. These challenges arise from the complex interplay between electrode montage, current intensity, stimulation duration, and individual brain anatomy and physiology. Recent research has demonstrated how AI approaches can address these challenges, potentially enhancing the efficacy and reliability of tDCS through personalized parameter optimization.

4.3.1 Understanding tDCS Mechanisms and Challenges

Transcranial direct current stimulation operates by applying a constant, low-intensity electrical current (typically 1–2 mA) between electrodes placed on the scalp. This current partially penetrates the skull and modulates the resting membrane potential of neurons in the underlying cortex, with anodal stimulation generally increasing excitability and cathodal stimulation generally decreasing it [25].

These effects can persist beyond the stimulation period through mechanisms that may involve long-term potentiation (LTP), long-term depression (LTD), and alterations in neurotransmitter concentrations.

Despite its promise, tDCS faces several challenges that have limited its clinical adoption. First, the effects of stimulation show significant interindividual variability, with some subjects showing robust responses while others show minimal or even opposite effects [26]. This variability may reflect differences in skull thickness, brain anatomy, genetics, age, and other factors that influence current flow and neural responsiveness. Second, the relationship between electrode placement and the resulting electric field distribution in the brain is complex and nonintuitive, making it difficult to target specific brain regions precisely [27]. Third, the optimal stimulation parameters may vary depending on the specific condition being treated, the individual patient characteristics, and even the current state of the brain at the time of stimulation.

Traditional approaches to tDCS have typically relied on standardized electrode montages and stimulation parameters based on population averages, with limited personalization beyond occasional adjustments for comfort or safety. While this approach has shown efficacy in some applications, the substantial interindividual variability and targeting challenges have likely contributed to the mixed results observed in clinical trials. These limitations have motivated increasing interest in more personalized approaches to tDCS that can optimize stimulation parameters for individual patients based on their specific characteristics and therapeutic goals.

4.3.2 Computational Modeling for Individualized tDCS

Computational models of current flow represent an important first step toward personalized tDCS. These models, typically based on finite element methods, simulate how electrical currents propagate through the various tissues of the head and brain, allowing researchers to predict the electric field distribution for different electrode configurations [28]. By incorporating individual MRI scans to create subject-specific head models, these approaches can account for anatomical variations that might influence current flow, such as differences in skull thickness, cerebrospinal fluid distribution, or cortical folding patterns.

Advanced modeling approaches have further refined these techniques by incorporating information about tissue conductivity, the anisotropy of white matter tracts, and even individual functional connectivity patterns derived from fMRI or EEG [29]. These enhancements aim to provide more accurate predictions of how tDCS will affect specific neural circuits in individual patients, potentially enabling more precise targeting of stimulation.

While computational modeling represents an important advance toward personalized tDCS, it faces several limitations. First, many models rely on simplified assumptions about tissue properties and boundary conditions that may not fully capture the complexity of individual heads. Second, the relationship between electric field distribution and physiological or behavioral effects remains incompletely understood, making it difficult to translate model predictions into optimal stimulation parameters. Third, creating and solving subject-specific models can be computationally intensive and require specialized expertise, limiting their practical implementation in routine clinical settings.

4.3.3 AI Approaches to tDCS Optimization

Artificial intelligence approaches offer promising solutions to overcome the limitations of traditional computational modeling while enhancing the personalization of tDCS. By leveraging data from previous stimulation sessions, neuroimaging, neurophysiological measures, and clinical outcomes, AI algorithms can learn complex relationships between individual characteristics, stimulation parameters, and therapeutic effects, potentially enabling more effective and reliable neuromodulation.

4.3.4 Predictive Modeling of tDCS Response

Machine learning models can predict individual responses to tDCS based on various factors, including demographic characteristics, neuroimaging features, baseline clinical measures, and neurophysiological markers. For example, Paul et al. [30] developed a random forest model that predicted response to anodal tDCS in chronic stroke patients based on baseline functional connectivity patterns, achieving 72.5% accuracy in identifying responders versus nonresponders.

More advanced approaches have employed deep learning to extract predictive features from complex neuroimaging data. For instance, Li et al. [31] developed a convolutional neural network that predicted motor improvements following tDCS in stroke patients based on structural and functional MRI, outperforming traditional prediction methods. These approaches may eventually enable clinicians to identify patients most likely to benefit from tDCS and potentially tailor stimulation parameters to individual characteristics.

4.3.5 Optimizing Electrode Montages and Parameters

Traditional tDCS typically employs standardized electrode montages based on the 10–20 EEG system, which may not optimally target specific brain regions or networks in individual patients. AI approaches can optimize electrode placement and stimulation parameters based on individual anatomy, functional connectivity, and therapeutic goals, potentially enhancing stimulation precision and efficacy.

Reinforcement learning represents one promising approach for tDCS optimization. For example, Rudroff et al. [32] developed a reinforcement learning algorithm that optimizes tDCS electrode configurations for targeting specific brain networks implicated in fatigue in Long COVID patients. The algorithm explores different montages and learns from the resulting changes in brain activity measured with functional MRI, progressively refining its recommendations to maximize the desired network modulation while minimizing off-target effects.

Genetic algorithms offer another approach for exploring the large parameter space of possible tDCS configurations. Saturnino et al. [33] employed a genetic algorithm to optimize multielectrode tDCS for targeting specific cortical regions while minimizing current flow through other areas. The algorithm iteratively generated and evaluated thousands of possible electrode arrangements, evolving toward increasingly optimal solutions based on computational modeling of the resulting electric fields.

4.3.6 Closed-Loop Systems for Adaptive Stimulation

Traditional tDCS typically delivers stimulation with fixed parameters for the entire session, regardless of ongoing changes in brain state or activity. Closed-loop systems that monitor neural activity in real time and adjust stimulation parameters accordingly represent a frontier in personalized neuromodulation, potentially enhancing efficacy by delivering stimulation at optimal times and intensities.

Neurophysiological measures like EEG can provide real-time feedback for closed-loop tDCS. Saeidi et al. [34] provided recommendations for applying supervised machine learning and deep learning algorithms for the neural decoding of EEG signals in various tasks and experimental protocols, which might be addressed in future research efforts in this field.

Functional neuroimaging can also guide adaptive tDCS, albeit with practical limitations for real-time applications. Bergmann et al. [35] proposed a system combining functional near-infrared spectroscopy (fNIRS) with tDCS, where hemodynamic responses measured by fNIRS inform adjustments to stimulation parameters.

4.3.7 Clinical Applications of AI-Optimized tDCS

The integration of AI approaches with tDCS has shown promise across various neurological and psychiatric conditions, potentially expanding the therapeutic applications of this neuromodulation technique while improving its reliability and efficacy.

4.3.8 Depression and Mood Disorders

Major depressive disorder represents one of the most extensively studied applications for tDCS, with meta-analyses supporting its efficacy as a treatment option [36]. However, response rates vary considerably across studies and individuals, motivating interest in more personalized approaches. AI-optimized tDCS may enhance outcomes by tailoring stimulation to individual characteristics and monitoring treatment response.

Machine learning models have been developed to predict response to tDCS in depression based on baseline clinical features, neuroimaging markers, and EEG patterns. For example, Li et al. [37] identified EEG connectivity features that predicted symptomatic improvement following dorsolateral prefrontal cortex (DLPFC) stimulation, potentially enabling better patient selection for this intervention.

4.3.9 Cognitive Enhancement and Rehabilitation

Cognitive impairment represents a significant challenge across various neurological conditions, from stroke and traumatic brain injury to neurodegenerative disorders. tDCS has shown potential for enhancing specific cognitive functions and supporting rehabilitation efforts, with AI approaches potentially optimizing these effects through personalized stimulation protocols.

Work by Rudroff and colleagues [32] has explored the potential of AI-optimized tDCS for cognitive symptoms in Long COVID. Using a combination of machine learning and computational modeling, they developed personalized stimulation protocols based on individual functional connectivity patterns associated with cognitive fatigue. This approach showed promising results in a pilot trial, with significant improvements in both subjective fatigue ratings and objective cognitive performance compared to sham stimulation or standard montages.

4.3.10 Pain Management

Chronic pain conditions often respond poorly to conventional treatments, creating significant interest in neuromodulation approaches like tDCS. AI optimization may enhance the analgesic effects of tDCS by targeting specific pain networks more precisely and adapting stimulation to individual pain characteristics.

Machine learning models have been developed to predict response to tDCS for chronic pain based on various factors, including pain duration, psychological features, and baseline neurophysiological measures. Beyond prediction, AI approaches can optimize tDCS parameters for pain management by accounting for individual variations in pain processing networks and customizing stimulation accordingly, recognizing the heterogeneous nature of chronic pain conditions.

4.3.11 Future Directions and Challenges

While AI-optimized tDCS shows considerable promise for enhancing the precision and efficacy of neuromodulation, several challenges and future directions warrant consideration as this field continues to evolve. The integration of multiple biomarkers to provide a more comprehensive characterization of individual brains and more precise targeting of stimulation represents an important direction. Future developments will likely integrate multiple biomarkers to provide a more comprehensive characterization of individual brains and more precise targeting of stimulation.

The combination of tDCS with other interventions represents another promising direction for enhancing outcomes. AI algorithms could optimize not only tDCS parameters but also their temporal relationship with complementary approaches like cognitive training, physical therapy, or pharmacological treatments.

Most current approaches to tDCS optimization focus on single sessions or short-term protocols, whereas many neurological conditions require ongoing management. Future systems may implement continuous adaptation over extended treatment periods, learning from individual responses and adjusting parameters accordingly.

Despite its potential, AI-optimized tDCS faces practical challenges for widespread clinical implementation. Many approaches require specialized equipment like high-density EEG systems or advanced neuroimaging that may not be available in typical clinical settings. Future developments will need to balance sophistication with practicality, perhaps developing streamlined systems that can deliver personalized stimulation based on more accessible measures.

The integration of AI with tDCS illustrates a broader trend toward more personalized, precise, and adaptive neuromodulation approaches. By accounting for individual variability and optimizing stimulation parameters based on objective measures, these approaches may enhance the therapeutic potential of noninvasive brain stimulation across various neurological conditions.

4.4 AI for Personalized Treatment Approaches in Neurological Disorders

The concept of personalized or precision medicine has gained significant momentum across medical specialties. This approach tailors treatment to individual characteristics rather than population averages. In neurology, this approach holds particular promise given the heterogeneous nature of many neurological disorders. Patients with the same diagnosis often show markedly different symptoms, progression rates, and treatment responses. Artificial intelligence represents a powerful tool for advancing personalized neurology by identifying meaningful patterns in complex, multimodal data that can inform more targeted and effective interventions.

4.4.1 The Need for Personalization in Neurological Care

Traditional approaches to neurological disorders have typically followed a "one-size-fits-all" model, where diagnosis leads to standardized treatment protocols based on population-level evidence. While this approach has yielded important advances, it fails to account for the substantial heterogeneity within diagnostic categories and the multifactorial nature of treatment response. Several factors highlight the need for more personalized approaches:

Biological Heterogeneity: Conditions like multiple sclerosis, epilepsy, and Parkinson's disease encompass diverse pathophysiological mechanisms that may respond differently to various treatments [38]. For example, MS can manifest with predominantly inflammatory or neurodegenerative features, potentially requiring different therapeutic strategies.

Genetic Variation: Pharmacogenomic differences can significantly influence medication metabolism, efficacy, and side effect profiles [39]. For instance, genetic variants affecting the cytochrome P450 system may alter the effectiveness of antiepileptic drugs or increase the risk of adverse reactions.

Comorbidity Profiles: Neurological patients often present with multiple comorbidities that can influence treatment selection and outcomes [40]. The presence of psychiatric conditions, cognitive impairment, or systemic diseases like diabetes or hypertension may necessitate tailored treatment approaches that consider potential interactions and cumulative burden.

Individual Preferences and Goals: Patients vary in their treatment priorities, risk tolerance, and quality of life considerations. Some may prioritize symptom control, while others place greater emphasis on minimizing side effects or preserving specific functions important to their personal or professional lives.

Despite this clear need, implementing personalized neurology in clinical practice faces substantial challenges. The relationships between individual characteristics and treatment outcomes are often complex and nonlinear, making them difficult to discern through traditional statistical approaches. Moreover, the relevant data may be spread across multiple modalities and sources, from genetic profiles and neuroimaging to electronic health records and patient-reported outcomes. AI approaches offer promising solutions to these challenges by integrating diverse data types and identifying complex patterns that can guide personalized treatment decisions.

4.4.2 AI Approaches to Treatment Selection and Optimization

4.4.2.1 Predictive Modeling of Treatment Response

Machine learning models can predict individual responses to specific treatments based on patterns in patient characteristics, potentially enabling more informed selection among therapeutic options. These models typically analyze pretreatment data to identify features associated with positive or negative outcomes for particular interventions.

In epilepsy, algorithms have been developed to predict response to specific antiseizure medications based on clinical features, EEG patterns, and genetic markers [41]. Similar approaches have been applied to predict surgical outcomes in drug-resistant epilepsy, identifying candidates most likely to achieve seizure freedom following temporal lobe resection.

For multiple sclerosis, machine learning models can predict response to disease-modifying therapies based on demographic, clinical, and imaging features [42]. These approaches may help optimize treatment selection, balancing efficacy against safety concerns like the risk of progressive multifocal leukoencephalopathy with certain medications.

In Parkinson's disease, AI approaches can predict response to levodopa therapy and the development of motor complications like dyskinesias [43]. These predictions may guide decisions about when to initiate pharmacological treatment and which medication to choose. For advanced therapies like deep brain stimulation, machine learning models can identify optimal candidates based on clinical phenotypes, cognitive profiles, and neuroimaging features.

4.4.2.2 Subtyping and Stratification

Beyond predicting response to existing treatments, AI approaches can identify meaningful subtypes within diagnostic categories, potentially revealing groups that share common mechanisms and may respond similarly to specific interventions. This data-driven approach to disease stratification represents an important step toward more personalized therapeutic strategies.

In Alzheimer's disease, unsupervised learning methods have identified distinct subtypes based on patterns of atrophy, tau deposition, and clinical features [44]. These subtypes show different progression rates and may respond differently to emerging therapies targeting specific pathological processes.

For Parkinson's disease, cluster analysis and machine learning approaches have revealed distinct clinical subtypes that may have different underlying mechanisms and treatment implications [45]. These distinctions may eventually guide more targeted approaches to both symptomatic therapy and neuroprotective interventions.

In multiple sclerosis, machine learning-based clustering has identified subtypes that transcend the traditional relapsing-remitting versus progressive categories [16]. By integrating clinical, imaging, and laboratory data, these approaches have revealed patterns that better predict long-term outcomes and may inform treatment selection.

4.4.2.3 Treatment Regimen Optimization

Beyond selecting among available treatments, AI approaches can optimize dosing, timing, and combinations of interventions for individual patients. These approaches recognize that even with the correct treatment selection, parameters must often be tailored to maximize efficacy while minimizing side effects.

In epilepsy, machine learning algorithms can optimize medication dosing based on patient characteristics, potentially enhancing seizure control while reducing adverse effects [46]. For example, reinforcement learning approaches have been developed that dynamically adjust antiepileptic drug doses based on seizure frequency and side effect reports, demonstrating improved outcomes compared to standard titration protocols.

For Parkinson's disease, AI approaches can optimize deep brain stimulation parameters based on clinical symptoms and neurophysiological recordings [47]. Closed-loop systems that adjust stimulation based on real-time measures of neural activity show promise for addressing fluctuating symptoms and reducing side effects compared to standard open-loop stimulation.

4.4.3 Incorporating Patient-Generated Data

The increasing availability of wearable devices, smartphone applications, and home monitoring systems has created new opportunities for capturing patient-generated data outside traditional clinical settings. AI approaches can analyze these continuous, real-world data streams to inform more personalized and responsive treatment approaches.

4.4.4 Digital Biomarkers and Remote Monitoring

Digital biomarkers derived from wearable sensors, smartphones, or smart home systems can provide objective, continuous measures of neurological function in real-world settings [48]. These measures may better reflect functional capacity and symptom burden than periodic clinical assessments, potentially enabling more personalized treatment adjustments.

In Parkinson's disease, machine learning algorithms can analyze data from wearable accelerometers and gyroscopes to quantify tremor, bradykinesia, dyskinesia, and gait disturbances [14]. These objective measures can track symptom fluctuations throughout the day and in response to medications, potentially informing more precise dosing schedules and combinations.

For epilepsy, AI approaches can analyze data from wearable EEG, heart rate monitors, accelerometers, and other sensors to detect seizures, predict seizure risk, and evaluate treatment effects [49]. Systems like the FDA-approved Embrace watch use machine learning to detect seizures based on autonomic nervous system changes and motion patterns, potentially enhancing safety and providing more comprehensive seizure counts than self-reporting.

In multiple sclerosis, smartphone-based assessments and wearable sensors can track gait, balance, vision, cognition, and fatigue in daily life [50]. Machine learning algorithms can analyze these data to detect subtle changes in function that might indicate disease activity or treatment effects, potentially enabling earlier intervention than would be possible with standard clinical follow-up.

4.4.5 Applications Across Neurological Disorders

While the underlying principles of AI-driven personalization apply broadly, specific applications vary across neurological conditions based on their unique characteristics, available treatments, and evidence base.

4.4.6 Epilepsy

Epilepsy exemplifies the need for personalized approaches, given its diverse etiologies, seizure types, and treatment responses. AI methods have demonstrated value across the epilepsy care continuum, from syndrome classification to medication selection and surgical planning.

Machine learning algorithms can analyze EEG patterns, clinical features, and genetic data to classify epilepsy syndromes more precisely than traditional approaches [51]. These refined classifications may better predict natural history and treatment response, potentially enabling more targeted interventions from the outset.

For medication selection, AI models can predict response to specific antiepileptic drugs based on clinical features, EEG patterns, and genetic markers. These predictions may help clinicians select optimal first-line therapy, potentially reducing the trial-and-error approach that currently characterizes medication choice.

In surgical planning, AI approaches can enhance the localization of seizure onset zones and predict surgical outcomes based on multimodal data [22]. Machine learning algorithms can integrate information from structural MRI, functional MRI, EEG, magnetoencephalography (MEG), and PET to identify surgical targets with greater precision than traditional methods.

4.4.7 Neurodegenerative Disorders

Neurodegenerative disorders like Alzheimer's disease and Parkinson's disease pose unique challenges for personalization given their progressive nature, complex pathophysiology, and limited disease-modifying options. Nevertheless, AI approaches show promise for enhancing both symptom management and emerging disease-modifying strategies.

In Alzheimer's disease, machine learning models can predict individual progression trajectories based on baseline clinical, cognitive, genetic, and neuroimaging features [17]. These predictions may guide decisions about when to initiate treatment and which interventions to prioritize.

With the emergence of disease-modifying therapies targeting specific pathological processes like amyloid accumulation, AI approaches can help identify patients most likely to benefit from these interventions [52]. Machine learning algorithms can analyze multimodal data to predict response to anti-amyloid antibodies or other targeted therapies, potentially enabling more precise patient selection for these expensive and sometimes risky treatments.

In Parkinson's disease, AI approaches can characterize individual symptom patterns and fluctuations, informing personalized medication regimens [14]. Machine learning algorithms can analyze data from wearable sensors to detect subtle motor changes throughout the day, potentially enabling more precise timing and dosing of dopaminergic medications.

For surgical interventions like deep brain stimulation (DBS), AI can enhance both patient selection and parameter optimization [23]. Machine learning models can predict DBS outcomes based on clinical features, connectivity patterns, and electrode locations, potentially identifying optimal candidates and targets.

4.4.8 Challenges and Future Directions

Despite its promise, implementing AI-driven personalized neurology in routine clinical practice faces several challenges that must be addressed through technological advances, validation studies, and implementation science.

4.4.9 Data Quality and Integration

The performance of AI models depends critically on the quality, quantity, and representativeness of training data. Current approaches often rely on data from specialized research centers or clinical trials, which may not fully represent the diversity of patients seen in routine practice [53]. Efforts to develop more inclusive datasets that capture the full spectrum of disease manifestations, comorbidities, and demographic characteristics will be essential for building generalizable and equitable models.

Integrating data across different sources, formats, and time points presents additional challenges. Electronic health records, imaging repositories, genetic databases, and patient-generated data often exist in siloed systems with limited interoperability [54]. Developing standardized data models and consent frameworks that enable secure data sharing while protecting privacy represents an important direction for advancing personalized neurology.

4.4.10 Validation and Implementation

Translating promising AI approaches from research to clinical practice requires rigorous validation and thoughtful implementation. Prospective studies demonstrating that AI-guided decisions improve patient outcomes compared to standard care are relatively rare but will be essential for clinical adoption and regulatory approval [53]. These studies should assess not only statistical performance metrics but also clinical utility, cost-effectiveness, and impacts on workflow and decision-making.

Implementation science approaches can help identify barriers and facilitators to adopting AI tools in neurological practice. These barriers may include technical factors like integration with electronic health records, human factors like clinician trust and understanding, organizational factors like workflow disruption, and policy factors like reimbursement and liability considerations.

4.4.11 Interpretability and Trust

The "black box" nature of many advanced AI methods raises concerns about interpretability and trustworthiness, particularly for high-stakes clinical decisions [55]. Clinicians may be reluctant to follow recommendations from systems they cannot understand, while patients may question decisions based on algorithms rather than human judgment.

Explainable AI techniques aim to make model predictions more transparent by identifying which features most strongly influence specific recommendations [56]. For example, attention mechanisms can highlight regions of a brain scan that drive diagnostic predictions, while feature importance measures can identify clinical variables that most strongly predict treatment response. These explanations may enhance trust and enable clinicians to evaluate the clinical plausibility of AI-generated recommendations.

The integration of AI approaches with personalized medicine frameworks holds significant promise for enhancing neurological care. By moving beyond one-size-fits-all approaches to more targeted interventions based on individual characteristics, these methods may improve outcomes while optimizing resource utilization. However, realizing this potential will require addressing substantial challenges related to data quality, validation, interpretability, and ethical considerations through collaborative efforts across disciplines.

4.5 Early Detection Systems for Alzheimer's Disease from Neuroimaging Biomarkers

Figure 4.3 presents a visual framework of how AI enhances the early detection of Alzheimer's disease across its progression stages, from preclinical changes to clinical manifestation. This visualization captures the relationship between neuroimaging biomarkers, AI detection capabilities, and disease progression, illustrating how advanced algorithms can identify subtle brain changes years before symptom onset. As we explore the specific techniques and applications throughout this section, this framework contextualizes the transformative potential of AI for early intervention and treatment planning in Alzheimer's care.

Alzheimer's disease (AD) represents one of the most significant healthcare challenges of our time, affecting over 35 million people worldwide with projections suggesting this number will triple by 2050 [57]. Current treatments offer limited symptomatic relief but do not alter the underlying disease course, partly because intervention typically begins after substantial neurodegeneration has already occurred [58]. Earlier detection, ideally during the preclinical or early mild cognitive impairment (MCI) stages, may offer a window for more effective intervention before irreversible damage accumulates [59]. Artificial intelligence approaches applied to neuroimaging biomarkers show particular promise for enabling this early detection, potentially identifying subtle brain changes years before clinical symptoms become apparent.

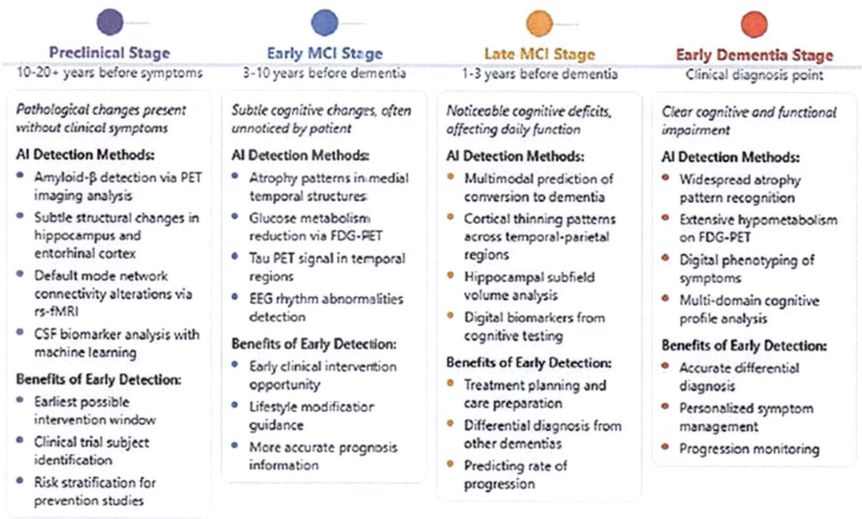

Fig. 4.3 Early detection timeline for Alzheimer's disease using AI

4.5.1 The Need for Early Detection

The pathophysiological processes of Alzheimer's disease begin decades before clinical symptoms emerge, creating a significant opportunity for early detection and intervention [60]. Several factors highlight the potential value of identifying AD in its earliest stages:

Therapeutic Window: Emerging disease-modifying therapies may be most effective when administered before extensive neuronal damage has occurred [61]. The recent FDA approval of aducanumab, which targets amyloid plaques, has intensified interest in identifying appropriate candidates for such interventions during early disease stages when amyloid burden might be more modifiable.

Clinical Trial Design: Identifying individuals with preclinical or early-stage AD can enhance clinical trial efficiency by enabling the enrollment of more homogeneous populations with a greater likelihood of measurable decline over the study period [62]. This approach may reduce the sample sizes needed to detect treatment effects and increase the probability of successful trials for disease-modifying therapies.

Planning and Support: Early diagnosis allows patients and families to plan for future care needs, engage in advance care planning, and access appropriate support services before crisis situations develop [63]. This preparation may reduce caregiver burden, delay institutionalization, and improve quality of life for both patients and care partners.

Despite these potential benefits, current approaches to AD diagnosis face significant limitations. Clinical assessments alone lack sensitivity for detecting subtle cognitive changes in high-functioning individuals, while traditional neuroimaging analysis may miss early pathological changes that do not yet cause obvious structural abnormalities [64]. Standard biomarker measurements such as cerebrospinal fluid (CSF) analysis for amyloid-β and tau require invasive procedures, limiting their applicability for widespread screening. These challenges have motivated interest in AI-enhanced analysis of neuroimaging data as a potentially more sensitive, noninvasive approach to early AD detection.

4.5.2 Neuroimaging Biomarkers for Alzheimer's Disease

Various neuroimaging modalities capture different aspects of AD pathophysiology, from structural changes to functional alterations, metabolic abnormalities, and molecular pathology. Each modality offers distinct biomarkers that may contribute to early detection when analyzed with appropriate AI methods.

4.5.3 Structural MRI Biomarkers

Structural magnetic resonance imaging (sMRI) provides detailed anatomical information about brain tissue and has been extensively studied for AD detection and monitoring. Key biomarkers derived from sMRI include the following:

Hippocampal and Medial Temporal Lobe Atrophy: The hippocampus and surrounding medial temporal lobe structures show early atrophy in AD, often preceding clinical symptoms [65]. Volumetric measurements of these regions have demonstrated value for distinguishing AD from normal aging and predicting progression from MCI to dementia. However, manual segmentation of these structures is time-consuming and subject to interrater variability, motivating automated approaches.

Cortical Thickness: Specific patterns of cortical thinning, particularly in temporal, parietal, and posterior cingulate regions, characterize early AD [66]. These patterns may differ from those observed in normal aging or other neurodegenerative conditions, potentially enhancing diagnostic specificity. Measuring cortical thickness across the entire brain surface generates high-dimensional data well suited for AI analysis.

4.5.4 Functional Neuroimaging Biomarkers

Functional neuroimaging techniques assess aspects of brain activity and can detect alterations in neural processing that may precede structural changes in AD.

Functional Connectivity: Resting-state functional MRI (rs-fMRI) measures temporal correlations in blood oxygen level-dependent (BOLD) signals between brain regions, reflecting functional connectivity within neural networks [67]. AD is associated with disrupted connectivity in specific networks, particularly the default mode network (DMN), which shows altered function early in the disease course. The complexity of these network relationships makes them well suited for machine learning analysis.

4.5.5 Metabolic and Molecular Biomarkers

Positron emission tomography (PET) can visualize various aspects of brain metabolism and molecular pathology using different radioactive tracers.

Glucose Metabolism: Fluorodeoxyglucose (FDG) PET measures regional glucose metabolism, which shows characteristic patterns of reduction in AD, particularly in posterior cingulate, precuneus, and temporoparietal regions [68]. These metabolic changes may appear before structural atrophy and correlate more closely with cognitive symptoms than some structural measures.

Amyloid Deposition: Amyloid PET tracers like Pittsburgh compound B (PiB) and florbetapir visualize amyloid-β plaque deposition, a core pathological feature of AD that begins years before symptom onset [69]. Amyloid positivity represents an important early biomarker of AD pathology, although its presence does not necessarily predict imminent cognitive decline without considering other factors.

4.5.6 AI Approaches for Early Detection

Artificial intelligence methods have demonstrated impressive capabilities for analyzing neuroimaging data, often outperforming traditional analytical approaches for early AD detection. Various AI techniques address different aspects of this challenge, from preprocessing and feature extraction to classification, prediction, and multimodal integration.

4.5.7 Deep Learning for Image Analysis

Convolutional neural networks (CNNs) and other deep learning architectures have transformed neuroimaging analysis by automatically learning relevant features from minimally processed images rather than relying on handcrafted features or predefined regions of interest.

Structural MRI Analysis: Deep learning approaches have achieved high accuracy in distinguishing AD from normal aging and predicting progression from MCI to dementia based on structural MRI [70]. For example, Wen et al. [71] developed a

3D convolutional neural network that classified AD, MCI, and normal controls with over 90% accuracy based on structural MRI alone, outperforming traditional volumetric methods.

Beyond classification, deep learning can automate the segmentation of brain structures relevant to AD diagnosis, such as the hippocampus and entorhinal cortex. These automated methods enhance the feasibility of volumetric analysis in clinical settings where specialized neuroimaging expertise may be limited.

PET Image Analysis: For metabolic and molecular imaging, deep learning techniques can extract subtle patterns that may elude visual inspection or standard quantitative analysis. Ding et al. [72] developed a CNN that predicted conversion from MCI to AD with 82% accuracy based on FDG-PET scans, identifying metabolic patterns that preceded clinical progression by up to 3 years.

4.5.8 Machine Learning for Feature-Based Analysis

While deep learning approaches often work directly with image data, traditional machine learning methods typically operate on extracted features such as regional volumes, cortical thickness measurements, or connectivity metrics. These approaches may offer advantages in interpretability and performance when working with smaller datasets.

Support Vector Machines (SVMs) have been widely applied for AD classification based on neuroimaging features. For example, Salvatore et al. [73] used SVMs to classify AD and MCI based on regional gray matter density from structural MRI, achieving over 85% accuracy in distinguishing MCI patients who later converted to AD from those who remained stable.

Random Forests and Ensemble Methods combine multiple decision trees to enhance classification performance and robustness. Moradi et al. [74] employed random forests to predict MCI-to-AD conversion based on structural MRI, FDG-PET, and cerebrospinal fluid biomarkers, achieving 91% accuracy and identifying key predictive features across modalities.

4.5.9 Multimodal Integration

Given that AD affects multiple aspects of brain structure and function, approaches that integrate information across imaging modalities often demonstrate superior performance compared to single-modality methods.

Feature-Level Fusion combines features extracted from different modalities into a single feature vector for subsequent classification. For example, Liu et al. [75] integrated regional volumes from structural MRI with metabolic measurements from FDG-PET and connectivity metrics from DTI, achieving 94% accuracy in distinguishing early AD from normal aging—significantly higher than any single modality alone.

Decision-Level Fusion applies separate classifiers to each modality and then combines their outputs through strategies like majority voting or weighted averaging. Venugopalan et al. [2] employed this approach for early AD detection, training specialized deep learning models for structural MRI, FDG-PET, and amyloid PET, then combining their predictions through a meta-learner.

4.5.10 Longitudinal Modeling

While many approaches focus on cross-sectional classification, modeling changes over time can enhance early detection by capturing disease trajectories rather than static measurements.

Recurrent Neural Networks (RNNs) and their variants like long short-term memory (LSTM) networks can model temporal sequences of imaging data, capturing patterns of change that may be more informative than single time points. For example, Cui et al. [76] employed LSTM networks to predict cognitive decline based on longitudinal structural MRI, demonstrating superior performance compared to cross-sectional models.

Event-Based Models reconstruct the temporal sequence of biomarker abnormalities in disease progression without requiring longitudinal data from individual patients [77]. These models can position individuals along a disease timeline based on their biomarker profiles, potentially enabling earlier detection by identifying the earliest biomarkers to become abnormal.

4.5.11 Key Examples: State-of-the-Art Systems

Several advanced AI systems for early AD detection have demonstrated particularly promising results in research settings and are progressing toward clinical translation.

4.5.12 Deep Learning for Multimodal Integration

The Alzheimer's disease prediction of longitudinal evolution (TADPOLE) Challenge provided a standardized framework for comparing different approaches to predicting AD progression using the Alzheimer's disease neuroimaging initiative (ADNI) dataset [78]. The winning approaches employed sophisticated AI methods for multimodal integration and longitudinal modeling:

Venugopalan et al. [2] developed a modular deep learning system that processes structural MRI, FDG-PET, and amyloid PET through specialized networks before combining their predictions. This approach achieved the highest accuracy in predicting MCI-to-AD conversion in the TADPOLE Challenge, with an area under the receiver operating characteristic curve (AUC) of 0.93.

4.5.13 Longitudinal Trajectory Modeling

Fisher et al. [17] developed a novel approach for modeling individualized trajectories of brain change using machine learning, demonstrating superior performance for early detection compared to cross-sectional analysis. Their system achieved an AUC of 0.89 for predicting cognitive decline within a 36-month window based on trends identified in the preceding 24 months.

4.5.14 Challenges and Limitations

Despite impressive advances in AI-based early detection systems for AD, several significant challenges and limitations must be addressed to realize their full potential in clinical practice.

4.5.15 Generalizability and External Validation

Many AI systems for AD detection have been developed and validated using research cohorts like ADNI, which may not fully represent the diversity of patients seen in routine clinical practice [79]. These cohorts often have strict inclusion criteria, standardized imaging protocols, and extensive quality control that exceed what is typical in clinical settings.

External validation studies have sometimes demonstrated lower accuracy than reported in initial development, highlighting the challenges of generalization across different populations and imaging protocols. For example, Klöppel et al. [80] found that a deep learning system for AD classification showed a 15–20% drop in accuracy when applied to an independent clinical cohort compared to its performance on research datasets.

4.5.16 Standardization and Harmonization

Variability in imaging acquisition parameters, scanner hardware, and reconstruction methods can significantly impact neuroimaging biomarkers, potentially introducing systematic biases that confound AI analysis [81]. This variability poses particular challenges for multisite implementation and longitudinal monitoring, where changes in equipment or protocols might be misinterpreted as biological changes.

Harmonization methods aim to address these challenges by minimizing systematic differences between scanners or protocols while preserving biological variability. Statistical approaches like ComBat have shown promise for harmonizing structural MRI measurements across sites [82], while deep learning-based methods like CycleGAN have been applied to generate standardized images from diverse acquisition parameters.

4.5.17 Clinical Integration and Workflow

Incorporating AI-based analysis into clinical workflows presents practical challenges related to data transfer, processing time, integration with electronic health records, and result presentation [83]. Many current neuroimaging AI systems require specialized expertise for implementation and interpretation, limiting their accessibility in general clinical settings where neuroradiology or neuroimaging expertise may be limited.

4.5.18 Future Directions

Looking ahead, several promising directions may enhance the impact and clinical utility of AI-based early detection systems for AD, addressing current limitations while expanding capabilities.

4.5.19 Multimodal Biomarker Integration

Future systems will likely integrate an increasing range of biomarkers beyond traditional neuroimaging, including blood-based markers like plasma amyloid-β and phosphorylated tau that have shown promise for AD detection [84]. Recent advances in fluid biomarkers offer less invasive alternatives to CSF analysis or PET imaging for detecting AD pathology, potentially enabling more accessible screening approaches when combined with AI analysis.

Digital biomarkers derived from smartphone-based cognitive testing, voice analysis, or the continuous monitoring of functional activities represent another emerging data source that could complement neuroimaging for early detection [85]. These measures capture different aspects of cognitive and functional status than traditional assessments and may detect subtle changes in everyday performance before clinical symptoms become apparent.

4.5.20 Personalized Risk Assessment and Progression Modeling

Beyond binary classification (AD vs. non-AD), future systems will increasingly focus on personalized risk assessment and progression modeling tailored to individual characteristics. These approaches recognize the heterogeneous nature of AD and aim to provide more granular information about individual disease trajectories, potential complications, and expected rates of progression.

Incorporating genetic information, comorbidities, lifestyle factors, and cognitive reserve into these models may enhance their precision and clinical relevance [86]. For example, the APOE genotype significantly influences AD risk and progression rate, while cardiovascular comorbidities may accelerate cognitive decline in individuals with AD pathology.

The early detection of Alzheimer's disease through AI-enhanced neuroimaging analysis represents a rapidly advancing field with significant potential to transform AD diagnosis and management. By identifying subtle brain changes years before clinical symptoms emerge, these approaches may expand the window for intervention, enhance clinical trial efficiency, and eventually enable more effective disease management. However, realizing this potential will require addressing substantial challenges related to generalizability, standardization, clinical integration, regulation, and ethical considerations.

4.6 Challenges in Clinical Implementation and Validation

Artificial intelligence shows tremendous promise for enhancing neurological care across the diagnostic, prognostic, and therapeutic domains. However, translating these technologies from research environments to routine clinical practice presents significant challenges. This section examines key barriers to implementation and validation, along with potential approaches for addressing them to accelerate responsible clinical adoption.

4.6.1 Technical and Infrastructural Challenges

4.6.1.1 Data Quality and Standardization

Clinical neuroimaging data often differ substantially from carefully curated research datasets. Most AI algorithms are developed and validated using these research datasets [54]. Variations in acquisition parameters, scanner hardware, reconstruction methods, and quality control procedures can significantly impact image characteristics and derived measurements. These variations potentially reduce the performance of AI systems when applied in routine clinical settings [81].

Harmonization methods represent one approach for addressing data heterogeneity, aiming to minimize technical variations while preserving biological information. Statistical techniques like ComBat have shown promise for reducing scanner effects in structural MRI measurements [82], while deep learning approaches like CycleGAN can generate standardized images from diverse acquisition parameters. However, these methods typically require representative data from each scanner or protocol, creating implementation challenges for widely distributed systems.

Data standardization initiatives offer complementary solutions by promoting consistent acquisition and processing methods across sites. The brain imaging data structure (BIDS) provides standardized formats for organizing and describing neuroimaging data, facilitating sharing and interoperability [87]. Similarly, consensus protocols like the Alzheimer's disease neuroimaging initiative (ADNI) MRI protocol establish standardized acquisition parameters that enhance comparability across sites. Wider adoption of these standards in clinical settings could reduce technical variability and enhance the robustness of AI applications.

4.6.1.2 IT Infrastructure and Integration

Implementing neuroimaging AI systems in clinical practice requires appropriate IT infrastructure for data storage, processing, and integration with existing clinical systems [83]. Many healthcare institutions face challenges related to legacy systems, limited computational resources, restricted network connectivity, and complex security requirements that can impede the adoption of advanced AI tools.

Cloud-based solutions offer potential advantages for computational scalability and centralized updates, but raise additional considerations regarding data privacy, security, and transfer speeds. Regulatory frameworks like HIPAA in the United States and GDPR in Europe impose specific requirements for protected health information that must be addressed in cloud-based implementations.

Integration with electronic health records (EHRs) and picture archiving and communication systems (PACS) represents another crucial challenge for clinical implementation. While standards like HL7 FHIR and DICOM provide frameworks for interoperability, practical implementation often requires custom interfaces and workflows adapted to specific institutional systems.

4.6.2 Clinical Validation and Evaluation Challenges

4.6.2.1 Performance Metrics and Benchmarks

Evaluating the clinical utility of neuroimaging AI systems requires appropriate performance metrics and benchmarks that align with clinical needs and decision-making processes [88]. While technical metrics like accuracy, sensitivity, specificity, and area under the receiver operating characteristic curve (AUC) provide important information about algorithmic performance, they may not fully capture clinical impact or value.

For diagnostic applications, metrics should reflect the consequences of different error types in specific clinical contexts. For example, in acute stroke triage, high sensitivity for large vessel occlusions may be prioritized over specificity to avoid missing patients who could benefit from thrombectomy. Conversely, in Alzheimer's disease screening, higher specificity might be valued to avoid unnecessary anxiety and follow-up testing.

For prognostic applications, calibration (how well predicted probabilities match actual outcomes) may be as important as discrimination (ability to distinguish between outcome groups). Well-calibrated risk predictions enable more informed decision-making about treatment intensity, monitoring frequency, and resource allocation.

Beyond technical performance, usability metrics that assess integration into clinical workflows, impact on decision-making, and user satisfaction provide crucial information about real-world utility. Time savings, a reduction in interreader variability, improved confidence in difficult cases, and the enhanced detection of subtle findings represent potential benefits that should be systematically evaluated in clinical validation studies.

4.6.3 Study Design and Population Representativeness

The design of clinical validation studies significantly influences their generalizability and relevance for real-world implementation [89]. Many published studies employ retrospective designs with selected cases, which may overestimate performance compared to prospective implementation in consecutive, unselected patients. Selection bias can occur when cases with technical limitations, rare presentations, or ambiguous findings are excluded, leading to inflated performance estimates that do not reflect routine clinical practice.

Population represents another critical consideration, as many AI systems are developed and validated using data from specific demographic groups or clinical settings that may not reflect the diversity of target populations [90]. Performance disparities across age, sex, race, ethnicity, and socioeconomic factors have been documented in various clinical AI applications, highlighting the importance of diverse training and validation data.

Multicenter validation studies offer particularly valuable evidence for generalizability, as they can assess performance across different patient populations, imaging equipment, and clinical workflows. Such studies may reveal performance variations that were not apparent in single-center evaluations and identify factors that influence real-world effectiveness.

4.6.4 Regulatory and Ethical Challenges

4.6.4.1 Evolving Regulatory Frameworks

Regulatory frameworks for AI-based medical devices continue to evolve as agencies like the U.S. Food and Drug Administration (FDA), the European Medicines Agency (EMA), and others develop approaches for evaluating these novel technologies [91]. Traditional regulatory pathways designed for conventional medical devices may not fully address the unique characteristics of AI systems, including their adaptability, opacity, and potential for continuous learning and updating after deployment.

The FDA's proposed regulatory framework for AI/ML-based Software as a Medical Device (SaMD) represents one approach to addressing these challenges, introducing the concept of "predetermined change control plans" that specify how systems may be updated without requiring completely new regulatory reviews [92]. This approach aims to balance innovation and safety by establishing bounds within which systems can evolve while maintaining regulatory oversight for significant changes that might affect clinical performance or safety.

Demonstrating clinical validation for regulatory approval typically requires substantial evidence from well-designed studies, but the specific requirements may vary depending on the intended use, risk level, and novelty of the technology. AI systems intended for triage or prioritization may face different evidentiary standards than those providing definitive diagnoses or treatment recommendations.

4.6.4.2 Explainability and Transparency

The "black box" nature of many advanced AI methods, particularly deep learning approaches, raises concerns about explainability and transparency in clinical applications [55]. Clinicians may be reluctant to incorporate AI recommendations into their decision-making without understanding the basis for these recommendations, particularly when they contradict clinical impressions or established practices.

Explainable AI (XAI) techniques aim to address these concerns by making model decisions more transparent and interpretable without necessarily sacrificing performance [56]. These approaches range from inherently interpretable models like sparse linear classifiers to post hoc explanation methods for complex models, such as attention maps, feature importance rankings, or example-based explanations.

The appropriate level of explainability may vary depending on the specific application, user needs, and risk level. For high-risk applications like treatment selection or surgical planning, detailed explanations with mechanistic plausibility may be essential for clinical acceptance. For lower-risk applications like image enhancement or worklist prioritization, more general performance metrics might suffice if the ultimate clinical decision remains with human experts.

4.6.5 Implementation Science and Adoption Challenges

4.6.5.1 Workflow Integration and User Acceptance

Successful clinical implementation requires careful attention to workflow integration and user acceptance, regardless of algorithmic performance [93]. AI systems that disrupt established workflows, increase documentation burden, or generate alerts that do not align with clinical priorities are unlikely to achieve widespread adoption despite potential benefits.

Understanding the clinical context and decision-making environment represents an essential first step in effective implementation. Different clinical settings—emergency departments, inpatient units, outpatient clinics, or remote consultations—have distinct workflows, time constraints, and information needs that should inform AI system design and integration.

User acceptance depends on multiple factors beyond technical performance, including perceived usefulness, ease of use, compatibility with existing practices, and professional autonomy considerations. Engaging end-users throughout the development and implementation process, from initial needs assessment to iterative refinement based on feedback, can enhance both system design and eventual adoption.

4.6.6 Clinical and Economic Impact Assessment

Demonstrating meaningful clinical and economic impact represents a crucial challenge for neuroimaging AI implementation, particularly in resource-constrained healthcare environments. While technical performance metrics provide important information about algorithmic capabilities, healthcare decision-makers increasingly require evidence of improved patient outcomes, enhanced efficiency, or cost-effectiveness to justify investment in new technologies.

Clinical impact assessment should examine effects on diagnostic accuracy, treatment decisions, workflow efficiency, and, ultimately, patient outcomes. Relevant metrics might include changes in diagnostic sensitivity and specificity, reductions in reading time or interreader variability, modifications in treatment plans, or improvements in clinical endpoints like time to treatment, complication rates, or functional outcomes.

Economic evaluations should consider both direct costs (software licensing, hardware requirements, and implementation support) and indirect effects on healthcare utilization, workforce requirements, and reimbursement. Cost-effectiveness analyses comparing AI implementation to standard care can provide valuable information for coverage and reimbursement decisions, while budget impact models assess affordability and resource implications for specific healthcare systems or payers.

4.6.7 Training and Education

The effective use of AI systems in clinical neuroscience requires appropriate training and education for healthcare professionals, many of whom may have limited background in artificial intelligence, data science, or advanced statistics. Understanding AI capabilities, limitations, appropriate use cases, and potential biases is essential for responsible clinical implementation and interpretation of AI-generated information.

Educational initiatives should address technical aspects of AI systems while emphasizing clinical context and integration with existing knowledge and workflows. Training should cover system operation, result interpretation, appropriate override procedures for questionable recommendations, and recognition of potential failure modes or limitations.

Professional societies and certification bodies have important roles in establishing standards and competencies for AI use in clinical neuroscience. Organizations like the American College of Radiology, the American Academy of Neurology, and the European Society of Radiology have begun developing guidelines, use cases, and educational resources for AI applications in their respective domains.

4.6.8 Implementation of Strategies and Solutions

4.6.8.1 Phased Implementation and Continuous Evaluation

Phased implementation approaches that begin with limited clinical scenarios and gradually expand based on demonstrated performance and value can mitigate risks while enabling systematic evaluation and refinement. Initial implementation might focus on well-defined use cases with clear metrics for success, strong evidence of benefit, and minimal disruption to existing workflows.

Continuous monitoring and evaluation throughout the implementation process enable the early detection of performance issues, workflow challenges, or unintended consequences that might not have been apparent during initial validation. Establishing clear metrics, reporting mechanisms, and review processes helps ensure that problems are identified and addressed promptly while providing data for ongoing improvement.

Learning health system frameworks that integrate research and practice can facilitate this continuous evaluation and improvement cycle [94]. By systematically collecting data on AI system performance, clinical decisions, and patient outcomes during routine care, these frameworks enable ongoing refinement of algorithms and implementation approaches based on real-world experience.

4.6.9 Multidisciplinary Collaboration and Implementation Teams

Successful implementation typically requires multidisciplinary collaboration among various stakeholders, including clinicians, data scientists, IT professionals, administrators, and patient representatives. Each group brings unique perspectives and expertise regarding clinical needs, technical capabilities, operational constraints, and patient perspectives that collectively inform effective implementation strategies.

Dedicated implementation teams with representation from these diverse stakeholders can coordinate the complex process of integrating AI systems into clinical practice. These teams should have clear leadership, defined roles and responsibilities, appropriate resources, and executive support to address the technical, clinical, operational, and cultural aspects of implementation.

Clinical champions who understand both the technology and clinical context can play particularly valuable roles in implementation efforts. These individuals help bridge the gap between technical developers and clinical users, translate between different professional languages and perspectives, and address concerns or resistance that might emerge during implementation.

The clinical implementation and validation of AI in neuroscience practice represents a complex sociotechnical challenge that extends far beyond algorithmic performance. Addressing the multifaceted technical, clinical, regulatory, and organizational aspects of implementation requires coordinated efforts across disciplines, institutions, and stakeholders. By acknowledging these challenges explicitly and developing systematic approaches to address them, the field can accelerate responsible translation of promising research into clinical impact while establishing sustainable frameworks for ongoing innovation and improvement.

4.6.10 Conclusion

Artificial intelligence is transforming clinical neuroscience practice across the care continuum, from diagnosis and prognosis to treatment planning and monitoring. Advanced algorithms can now analyze complex neuroimaging data with accuracy approaching or exceeding human experts in specific tasks, potentially enhancing consistency, efficiency, and early detection of subtle abnormalities. Machine learning models can integrate diverse clinical data to predict individual disease trajectories and treatment responses, potentially enabling more personalized approaches to neurological care. AI-optimized neuromodulation techniques like transcranial direct current stimulation show promise for enhancing treatment efficacy through individualized targeting and parameter selection. Early detection systems for conditions like Alzheimer's disease may identify subtle brain changes years before clinical symptoms emerge, potentially expanding the window for intervention.

Despite these promising advances, significant challenges remain for widespread clinical implementation. Data quality, standardization, and interoperability issues complicate algorithm development and generalization across diverse settings. Regulatory frameworks continue to evolve for AI-based medical applications, creating uncertainty about validation requirements and approval pathways. Questions about explainability, transparency, and liability pose challenges for clinical integration, particularly for high-stakes decisions. Practical implementation requires attention to workflow integration, user acceptance, and economic considerations beyond technical performance.

Looking ahead, several promising directions may enhance the clinical impact of AI in neuroscience. Multimodal approaches that integrate diverse data types—imaging, electrophysiology, genetics, and digital biomarkers—show particular promise for comprehensive characterization of complex neurological conditions. Explainable AI techniques that make algorithmic decisions more transparent and interpretable may enhance clinician trust and facilitate responsible implementation. Federated learning and privacy-preserving techniques enable collaborative model development without centralizing sensitive data, potentially accelerating progress while protecting patient privacy. Implementation science approaches that systematically address workflow, training, and organizational factors will be crucial for translating promising research into routine clinical practice.

The integration of AI into clinical neuroscience represents a significant opportunity to enhance the precision, accessibility, and effectiveness of neurological care. Realizing this potential will require continued collaboration among neuroscientists, clinicians, computer scientists, patients, and policymakers to develop systems that augment human capabilities while addressing ethical, practical, and technical challenges. With thoughtful development and implementation, AI technologies may ultimately transform our approach to neurological disorders, enabling earlier detection, more accurate diagnosis, and more personalized treatment for patients worldwide.

4.6.11 Summary

This chapter explores the transformative potential of artificial intelligence in clinical neuroscience, examining how AI technologies are revolutionizing the diagnosis, treatment, and monitoring of neurological disorders. This chapter presents a comprehensive overview of AI applications across the entire neurology care continuum, from initial assessment to long-term management.

The opening sections detail real-world AI applications in neuroimaging analysis, where deep learning algorithms now match or exceed human expert performance in detecting conditions like stroke, brain tumors, and neurodegenerative diseases. This chapter describes how AI enhances EEG interpretation for epilepsy diagnosis, supports clinical decision-making through integrated data analysis, and enables more accurate prediction of disease progression and treatment outcomes across conditions including multiple sclerosis, Parkinson's disease, and Alzheimer's disease.

A significant portion focuses on AI-optimized transcranial direct current stimulation (tDCS), illustrating how machine learning approaches can personalize neuromodulation therapy. This chapter explains how AI addresses the substantial interindividual variability in treatment response by optimizing electrode placement, stimulation parameters, and timing based on individual brain anatomy and functional patterns.

The discussion of personalized treatment approaches demonstrates how AI moves beyond "one-size-fits-all" protocols to enable truly individualized care. This includes treatment selection based on predicted response patterns, disease subtyping through unsupervised learning, and optimization of medication regimens and rehabilitation protocols tailored to individual characteristics and preferences.

This chapter extensively covers early detection systems for Alzheimer's disease, showing how AI analysis of neuroimaging biomarkers can identify subtle brain changes years before clinical symptoms emerge. This section details various AI approaches across multiple imaging modalities and discusses the potential for earlier intervention when treatments may be most effective.

Throughout, this chapter emphasizes the complementary relationship between AI systems and clinical expertise, positioning AI as a tool that augments rather than replaces human judgment. The concluding section addresses the substantial challenges in clinical implementation, including data quality issues, regulatory frameworks, workflow integration, and the need for transparent, interpretable AI systems that clinicians and patients can trust.

This chapter ultimately presents AI as a powerful enabler of more precise, personalized, and effective neurological care while acknowledging that realizing this potential requires careful attention to validation, implementation science, and ethical considerations.

References

1. Durstewitz D, Koppe G, Meyer-Lindenberg A. Deep neural networks in psychiatry. Mol Psychiatry. 2019;24(11):1583–98.
2. Venugopalan J, Tong L, Hassanzadeh HR, Wang MD. Multimodal deep learning models for early detection of Alzheimer's disease stage. Sci Rep. 2021;11(1):1–13.
3. Cinel C, Valeriani D, Poli R. Neurotechnologies for human cognitive augmentation: current state of the art and future prospects. Front Hum Neurosci. 2019;13:13.
4. Akkus Z, Galimzianova A, Hoogi A, Rubin DL, Erickson BJ. Deep learning for brain MRI segmentation: state of the art and future directions. J Digit Imaging. 2017;30(4):449–59.
5. Goodkin O, Pemberton H, Vos SB, Prados F, Sudre CH, Moggridge J, et al. The quantitative neuroradiology initiative framework: application to dementia. Br J Radiol. 2019;92(1101):20190365.
6. Chilamkurthy S, Ghosh R, Tanamala S, Biviji M, Campeau NG, Venugopal VK, et al. Deep learning algorithms for detection of critical findings in head CT scans: a retrospective study. Lancet. 2018;392(10162):2388–96.
7. Murray NM, Unberath M, Hager GD, Hui FK. Artificial intelligence to diagnose ischemic stroke and identify large vessel occlusions: a systematic review. J Neurointerv Surg. 2020;12(2):156–64.
8. U.S. Food and Drug Administration. 2018. FDA Press Announcement. Retrieved from https://www.fda.gov/news-events/press-announcements/fda-permits-marketing-clinical-decision-support-software-alerting-providers-potential-stroke.
9. Hao W, Zhao X, Zhang S, Yang G, Zhang Y. Multi-modal neuroimaging feature selection with consistent metric constraint for diagnosis of Alzheimer's disease. Med Image Anal. 2020;60:101625.
10. Tjepkema-Cloostermans MC, de Carvalho RC, van Putten MJ. Deep learning for detection of focal epileptiform discharges from scalp EEG recordings. Clin Neurophysiol. 2018;129(10):2191–6.
11. Kiral-Kornek I, Roy S, Nurse E, Mashford B, Karoly P, Carroll T, et al. Epileptic seizure prediction using big data and deep learning: toward a mobile system. EBioMedicine. 2018;27:103–11.
12. Struck AF, Ustun B, Ruiz AR, Lee JW, LaRoche SM, Hirsch LJ, et al. Association of an electroencephalography-based risk score with seizure probability in hospitalized patients. JAMA Neurol. 2017;74(12):1419–24.
13. Bhagwat N, Pipitone J, Winterburn JL, Guo T, Duerden EG, Voineskos AN, et al. An artificial neural network model for clinical score prediction in Alzheimer disease using structural neuroimaging measures. J Psychiatry Neurosci. 2018;43(1):16.
14. Espay AJ, Bonato P, Nahab FB, Maetzler W, Dean JM, Klucken J, et al. Technology in Parkinson's disease: challenges and opportunities. Mov Disord. 2016;31(9):1272–82.
15. Zhao Y, Healy BC, Rotstein D, Guttmann CR, Bakshi R, Weiner HL, et al. Exploration of machine learning techniques in predicting multiple sclerosis disease course. PLoS One. 2017;12(4):e0174866.
16. Eshaghi A, Young AL, Wijeratne PA, Prados F, Arnold DL, Narayanan S, et al. Identifying multiple sclerosis subtypes using unsupervised machine learning and MRI data. Nat Commun. 2021;12(1):1–12.
17. Fisher CK, Smith AM, Walsh JR, Coalition Against Major Diseases (CAMD). Machine learning for comprehensive forecasting of Alzheimer's disease progression. Sci Rep. 2019;9(1):1–14.
18. Goldenholz DM, Goldenholz SR, Romero J, Moss R, Sun H, Westover B. Development and validation of forecasting next reported seizure using e-diaries. Ann Neurol. 2020;88(3):588–95.
19. Chen Z, Brodie MJ, Liew D, Kwan P. Treatment outcomes in patients with newly diagnosed epilepsy treated with established and new antiepileptic drugs: a 30-year longitudinal cohort study. JAMA Neurol. 2018;75(3):279–86.

20. Durso-Finley J, Falet J-PR, Nichyporuk B, Arnold DL, Arbel T. Personalized prediction of future lesion activity and treatment effect in multiple sclerosis from baseline MRI. In: Proceedings of the 5th international conference on medical imaging with deep learning. PMLR; 2022. p. 172.
21. Bakas S, Reyes M, Jakab A, Bauer S, Rempfler M, Crimi A, et al. Identifying the best machine learning algorithms for brain tumor segmentation, progression assessment, and overall survival prediction in the BRATS challenge. 2018. arXiv preprint arXiv:1811.02629.
22. Varatharajah Y, Berry B, Cimbalnik J, Kremen V, Van Gompel J, Stead M, et al. Integrating artificial intelligence with real-time intracranial EEG monitoring to automate interictal identification of seizure onset zones in focal epilepsy. J Neural Eng. 2018;15(4):046035.
23. Horn A, Reich M, Vorwerk J, Li N, Wenzel G, Fang Q, et al. Connectivity predicts deep brain stimulation outcome in Parkinson disease. Ann Neurol. 2019;82(1):67–78.
24. Nitsche MA, Cohen LG, Wassermann EM, Priori A, Lang N, Antal A, et al. Transcranial direct current stimulation: state of the art 2008. Brain Stimul. 2008;1(3):206–23.
25. Stagg CJ, Nitsche MA. Physiological basis of transcranial direct current stimulation. Neuroscientist. 2011;17(1):37–53.
26. López-Alonso V, Cheeran B, Río-Rodríguez D, Fernández-del-Olmo M. Inter-individual variability in response to non-invasive brain stimulation paradigms. Brain Stimul. 2014;7(3):372–80.
27. Datta A, Bansal V, Diaz J, Patel J, Reato D, Bikson M. Gyri-precise head model of transcranial direct current stimulation: improved spatial focality using a ring electrode versus conventional rectangular pad. Brain Stimul. 2009;2(4):201–7.
28. Bikson M, Bulow P, Stiller JW, Datta A, Battaglia F, Karnup SV, Postolache TT. Transcranial direct current stimulation for major depression: a general system for quantifying transcranial electrotherapy dosage. Curr Treat Options Neurol. 2008;10(5):377–85.
29. Ruffini G, Fox MD, Ripolles O, Miranda PC, Pascual-Leone A. Optimization of multifocal transcranial current stimulation for weighted cortical pattern targeting from realistic modeling of electric fields. NeuroImage. 2014;89:216–25.
30. Paul AK, Bose A, Kalmady SV, Shivakumar V, Sreeraj VS, Parlikar R, Narayanaswamy JC, Dursun SM, Greenshaw AJ, Greiner R, Venkatasubramanian G. Superior temporal gyrus functional connectivity predicts transcranial direct current stimulation response in schizophrenia: a machine learning study. Front Psychiatry. 2022;13:923938.
31. Li H, Habes M, Fan Y. Deep ordinal ranking for multi-category diagnosis of Alzheimer's disease using hippocampal MRI data. 2020. arXiv preprint arXiv:1709.01599.
32. Rudroff T, Rainio O, Klen R. Leveraging artificial intelligence to optimize transcranial direct current stimulation for Long COVID management: a forward-looking perspective. Brain Sci. 2024;14(8):831.
33. Saturnino GB, Madsen KH, Thielscher A. Electric field simulations for transcranial brain stimulation using FEM: an efficient implementation and error analysis. J Neural Eng. 2019;16(6):066032.
34. Saeidi M, Karwowski W, Farahani FV, Fick K, Taiar R, Hancock PA, Al-Juaid A. Neural decoding of EEG signals with machine learning: a systematic review. Brain Sci. 2021;11(11):1525.
35. Bergmann TO, Groppa S, Seeger M, Mölle M, Marshall L, Siebner HR. Acute changes in motor cortical excitability during slow oscillatory and constant anodal transcranial direct current stimulation. J Neurophysiol. 2016;102(5):2303–11.
36. Brunoni AR, Moffa AH, Fregni F, Palm U, Padberg F, Blumberger DM, et al. Transcranial direct current stimulation for acute major depressive episodes: meta-analysis of individual patient data. Br J Psychiatry. 2016;208(6):522–31.
37. Li X, Chen M, Liu Q, Zheng C, Yu C, Hou G, Chen Z, Chen Y, Chen Y, Zhu G, Zhou D, Xu W. TMS-evoked potential in the dorsolateral prefrontal cortex to assess the severity of depression disease: a TMS-EEG study. Front Pharmacol. 2023;14:1207020.
38. Thompson AJ, Baranzini SE, Geurts J, Hemmer B, Ciccarelli O. Multiple sclerosis. Lancet. 2018;391(10130):1622–36.

39. Cavalleri GL, McCormack M, Alhusaini S, Chaila E, Delanty N. Pharmacogenomics and epilepsy: the road ahead. Pharmacogenomics. 2017;12(10):1429–47.
40. Barnett K, Mercer SW, Norbury M, Watt G, Wyke S, Guthrie B. Epidemiology of multimorbidity and implications for health care, research, and medical education: a cross-sectional study. Lancet. 2012;380(9836):37–43.
41. Sinha N, Wang Y, Dauwels J, Kaiser M, Thesen T, Feltenberger G, et al. Computer modelling of connectivity change suggests epileptogenesis mechanisms in idiopathic generalised epilepsy. Neuroimage Clin. 2021;29:102564.
42. Bejarano B, Bianco M, Gonzalez-Moron D, Sepulcre J, Goñi J, Arcocha J, et al. Computational classifiers for predicting the short-term course of multiple sclerosis. BMC Neurol. 2019;11:67.
43. Johnston TH, Lacoste AMB, Ravenscroft P, Su J, Tamadon S, Seifi M, Lang AE, Fox SH, Brotchie JM, Visanji NP. Using artificial intelligence to identify drugs for repurposing to treat l-DOPA-induced dyskinesia. Neuropharmacology. 2024;248:109880.
44. Habes M, Grothe MJ, Tunc B, McMillan C, Wolk DA, Davatzikos C. Disentangling heterogeneity in Alzheimer's disease and related dementias using data-driven methods. Biol Psychiatry. 2020;88(1):70–82.
45. Fereshtehnejad SM, Zeighami Y, Dagher A, Postuma RB. Clinical criteria for subtyping Parkinson's disease: biomarkers and longitudinal progression. Brain. 2017;140(7):1959–76.
46. Chakraborty S, Tomsett R, Raghavendra R, Harborne D, Alzantot M, Cerutti F, et al. Interpretability of deep learning models: a survey of results. In: 2019 IEEE smartworld, ubiquitous intelligence & computing, advanced & trusted computing, scalable computing & communications, cloud & big data computing, internet of people and smart city innovation (SmartWorld/SCALCOM/UIC/ATC/CBDCom/IOP/SCI). IEEE; 2019. p. 1294–303.
47. Grado LL, Johnson MD, Netoff TI. Bayesian adaptive dual control of deep brain stimulation in a computational model of Parkinson's disease. PLoS Comput Biol. 2018;14(12):e1006606.
48. Dorsey ER, Glidden AM, Holloway MR, Birbeck GL, Schwamm LH. Teleneurology and mobile technologies: the future of neurological care. Nat Rev Neurol. 2017;14(5):285–97.
49. Beniczky S, Ryvlin P. Standards for testing and clinical validation of seizure detection devices. Epilepsia. 2018;59:9–13.
50. Midaglia L, Mulero P, Montalban X, Graves J, Hauser SL, Julian L, et al. Adherence and satisfaction of smartphone-and smartwatch-based remote active testing and passive monitoring in people with multiple sclerosis: nonrandomized interventional feasibility study. J Med Internet Res. 2019;21(8):e14863.
51. Mei X, Lee HC, Diao KY, Huang M, Lin B, Liu C, et al. Artificial intelligence-enabled rapid diagnosis of patients with COVID-19. Nat Med. 2020;26(8):1224–8.
52. Gauthier S, Albert M, Fox N, Goedert M, Kivipelto M, Mestre-Ferrandiz J, Middleton LT. Why has therapy development for dementia failed in the last two decades? Alzheimers Dement. 2021;12(1):e12046.
53. Kelly CJ, Karthikesalingam A, Suleyman M, Corrado G, King D. Key challenges for delivering clinical impact with artificial intelligence. BMC Med. 2019;17(1):1–9.
54. Wilkinson MD, Dumontier M, Aalbersberg IJ, Appleton G, Axton M, Baak A, et al. The FAIR guiding principles for scientific data management and stewardship. Sci Data. 2016;3(1):1–9.
55. Rudin C. Stop explaining black box machine learning models for high stakes decisions and use interpretable models instead. Nat Mach Intell. 2019;1(5):206–15.
56. Tjoa E, Guan C. A survey on explainable artificial intelligence (XAI): toward medical XAI. IEEE Trans Neural Netw Learn Syst. 2020;32(11):4793–813.
57. Alzheimer's Disease International. World Alzheimer report 2021: journey through the diagnosis of dementia. London: Alzheimer's Disease International; 2021.
58. Long JM, Holtzman DM. Alzheimer disease: an update on pathobiology and treatment strategies. Cell. 2019;179(2):312–39.
59. Sperling RA, Aisen PS, Beckett LA, Bennett DA, Craft S, Fagan AM, et al. Toward defining the preclinical stages of Alzheimer's disease: recommendations from the National Institute on Aging-Alzheimer's Association workgroups on diagnostic guidelines for Alzheimer's disease. Alzheimers Dement. 2011;7(3):280–92.

60. Jack CR, Knopman DS, Jagust WJ, Petersen RC, Weiner MW, Aisen PS, et al. Tracking pathophysiological processes in Alzheimer's disease: an updated hypothetical model of dynamic biomarkers. Lancet Neurol. 2013;12(2):207–16.
61. Cummings J, Aisen P, Lemere C, Atri A, Sabbagh M, Salloway S. Aducanumab produced a clinically meaningful benefit in association with amyloid lowering. Alzheimer's Res Ther. 2021;13(1):1–3.
62. Cummings J, Lee G, Ritter A, Sabbagh M, Zhong K. Alzheimer's disease drug development pipeline: 2018. Alzheimer's Dement Transl Res Clin Interv. 2018;4:195–214.
63. de Vugt ME, Verhey FR. The impact of early dementia diagnosis and intervention on informal caregivers. Prog Neurobiol. 2013;110:54–62.
64. Beach TG, Monsell SE, Phillips LE, Kukull W. Accuracy of the clinical diagnosis of Alzheimer disease at National Institute on Aging Alzheimer Disease Centers, 2005–2010. J Neuropathol Exp Neurol. 2012;71(4):266–73.
65. Schuff N, Woerner N, Boreta L, Kornfield T, Shaw LM, Trojanowski JQ, et al. MRI of hippocampal volume loss in early Alzheimer's disease in relation to ApoE genotype and biomarkers. Brain. 2009;132(4):1067–77.
66. Dickerson BC, Bakkour A, Salat DH, Feczko E, Pacheco J, Greve DN, et al. The cortical signature of Alzheimer's disease: regionally specific cortical thinning relates to symptom severity in very mild to mild AD dementia and is detectable in asymptomatic amyloid-positive individuals. Cereb Cortex. 2009;19(3):497–510.
67. van den Heuvel MP, Hulshoff Pol HE. Exploring the brain network: a review on resting-state fMRI functional connectivity. Eur Neuropsychopharmacol. 2010;20(8):519–34.
68. Landau SM, Harvey D, Madison CM, Reiman EM, Foster NL, Aisen PS, et al. Comparing predictors of conversion and decline in mild cognitive impairment. Neurology. 2011;75(3):230–8.
69. Villemagne VL, Burnham S, Bourgeat P, Brown B, Ellis KA, Salvado O, et al. Amyloid β deposition, neurodegeneration, and cognitive decline in sporadic Alzheimer's disease: a prospective cohort study. Lancet Neurol. 2013;12(4):357–67.
70. Jo T, Nho K, Saykin AJ. Deep learning in Alzheimer's disease: diagnostic classification and prognostic prediction using neuroimaging data. Front Aging Neurosci. 2019;11:220.
71. Wen J, Thibeau-Sutre E, Diaz-Melo M, Samper-González J, Routier A, Bottani S, et al. Convolutional neural networks for classification of Alzheimer's disease: overview and reproducible evaluation. Med Image Anal. 2020;63:101694.
72. Ding Y, Sohn JH, Kawczynski MG, Trivedi H, Harnish R, Jenkins NW, et al. A deep learning model to predict a diagnosis of Alzheimer disease by using 18F-FDG PET of the brain. Radiology. 2019;290(2):456–64.
73. Salvatore C, Cerasa A, Battista P, Gilardi MC, Quattrone A, Castiglioni I. Magnetic resonance imaging biomarkers for the early diagnosis of Alzheimer's disease: a machine learning approach. Front Neurosci. 2015;9:307.
74. Moradi E, Pepe A, Gaser C, Huttunen H, Tohka J, Alzheimer's Disease Neuroimaging Initiative. Machine learning framework for early MRI-based Alzheimer's conversion prediction in MCI subjects. Neuroimage. 2015;104:398–412.
75. Liu M, Zhang J, Yap PT, Shen D. View-aligned hypergraph learning for Alzheimer's disease diagnosis with incomplete multi-modality data. Med Image Anal. 2027;36:123–34.
76. Cui R, Liu M, Alzheimer's Disease Neuroimaging Initiative. RNN-based longitudinal analysis for diagnosis of Alzheimer's disease. Comput Med Imaging Graph. 2019;73:1–10.
77. Young AL, Oxtoby NP, Daga P, Cash DM, Fox NC, Ourselin S, et al. A data-driven model of biomarker changes in sporadic Alzheimer's disease. Brain. 2014;137(9):2564–77.
78. Marinescu RV, Oxtoby NP, Young AL, Bron EE, Toga AW, Weiner MW, et al. The Alzheimer's disease prediction of longitudinal evolution (TADPOLE) challenge: results after 1 year follow-up. 2020. arXiv preprint arXiv:2002.03419.
79. Bron EE, Smits M, Van Der Flier WM, Vrenken H, Barkhof F, Scheltens P, et al. Standardized evaluation of algorithms for computer-aided diagnosis of dementia based on structural MRI: the CADDementia challenge. Neuroimage 2015;111:562–79.

80. Klöppel S, Mangin JF, Vongerichten A, Frackowiak RS, Siebner HR. Nurture versus nature: long-term impact of forced right-handedness on structure of pericentral cortex and basal ganglia. J Neurosci. 2018;30(9):3271–5.
81. Glocker B, Robinson R, Castro DC, Dou Q, Konukoglu E. Machine learning with multi-site imaging data: an empirical study on the impact of scanner effects. 2019. arXiv preprint arXiv:1910.04597.
82. Fortin JP, Parker D, Tunç B, Watanabe T, Elliott MA, Ruparel K, et al. Harmonization of multi-site diffusion tensor imaging data. Neuroimage. 2018;161:149–70.
83. Topol EJ. High-performance medicine: the convergence of human and artificial intelligence. Nat Med. 2019;25(1):44–56.
84. Hampel H, O'Bryant SE, Molinuevo JL, Zetterberg H, Masters CL, Lista S, et al. Blood-based biomarkers for Alzheimer disease: mapping the road to the clinic. Nat Rev Neurol. 2018;14(11):639–52.
85. Kourtis LC, Regele OB, Wright JM, Jones GB. Digital biomarkers for Alzheimer's disease: the mobile/wearable devices opportunity. NPJ Digit Med. 2019;2(1):1–9.
86. Vemuri P, Lesnick TG, Przybelski SA, Knopman DS, Preboske GM, Kantarci K, et al. Vascular and amyloid pathologies are independent predictors of cognitive decline in normal elderly. Brain. 2015;138(3):761–71.
87. Gorgolewski KJ, Auer T, Calhoun VD, Craddock RC, Das S, Duff EP, et al. The brain imaging data structure, a format for organizing and describing outputs of neuroimaging experiments. Sci Data. 2016;3(1):1–9.
88. Park SH, Han K. Methodologic guide for evaluating clinical performance and effect of artificial intelligence technology for medical diagnosis and prediction. Radiology. 2018;286(3):800–9.
89. Shao L, Wang Z, Xie X, Xiao L, Shi Y, Wang ZA, Zhang JE. Development and external validation of a machine learning-based fall prediction model for nursing home residents: a prospective cohort study. J Am Med Dir Assoc. 2024;25(9):105169.
90. Gianfrancesco MA, Tamang S, Yazdany J, Schmajuk G. Potential biases in machine learning algorithms using electronic health record data. JAMA Intern Med. 2018;178(11):1544–7.
91. Gerke S, Minssen T, Cohen IG. Ethical and legal challenges of artificial intelligence-driven healthcare. In: Artificial intelligence in healthcare. Academic Press; 2020. p. 295–336.
92. U.S. Food and Drug Administration. 2019. Proposed regulatory framework for modifications to artificial intelligence/machine learning (AI/ML)-based software as a medical device (SaMD): discussion paper and request for feedback. https://www.fda.gov/files/medicaldevices/published/US-FDA-Artificial-Intelligence-and-Machine-Learning-Discussion-Paper.pdf.
93. Yang Q, Steinfeld A, Zimmerman J. Unremarkable AI: Fitting intelligent decision support into critical, clinical decision-making processes. In Proceedings of the 2019 CHI conference on human factors in computing systems, 2019. p. 1–11. arXiv:1904.09612v1 [cs.HC].
94. Foley T, Fairmichael F. The potential of learning healthcare systems. Newcastle: The Learning Healthcare Project; 2015.

Part II

Ethical Dimensions and Challenges

Decoding Thoughts, Encoding Ethics

5

If we lose our mental privacy, what else is there to lose? That's it, we lose the essence of who we are.

—Rafael Yuste

5.1 Introduction

The intersection of neuroscience, artificial intelligence, and brain-computer interfaces (BCIs) represents one of the most profound technological frontiers of our time. As these technologies advance, we gain unprecedented abilities to decode neural activity, interpret cognitive processes, and even influence brain function. With these capabilities comes a host of complex ethical challenges that extend beyond traditional bioethics frameworks. The ability to access and interpret neural data raises fundamental questions about mental privacy, cognitive liberty, and the very nature of human autonomy. These data potentially reveal thoughts, intentions, emotions, and memories.

The past decade has witnessed remarkable progress in neural decoding technologies. Machine learning algorithms can now reconstruct visual experiences from brain activity [1], identify semantic representations of speech from neural signals [2], and recognize emotional states from patterns of brain activation [3]. Brain-computer interfaces have progressed from laboratory curiosities to clinical interventions and consumer products, enabling direct communication between neural tissue and external devices [4]. Meanwhile, neurostimulation technologies are increasingly capable of modulating cognitive processes, from enhancing memory [5] to altering decision-making [6]. These advances suggest a future where the boundary between mind and machine becomes increasingly permeable, with profound implications for personal identity, human relationships, and societal institutions.

However, the rapid development of these technologies has outpaced the evolution of ethical frameworks and regulatory mechanisms designed to govern their use.

T. Rudroff, *AI for the Healthy Brain and Neurological Disorders*, Neurocultural Health and Wellbeing, https://doi.org/10.1007/978-3-032-11596-6_5

Traditional bioethical principles—beneficence, nonmaleficence, autonomy, and justice—provide important foundations but require significant expansion to address novel questions raised by neural engineering [7]. What constitutes appropriate consent for technologies that may decode private mental states? How should we balance the benefits of neural data collection against risks to mental privacy? What protections should be established against unauthorized access to or manipulation of neural activity? These questions take on additional urgency as neurotechnology applications expand beyond medical contexts into consumer, commercial, military, and legal domains [8].

This chapter explores the ethical challenges posed by advancing neurotechnology, with particular focus on neural decoding and brain-computer interfaces. We begin by examining the current technical reality of "mind reading" technologies, distinguishing between scientific capabilities and public perceptions that may alternately underestimate or exaggerate current abilities. We then explore emerging ethical frameworks specifically designed for neurotechnology, considering how principles like cognitive liberty and mental privacy extend traditional bioethical approaches. Next, we address the unique privacy considerations associated with neural data, examining both technical and policy approaches to protecting this exceptionally sensitive information. We then explore the emerging concept of "neural rights" within legal and philosophical contexts, considering how basic human rights frameworks might expand to incorporate neurotechnology-specific protections. Finally, we examine potential governance mechanisms for responsible neurotechnology development, including regulatory approaches, professional guidelines, and international agreements.

Throughout this chapter, we emphasize that ethical analysis must be grounded in an accurate understanding of both technical capabilities and limitations while remaining forward-looking enough to anticipate future developments. Rather than adopting either uncritical techno-optimism or reflexive techno-pessimism, we aim to chart a course for responsible innovation that harnesses the transformative potential of neurotechnology while protecting fundamental human values. The decisions we make today about how to develop, regulate, and implement these technologies will shape not just their societal impact but potentially the future of human cognition itself.

5.2 Mind Reading Technologies: Technical Reality Versus Public Perception

The notion of "mind reading" has long captivated human imagination, appearing in science fiction, mythology, and philosophical thought experiments across cultures and throughout history. Today, advances in neural decoding have begun to transform this concept from fiction toward scientific reality. Neural decoding uses neuroimaging and computational methods to infer mental content from brain activity. However, a significant gap exists between current technical capabilities and public perceptions. Media coverage often either overstates present abilities or underestimates future potential. Understanding this gap is essential for developing balanced ethical frameworks and appropriate regulatory approaches.

5.2.1 Current Capabilities in Neural Decoding

Neural decoding encompasses a spectrum of approaches that vary in their invasiveness, spatial and temporal resolution, and the types of mental content they can access. These approaches can be broadly categorized by their underlying methodologies and the nature of information they extract:

5.2.1.1 Perceptual Decoding

Some of the most advanced neural decoding applications focus on reconstructing perceptual experiences, particularly visual information. Pioneering work by Nishimoto et al. [9] demonstrated the ability to reconstruct movie clips that participants viewed during functional magnetic resonance imaging (fMRI). The researchers developed encoding models that mapped brain activity patterns to features in the visual stimuli and then inverted these models to reconstruct the original videos from new brain activity patterns. While initial reconstructions were blurry and imprecise, they nonetheless captured significant semantic content from the original videos.

More recent work has leveraged deep learning approaches to achieve higher-fidelity reconstructions. Shen et al. [10] used generative adversarial networks to reconstruct perceived images from fMRI data with remarkable detail, including recognizable faces and objects. Similarly, Horikawa and Kamitani [11] demonstrated the reconstruction of visual imagery during dreams, albeit with lower fidelity than waking perception. These advances suggest that perceptual experiences leave distinct neural signatures that can be increasingly interpreted by sophisticated algorithms.

Beyond vision, neural decoding of auditory experiences has also progressed significantly. Moses et al. [2] demonstrated decoding of speech directly from cortical activity using intracranial electrodes in epilepsy patients, achieving intelligible reconstruction of spoken sentences. These approaches leverage the hierarchical organization of sensory systems, where low-level features combine to form more complex representations that correspond to perceptual experiences.

5.2.2 Semantic and Cognitive Decoding

Beyond perceptual information, researchers have made progress in decoding more abstract semantic and cognitive processes. Mitchell et al. [12] demonstrated that patterns of brain activity could predict which words participants were thinking about, based on the semantic features of those words. This approach has been extended to decode more complex linguistic constructs, including phrases and sentences.

More recently, Willett et al. [13] demonstrated decoding of attempted handwriting movements directly from motor cortex activity in a paralyzed individual, achieving typing speeds comparable to able-bodied smartphone users. This work highlights how motor intentions can be decoded and translated into functional outputs even when physical movement is impossible. In the domain of decision-making, Haynes et al. [14] showed that simple binary decisions could be predicted from prefrontal cortex activity several seconds before participants reported making their choice, raising intriguing questions about the neural basis of intention and the relationship between neural activity and conscious awareness.

5.2.3 Emotional and Affective Decoding

Emotional states represent another domain where neural decoding has shown promise. Kragel and LaBar [3] demonstrated that distinct emotional states evoke unique patterns of brain activity that can be classified using machine learning approaches. Their work identified reliable neural signatures for emotions like fear, anger, disgust, and happiness across individuals. These approaches suggest potential applications in mental health assessment, although they also raise concerns about emotional privacy.

5.2.4 Limitations and Technical Challenges

Despite impressive advances, current neural decoding technologies face significant limitations that constrain their real-world applications and ethical implications:

5.2.4.1 Resolution and Signal Quality Constraints

Noninvasive neuroimaging methods like fMRI and electroencephalography (EEG) provide relatively coarse measurements of neural activity. FMRI measures blood oxygen level-dependent (BOLD) signals as indirect indicators of neural activity with excellent spatial resolution (2–3 mm) but poor temporal resolution (seconds), while EEG offers millisecond temporal precision but limited spatial specificity (centimeters). These constraints fundamentally limit the granularity of information that can be decoded noninvasively [15].

Invasive approaches using electrocorticography (ECoG) or intracortical microelectrodes offer superior resolution but are primarily limited to clinical contexts like epilepsy monitoring or experimental BCI applications for severe paralysis. Even the most advanced invasive techniques record from only a small fraction of neurons in targeted brain regions, providing limited coverage of the approximately 86 billion neurons in the human brain.

5.2.5 Idiosyncratic Neural Representations

Individual brains differ substantially in their functional organization, with significant variability in how information is represented across people. While some aspects of neural processing are relatively consistent across individuals—like primary sensory cortex organization—higher cognitive functions show much greater idiosyncrasy [16]. This variability necessitates subject-specific training for most neural decoding applications, limiting the development of universal "mind reading" technologies that would work across individuals without calibration.

Neural representations also change over time within individuals due to learning, development, and neural plasticity. Decoders trained on one day's data may perform poorly on data collected weeks or months later, requiring periodic recalibration to maintain accuracy. This temporal instability presents challenges for long-term applications and raises questions about the reliability of neural evidence in contexts like legal proceedings.

5.2.6 Restricted Decoding Contexts

Most successful neural decoding studies operate within highly constrained experimental paradigms with limited stimulus sets, controlled environments, and cooperative participants. Participants typically view specific images from a predefined set, think about items within a restricted category, or perform simple binary tasks while minimizing movement and maintaining stable attention. These constraints facilitate decoding by reducing the dimensionality of the problem but limit generalization to real-world scenarios where mental content is unconstrained and context dependent.

Additionally, most current decoding approaches require comparison against a baseline or reference condition, making it difficult to identify truly novel thoughts or experiences not represented in training data. This requirement fundamentally constrains the "open-ended" nature of potential mind reading technologies, as they can generally only recognize patterns similar to those they have previously encountered.

5.2.7 The Inner Speech Problem

A particularly significant limitation involves decoding internal dialogue or "inner speech." Despite popular conceptions of mind reading as accessing a person's internal monologue, current technologies cannot reliably decode the specific content of inner speech in open-ended contexts. While some studies have shown promising results in classifying which of several predefined words or phrases a person is thinking about, these approaches do not generalize to the infinite variety of potential inner speech content that characterizes human thought. This limitation is particularly important for ethical discussions, as inner speech represents a core aspect of mental privacy that many people would be concerned about protecting.

5.2.8 Public Perceptions and Media Representations

Public understanding of neural decoding capabilities is shaped significantly by media coverage, which often fails to communicate technical nuances and limitations. Analysis of media reporting on neurotechnology reveals several common patterns that contribute to misperceptions:

5.2.8.1 Hype Cycles and Overstatement

Media coverage of neural decoding advances frequently employs sensationalist framing, with headlines like "Scientists Can Now Read Your Mind" or "Brain Scanning Can Reveal Your Thoughts." Such coverage typically focuses on the most successful aspects of studies while downplaying limitations, experimental constraints, and the gap between laboratory demonstrations and practical applications. This pattern contributes to public misconceptions about current capabilities and potentially raises unwarranted fears about imminent threats to mental privacy.

Conversely, some reporting understates the significance of incremental advances, failing to recognize how multiple modest improvements may combine to enable transformative applications. The iterative nature of scientific progress means that capabilities dismissed as limited today may evolve into more powerful technologies through sustained research and development. This dynamic creates challenges for anticipatory ethics and governance, which must balance responding to current realities with preparing for emerging possibilities.

5.2.9 The Influence of Science Fiction

Popular culture significantly shapes public perceptions of neural decoding through science fiction depictions in films, television, and literature. Works like "Inception," "Black Mirror," and "The Matrix" portray mind reading and neural manipulation as seamless, high-fidelity processes that can extract or implant complex memories, emotions, and thoughts. These fictional representations rarely acknowledge the technical challenges and constraints faced by actual neurotechnology, leading to unrealistic expectations and concerns.

Science fiction can serve valuable purposes in exploring ethical implications of emerging technologies before they fully materialize, providing "anticipatory ethics" frameworks that help society prepare for potential futures [17]. However, when fictional capabilities are confused with current realities, this can complicate public discourse and policy development. Survey research indicates that individuals familiar with neurotechnology primarily through science fiction tend to have less accurate understandings of actual capabilities and more polarized attitudes regarding regulation.

5.2.10 Bridging the Reality Gap

Addressing the gap between technical reality and public perception requires multifaceted approaches involving scientists, journalists, policymakers, and educators:

5.2.10.1 Responsible Research Communication

The neuroscience community has increasingly recognized its responsibility to communicate research accurately and contextually. Organizations like the International Neuroethics Society have developed guidelines for responsible communication of neuroscience research, emphasizing the importance of acknowledging limitations, avoiding deterministic language, and contextualizing findings within broader scientific understanding [18].

5.2.11 Anticipatory Ethics and Governance

Despite current limitations, the rapid pace of advancement in neural decoding technologies necessitates forward-looking ethical and governance frameworks. Approaches like "responsible research and innovation" (RRI) emphasize the early integration of ethical considerations into technology development processes rather than treating ethics as an afterthought [19]. This approach acknowledges that while current capabilities may be limited, responsible development requires anticipating potential futures and establishing adaptive governance mechanisms.

Multistakeholder dialogues involving neuroscientists, ethicists, policymakers, industry representatives, and diverse public perspectives can help develop a shared understanding of both current capabilities and potential trajectories. These conversations can inform governance approaches that remain responsive to technological developments while protecting core ethical values [7].

The gap between technical reality and public perception of neural decoding technologies creates both challenges and opportunities for ethical governance. Unrealistic fears may lead to excessive regulation that impedes beneficial applications, while complacency about limitations could leave important values unprotected as capabilities advance. Navigating this tension requires ongoing dialogue grounded in an accurate understanding of current capabilities, reasonable projections of future developments, and careful consideration of the values at stake.

5.3 Ethical Frameworks for Brain-Computer Interfaces

Brain-computer interfaces (BCIs) represent a particularly significant neurotechnology application. They encompass diverse implementation approaches and use cases. BCIs range from medical applications for communication in locked-in syndrome to consumer devices for gaming and productivity. They demonstrate both the promise and ethical complexity of direct brain-machine interaction. This section explores ethical frameworks specifically developed for BCI technologies, considering how traditional bioethical principles apply to this domain while identifying novel ethical considerations that arise in BCI development and deployment.

5.3.1 The Evolving Landscape of BCI Applications

Brain-computer interfaces encompass a spectrum of technologies that enable direct communication between neural tissue and external devices. These technologies vary significantly in their invasiveness, capabilities, target users, and intended applications:

5.3.1.1 Medical and Assistive BCIs

The most established BCI applications focus on restoring function or enabling communication for individuals with severe motor impairments. Invasive BCIs using implanted electrodes have enabled individuals with paralysis to control robotic limbs [20], type text through imagined handwriting [13], and operate computer cursors with neurofeedback. Noninvasive approaches using EEG or functional near-infrared spectroscopy (fNIRS) have enabled basic communication for individuals with locked-in syndrome, allowing for the expression of yes/no responses or the selection of letters through mental imagery.

More recently, bidirectional BCIs that both record neural activity and deliver stimulation have shown promise for treating neurological conditions like epilepsy and movement disorders [21]. These systems can detect abnormal neural activity patterns and deliver targeted stimulation to normalize brain function, potentially offering more precise and adaptive treatment than traditional approaches. The medical focus of these applications generally situates them within established clinical ethics frameworks, although novel questions arise regarding device control, neural data ownership, and long-term brain-machine integration.

5.3.2 Consumer and Commercial BCIs

Beyond clinical applications, consumer-oriented BCIs have expanded significantly in recent years. Companies like Neurable, Emotiv, Neuralink, and Kernel have developed various noninvasive and (prospectively) minimally invasive BCI systems for applications ranging from gaming and entertainment to productivity and self-monitoring. These devices typically offer lower signal quality than medical-grade systems but provide accessibility to broader populations without medical necessity.

Commercial applications raise distinct ethical questions from medical contexts, particularly regarding informed consent, data privacy, and marketing claims. Unlike medical devices, consumer BCIs may operate with less regulatory oversight and fewer institutional protections for users while collecting potentially sensitive neural data for commercial purposes [22]. Additionally, questions about enhancement versus therapy become more pronounced in consumer contexts, where devices may be marketed for improving normal cognitive function rather than addressing medical conditions.

5.3.3 Traditional Bioethical Principles in BCI Contexts

Established bioethical principles provide important foundations for BCI ethics but require adaptation to address novel challenges in this domain:

5.3.3.1 Beneficence and Nonmaleficence

The principles of promoting benefit and avoiding harm remain central to BCI ethics, although defining "benefit" and "harm" becomes increasingly complex as technologies interface directly with neural systems. Beyond physical safety, considerations must include psychological impacts, effects on identity and agency, and potential unintended consequences of altering neural function.

Risk-benefit assessments for invasive BCIs must consider not only immediate surgical risks but also long-term implications of having technological components integrated with neural tissue. These include potential infection risks, device degradation, compatibility with future technologies, and psychological adaptation to brain-machine integration. Noninvasive BCIs present different risk profiles, with generally lower physical risks but potential concerns about the long-term effects of modulating neural activity through techniques like transcranial magnetic stimulation or transcranial direct current stimulation.

The principle of nonmaleficence also raises important questions about BCI security and potential misuse. Systems that can record or stimulate neural activity create novel vulnerability vectors, including potential for "brainjacking" (unauthorized access to neural devices), neurological data theft, or intentional manipulation of cognitive or emotional states [23]. Protecting against these harms requires robust security measures throughout BCI design and implementation.

5.3.4 Autonomy and Informed Consent

Respect for autonomy in BCI contexts involves complex considerations about agency, identity, and decision-making capacity. For medical BCIs, obtaining informed consent may be complicated by the cognitive impairments or communication challenges that necessitate the intervention. Surrogate decision-making processes may be required, raising questions about how to properly respect the autonomy of individuals who cannot directly express their preferences.

The informational requirements for truly informed consent to BCI use are substantial and evolving. Potential users need to understand not only immediate risks and benefits but also long-term implications for neural plasticity, identity, data privacy, and potential dependencies on specific technological ecosystems. The novelty of many BCI applications means that some long-term implications remain speculative, creating challenges for comprehensive informed consent processes.

Consumer BCIs raise additional autonomy concerns related to marketing claims, subtle coercion, and user understanding. When devices are promoted with promises of cognitive enhancement, productivity improvement, or emotional well-being, users may make decisions based on exaggerated or unsubstantiated claims. The technical complexity of these systems may also impede user understanding of how their neural data is collected, processed, and utilized, potentially undermining informed choice.

5.3.5 Emerging Ethical Principles for Neurotechnology

Beyond traditional bioethical principles, several emerging concepts address neurotechnology-specific ethical considerations:

5.3.5.1 Cognitive Liberty

Cognitive liberty encompasses the right to control one's own cognitive processes, mental life, and consciousness without external interference. This concept extends traditional notions of autonomy to specifically protect mental self-determination in contexts where technologies can monitor or influence neural function [24]. Cognitive liberty has both negative dimensions (freedom from unwanted neural interference) and positive dimensions (freedom to modify one's own cognitive function through consensual neurotechnological intervention).

In BCI contexts, cognitive liberty raises questions about appropriate boundaries for neural monitoring and modulation, particularly in nonmedical applications. When employers deploy BCIs to monitor worker attention or emotional states, or when educational institutions use neurotechnology to track student engagement, cognitive liberty concerns arise about potential coercion and the right to mental privacy. Similarly, entertainment or marketing applications that influence emotional responses through neural feedback raise questions about appropriate limits on such influence, even when initially consensual.

5.3.6 Neural Privacy

Neural privacy refers specifically to protecting information generated by or inferred from brain activity from unauthorized access, use, or disclosure. This concept recognizes that neural data represent a uniquely sensitive category of personal information that may reveal aspects of cognition, emotion, and identity not consciously disclosed or otherwise observable [25]. Neural privacy extends traditional privacy concepts to address the specific challenges of brain data, which may contain information individuals are unaware of revealing and may not fully understand.

For BCIs that continuously record neural activity, privacy considerations include what data is collected, how long it is retained, who can access it, how it can be used, and what inferences can be drawn from it. These questions become particularly significant as advances in neural decoding enable increasingly detailed inferences from brain activity patterns. Core neural privacy principles include providing transparency about data collection and use, obtaining specific consent for different data applications, minimizing data collection to what is necessary for device function, and giving users meaningful control over their neural information.

5.3.7 Neural Agency and Identity

Brain-computer interfaces raise novel questions about agency, identity, and the boundaries between self and technology. As BCIs become more integrated with neural function, particularly bidirectional systems that both record and stimulate, questions arise about how these technologies affect experiences of self, sense of agency, and personal identity over time.

Users of advanced BCIs may experience "distributed agency" where actions result from complex interactions between neural processes, device algorithms, and external systems. This distribution complicates traditional notions of individual responsibility and control, particularly when systems incorporate machine learning components that adapt and evolve over time. Questions arise about appropriate attribution of actions, responsibility for outcomes, and maintaining meaningful human control in brain-machine systems.

5.3.8 Toward Comprehensive Ethical Frameworks

Addressing the complex ethical challenges of BCI development and deployment requires integrative frameworks that combine traditional bioethical principles with neurotechnology-specific considerations:

5.3.8.1 Responsible Research and Innovation

The responsible research and innovation (RRI) framework emphasizes the early integration of ethical, legal, and social considerations into technology development processes rather than treating them as downstream constraints. For BCIs, this approach involves engaging diverse stakeholders—including potential users, clinicians, ethicists, and regulators—throughout the research and development process to identify concerns, establish design requirements, and develop appropriate governance mechanisms [19].

5.3.9 Value-Sensitive Design

Value-sensitive design (VSD) offers a methodological framework for incorporating ethical values directly into technology design processes. This approach identifies stakeholder values, translates them into specific design requirements, and evaluates how well technologies embody and support these values in practice [26].

For BCIs, VSD might involve identifying core values like autonomy, privacy, identity, and well-being and then developing specific design approaches that protect and enhance these values. Examples include creating granular privacy controls that give users meaningful choices about neural data collection, designing transparent user interfaces that support informed decision-making, and implementing security measures that protect against unauthorized access or influence.

Brain-computer interfaces represent one of the most transformative applications of neurotechnology, offering unprecedented opportunities for treating neurological conditions, enhancing human capabilities, and establishing new forms of human-machine interaction. The ethical frameworks governing these technologies must evolve alongside technical capabilities, integrating traditional bioethical principles with novel considerations specific to neural interfaces. By developing comprehensive, forward-looking approaches to BCI ethics, we can work toward technologies that respect fundamental human values while realizing the transformative potential of direct brain-machine communication.

5.4 Privacy Considerations Specific to Neural Data

As neurotechnology advances, neural data raises unprecedented privacy challenges. The collection, analysis, and storage of these data extend beyond traditional data protection frameworks. Neural data represent a uniquely sensitive category of personal information with distinct characteristics and vulnerabilities. This information is generated by or derived from monitoring brain activity. This section explores the special privacy considerations associated with neural data and examines potential approaches for protecting this information while enabling beneficial applications.

5.4.1 The Unique Character of Neural Data

Neural data differ from other forms of personal information in several fundamental ways that necessitate specialized privacy approaches:

5.4.1.1 Exceptional Intimacy and Sensitivity

Neural data captures the biological basis of thoughts, emotions, perceptions, and intentions—aspects of human experience traditionally considered most private and internal. While conventional digital traces reveal what people do, neural data have the potential to reveal what people think, feel, and experience subjectively [22].

This exceptional intimacy suggests that neural data warrants heightened privacy protections beyond those applied to other personal information categories.

The sensitivity of neural data stems partly from its potential to reveal information individuals themselves may not be conscious of or intend to disclose. Neural signals might indicate reactions, biases, or responses that occur below the threshold of awareness or contrary to conscious intentions. For example, neural markers might reveal implicit associations or emotional responses that contradict explicitly stated preferences or beliefs. This creates a novel form of privacy vulnerability where individuals might inadvertently reveal more than they realize or intend.

Furthermore, neural data may contain information about highly sensitive conditions or traits, including mental health status, cognitive abilities, personality characteristics, and potentially neurological or psychiatric conditions not yet diagnosed. The diagnostic potential of neural data creates complex questions about incidental findings and their management in nonclinical contexts.

5.4.2 Difficulty of Anonymization

Neural activity patterns contain distinctive individual characteristics that function similarly to biometric identifiers, making true anonymization particularly challenging. Emerging research suggests that neural signatures can uniquely identify individuals with high accuracy across different tasks and time periods [27]. For example, functional connectivity patterns derived from fMRI data could identify specific individuals from a large dataset with over 95% accuracy, even when comparing scans taken months apart.

Similarly, EEG data contains individual-specific features that enable identification across sessions. These findings suggest that neural data might function effectively as a "brain fingerprint," complicating efforts to share or analyze such data while preserving anonymity. Traditional de-identification approaches like removing personal identifiers are likely insufficient for neural data, as the data itself contains inherently identifying patterns. This characteristic creates tensions between privacy protection, data sharing for scientific progress, and secondary uses of neural information.

5.4.3 Inferential Potential and Data Fusion

The inferential potential of neural data—what can be derived or predicted beyond what is directly measured—creates additional privacy challenges. Advanced analysis techniques can extract information not explicitly present in the original recordings, potentially revealing aspects of cognition or experience that individuals did not consent to share [8].

This challenge intensifies when neural data is combined with other information sources through data fusion techniques. For example, integrating neural recordings with behavioral data, genetic information, or digital traces from applications and devices can enable more comprehensive profiling than would be possible from any single data source. This combinatorial approach might reveal patterns and predictions not apparent in isolated datasets, potentially circumventing privacy protections applied to individual data categories.

5.4.4 The Neural Privacy Decision Dilemma

The complexities surrounding neural data protection become particularly apparent when examining the decisions individual users must make about sharing their brain activity information. Unlike other forms of personal data, neural information presents users with a profound paradox: the very characteristics that make these data valuable for beneficial applications—its intimacy, uniqueness, and inferential richness—simultaneously make it exceptionally risky to share.

Figure 5.1 illustrates the complex decision tree that users face when considering whether to share neural data, highlighting the multifaceted nature of this choice and the inherent difficulties in obtaining truly informed consent. This "neural privacy paradox" demonstrates why traditional consent models prove inadequate for neurotechnology contexts and why new approaches to data governance are essential.

As the flowchart demonstrates, users confronting decisions about neural data sharing face several interconnected challenges that distinguish these choices from conventional privacy decisions. First, the temporal asymmetry problem: users must make immediate decisions about data sharing while the most significant implications may only emerge years or decades later as analytical capabilities advance. Second, the technical complexity barrier: the sophisticated nature of neural signal processing and machine learning analytics makes it difficult for users to fully comprehend what information might be extractable from their data, both now and in the future. Third, the irreversibility dilemma: once neural data are collected and analyzed, it may be technically impossible to fully "unlearn" the insights derived from it, making withdrawal of consent challenging to implement meaningfully.

5.4.5 Threats to Neural Privacy

Understanding the full spectrum of potential threats to neural privacy is essential for developing appropriate protections:

5.4.5.1 Corporate Collection and Commercial Exploitation

As consumer neurotechnology expands, companies are increasingly collecting neural data through wearable devices, mobile applications, and other consumer products. These data may be used for improving products, developing new applications, or generating insights for commercial purposes including targeted advertising or product development.

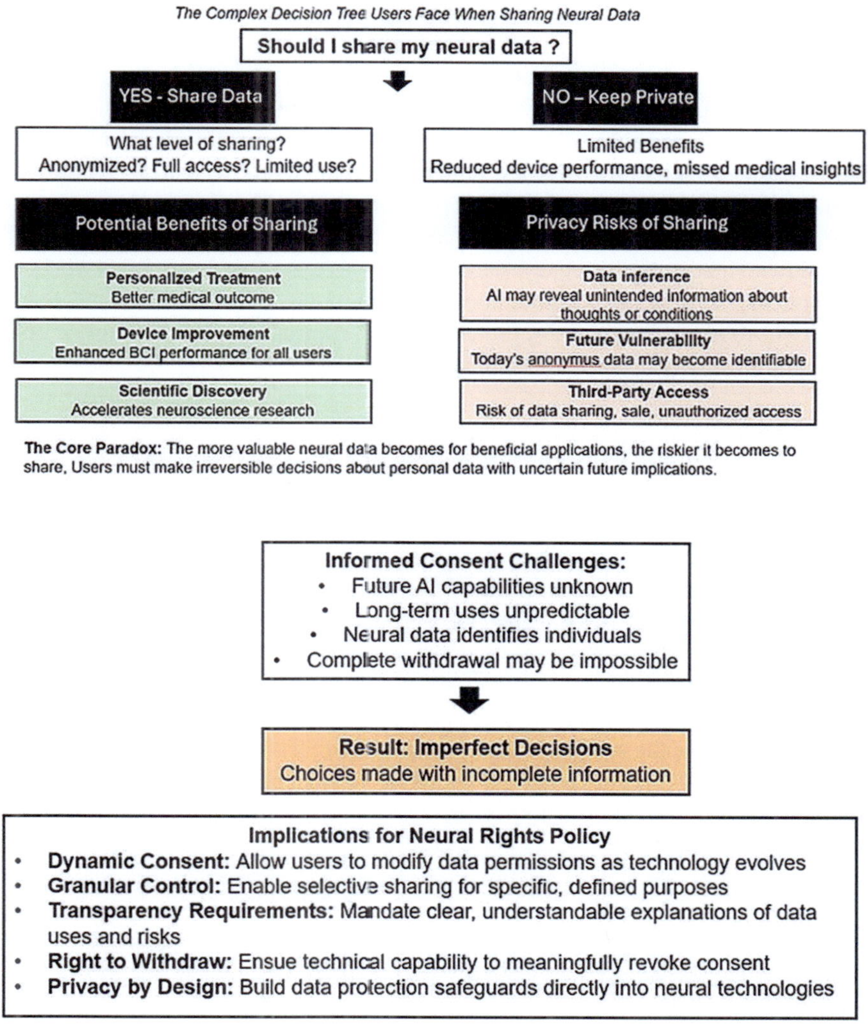

Fig. 5.1 Neural privacy paradox

Business models based on monetizing neural data may create incentives for excessive collection and secondary uses beyond users' expectations or preferences. Companies might analyze neural responses to advertisements, entertainment content, or product interactions to develop increasingly persuasive marketing approaches or personalized experiences designed to maximize engagement or purchasing behavior.

5.4.6 State Surveillance and Security Applications

Governmental entities may deploy neurotechnology for security, law enforcement, or intelligence purposes, creating potential for privacy violations through neural surveillance. Conceivable applications include monitoring for deception during interrogation, screening for security threats based on neural signatures, or tracking cognitive states of employees in sensitive positions.

The national security context may enable exceptional access to neural data with limited oversight or transparency, particularly where existing legal frameworks do not explicitly address this novel data category. Without specific protections, neural data might receive less stringent protection than communications content or other personal information in security contexts.

5.4.7 Unauthorized Access and Neural Security Violations

Beyond legitimate but potentially problematic uses by corporations or governments, neural data faces threats from malicious actors seeking unauthorized access. These threats include potential "brain hacking" of neural devices, interception of neural data during transmission, or breaches of databases containing stored neural information.

For implantable or wearable neural technologies, security vulnerabilities might enable not only data theft but also direct manipulation of device function, potentially allowing unauthorized influence over neural processing itself. This possibility, sometimes termed "brainjacking," represents a novel category of security threat that combines traditional cybersecurity concerns with direct risks to neural function and cognitive autonomy [23].

5.4.8 Technical Approaches to Neural Privacy Protection

Several technical approaches show promise for enhancing neural data privacy while enabling beneficial applications:

5.4.8.1 Privacy-Preserving Analysis Techniques

Differential privacy techniques add calibrated noise to data or analysis results to protect individual information while maintaining statistical utility for legitimate research or applications. These approaches provide mathematical guarantees about privacy protection regardless of an adversary's background knowledge or computational resources [28]. Applied to neural data, differential privacy could enable population-level insights without revealing individual neural patterns or characteristics.

Federated learning approaches keep data localized on individual devices while allowing distributed model training across many users. Rather than centralizing sensitive neural data, only model updates are shared, potentially reducing privacy risks compared to conventional centralized analysis. This approach seems particularly promising for consumer neurotechnology, where users might benefit from improved device performance without sharing raw neural data.

5.4.9 User-Controlled Data Management

Personal data stores or neural data cooperatives could give individuals greater control over their neural information, enabling selective sharing under specific conditions while maintaining ownership and oversight. These user-controlled repositories could implement granular permission systems allowing individuals to specify what data is shared, with whom, for what purposes, and for how long.

Progressive disclosure interfaces could help users understand the implications of sharing neural data at different levels of detail or for different applications. By presenting information about potential inferences and uses in accessible formats, such interfaces might enable more meaningful consent than conventional terms of service agreements.

5.4.10 Legal and Policy Approaches

Existing legal frameworks provide some protections for neural data, but generally were not designed with neurotechnology in mind. Developing appropriate legal approaches requires both adapting existing frameworks and considering novel protections:

5.4.10.1 Adaptation of Existing Data Protection Frameworks

Major data protection frameworks like the European Union's General Data Protection Regulation (GDPR) and the California Consumer Privacy Act (CCPA) contain provisions potentially applicable to neural data, including requirements for consent, purpose limitation, and data subject rights. However, these frameworks do not explicitly address the unique characteristics of neural information or provide neurotechnology-specific protections.

Some jurisdictions have begun considering neural data as a special category requiring enhanced protection. The GDPR's classification of biometric data as sensitive personal information provides one potential model, although questions remain about whether all neural data qualifies as biometric under current definitions. Clarifying this classification across jurisdictions would provide greater certainty about applicable protections.

5.4.11 Neurotechnology-Specific Legislation

Several jurisdictions have begun developing legislation specifically addressing neurotechnology and neural data. The Chilean constitution now includes explicit "neurorights" protecting mental privacy and cognitive liberty, representing the first constitutional-level protection specifically for neural information [29]. Similar legislative initiatives have emerged in Spain, Brazil, and other countries, suggesting an emerging recognition of neural data's exceptional status.

Emerging legal frameworks for neurotechnology typically address several key elements: the classification of neural data as uniquely sensitive information, requirements for specific and informed consent for neural data collection, limitations on secondary uses, enhanced security requirements for storage and transmission, and specific prohibitions on particularly problematic applications like involuntary neural monitoring [25].

5.4.12 Balancing Innovation and Protection

Developing appropriate neural privacy frameworks requires a careful balancing of multiple considerations:

5.4.12.1 Enabling Beneficial Applications While Preventing Harm

Overly restrictive approaches to neural privacy protection might impede beneficial applications in medicine, research, and assistive technology. For example, sharing neural data across research institutions enables larger, more diverse datasets that can advance understanding of neurological conditions and improve treatments. Privacy protections must enable such valuable uses while preventing problematic applications.

Similarly, many therapeutic neurotechnologies require detailed neural data collection and analysis to function effectively. Closed-loop deep brain stimulation systems for conditions like epilepsy or Parkinson's disease continuously monitor neural activity to deliver appropriate stimulation, necessarily involving substantial data collection and processing. Privacy frameworks must accommodate such beneficial medical applications while establishing appropriate safeguards.

The challenge involves developing contextual, purpose-sensitive approaches that apply different standards based on application domain, potential benefits, and risks. Medical and research applications might warrant different requirements than commercial or security uses, reflecting their distinctive risk-benefit profiles and ethical considerations.

5.4.13 User Autonomy and Meaningful Consent

Central to ethical neural data practices is respect for user autonomy through meaningful consent processes. Traditional informed consent models face significant challenges in neurotechnology contexts due to the technical complexity of these systems, the potential for unanticipated future uses, and difficulties communicating what information might be derived from neural data.

Dynamic consent models that allow individuals to modify permissions over time and across different use contexts show promise for neurotechnology applications.

Rather than one-time, all-or-nothing consent at initial use, these approaches enable ongoing control as users better understand technologies and as applications evolve. Digital platforms can facilitate such dynamic consent, allowing granular and adaptive permissions that better respect autonomy [30].

Neural data represents a uniquely intimate and sensitive information category that challenges existing privacy frameworks and necessitates specialized approaches. By developing comprehensive protection strategies that combine technical measures, legal frameworks, and ethical guidelines, we can harness the beneficial potential of neurotechnology while safeguarding fundamental values of mental privacy and cognitive liberty. The emerging field of neural privacy protection will continue evolving alongside technological capabilities, requiring ongoing attention, interdisciplinary collaboration, and inclusive deliberation about appropriate boundaries and safeguards.

5.5 The Concept of "Neural Rights" in an AI-Enhanced World

Artificial intelligence and neurotechnology are converging to create increasingly sophisticated interfaces between computational systems and human brains. This convergence means traditional conceptions of human rights may require expansion to address novel capabilities and vulnerabilities. The concept of "neural rights" represents an emerging framework for protecting human dignity, autonomy, and identity. This framework addresses contexts where AI systems can monitor, decode, and potentially influence neural activity. This section explores the theoretical foundations, current developments, and future implications of neural rights in an era of advancing AI and neurotechnology.

5.5.1 Philosophical Foundations of Neural Rights

Neural rights emerge from several philosophical traditions while addressing novel considerations specific to brain-computer interaction:

5.5.1.1 Extensions of Human Dignity and Personhood

The concept of human dignity—the inherent and inviolable worth of each person—provides a fundamental philosophical foundation for neural rights [31]. If dignity requires respecting persons as ends in themselves rather than merely as means, then protecting the neural substrates of personhood becomes essential when technologies can access and potentially manipulate these foundations [25].

As neuroscience increasingly identifies neural correlations of consciousness, decision-making, and personal identity, technologies that monitor or influence these processes raise profound questions about respecting human dignity. Neural rights frameworks extend traditional dignity protections to address these novel interfaces between technology and the biological basis of personhood.

5.5.2 Autonomy, Self-Determination, and Cognitive Liberty

Cognitive liberty—the right to control one's own consciousness and mental processes—represents a central component of most neural rights frameworks [24]. This concept extends traditional notions of autonomy to specifically address the right to maintain mental self-determination in contexts where technologies can monitor, decode, and potentially influence neural function.

Philosophical traditions emphasizing personal autonomy and self-determination provide important foundations for cognitive liberty, while new questions arise about what constitutes meaningful autonomy when technologies can potentially access thoughts, emotions, or intentions not consciously disclosed. If AI systems can derive information from neural activity that individuals themselves are unaware of revealing, traditional conceptions of autonomous choice may require reconsideration.

5.5.3 Privacy and Informational Self-Determination

Philosophical approaches to privacy provide another crucial foundation for neural rights. Traditional privacy concepts focused on physical spaces and explicit communications expand in neural rights contexts to include the internal mental domain traditionally considered entirely private and inaccessible to others.

Neural privacy involves not only protecting information explicitly shared but also safeguarding against the extraction of information individuals do not intend to reveal or may be unaware of disclosing through their neural activity. This raises complex questions about appropriate boundaries between legitimate information gathering and violations of mental privacy.

5.5.4 Articulating Specific Neural Rights

Building on these philosophical foundations, several specific neural rights have been proposed and increasingly recognized in academic, policy, and legal contexts:

5.5.4.1 Right to Mental Privacy

The right to mental privacy protects against unauthorized access to a person's neural information, creating a safeguarded domain for thought processes, emotional states, and internal experiences [25]. This right acknowledges the exceptional sensitivity of neural data as potentially revealing the most intimate aspects of human experience—thoughts, feelings, memories, and intentions traditionally considered entirely private.

Mental privacy protections might apply to various technologies and contexts, including neuroimaging in research or medical settings, consumer neurotechnology devices, ambient neural monitoring in public spaces, or workplace monitoring systems. The scope of protection may vary based on context, with particularly stringent safeguards for involuntary or covert neural monitoring.

5.5.5 Right to Cognitive Liberty

Cognitive liberty encompasses both negative freedoms (protection from unwanted neural interference) and positive freedoms (autonomy to modify one's own cognitive function through consensual neurotechnological intervention). This bidirectional concept protects both freedom from and freedom to in relation to neural modification [24].

The negative dimension protects against nonconsensual monitoring or modulation of brain activity, creating safeguards against coerced use of neurotechnology in contexts like employment, education, or legal proceedings. This aspect becomes particularly important as technologies like transcranial stimulation, focused ultrasound, or optogenetics develop increasing capabilities to influence neural function.

The positive dimension addresses individuals' autonomy to use neurotechnology for self-directed cognitive modification, enhancement, or exploration. This aspect engages with questions about rights to neuroenhancement, cognitive augmentation, or consciousness modification through consensual technological means.

5.5.6 Right to Mental Integrity

Mental integrity refers to protection against unauthorized harmful modification of neural function or cognitive processes. This right safeguards the continuity and coherence of one's mental life against technological disruption, manipulation, or damage [25].

Beyond physical harm to neural tissue, mental integrity encompasses protection against psychological harm through technological influence over cognitive or emotional processes. As technologies develop increasing capabilities to modulate mood, attention, memory, or decision-making, questions arise about appropriate boundaries for such influence and necessary safeguards against manipulation.

5.5.7 Right to Mental Agency

Mental agency encompasses the ability to maintain meaningful control over and responsibility for one's thoughts, decisions, and actions in contexts where AI systems might influence or share agency. This right addresses the distribution of control and responsibility in human-AI cognitive systems, particularly those involving direct brain-computer interfaces.

As AI systems increasingly function as cognitive partners rather than merely tools, questions arise about preserving appropriate human agency within these relationships. Advanced brain-computer interfaces might create "extended minds" where cognition spans biological and technological components, complicating traditional notions of individual agency and control.

5.5.8 Current Developments in Neural Rights Recognition

Neural rights have begun receiving formal recognition in various legal, policy, and institutional contexts:

5.5.9 Constitutional and Legislative Developments

Chile made history in 2021 by amending its constitution to explicitly protect neural rights, including mental privacy, personal identity, and freedom from neurotechnology manipulation without consent. This amendment represents the first constitutional-level protection specifically addressing neurotechnology and was influenced by the NeuroRights Initiative led by the neuroscientist Rafael Yuste [29].

Spain has developed draft legislation specifically addressing neurorights and protections for neural data, while Brazil, Mexico, and Argentina have initiated similar legislative processes. These developments suggest an emerging recognition of neurotechnology's unique implications and the need for specialized legal protections beyond conventional privacy or medical regulations.

5.5.10 International Declarations and Frameworks

The Organization for Economic Co-operation and Development (OECD) released its Recommendation on Responsible Innovation in Neurotechnology in 2019, establishing the first international standard addressing neural data and neurotechnology. While nonbinding, this framework emphasizes principles including promoting responsible innovation, protecting privacy, fostering inclusivity, and promoting international collaboration [32].

UNESCO's International Bioethics Committee has incorporated neurotechnology considerations into its work, addressing ethical implications of neural monitoring and modification in various reports and recommendations. These efforts situate neural rights within broader bioethical frameworks while acknowledging their distinctive characteristics.

5.5.11 Industry Standards and Self-Regulation

Various industry initiatives have begun developing standards and best practices for neurotechnology development and deployment. The IEEE Standards Association's Neuroethics Framework provides guidelines for responsible innovation in brain-machine intelligence, addressing concerns including privacy, accessibility, and autonomy [33].

Technology companies involved in neurotechnology development have established internal ethics committees and adopted various principles regarding neural data protection. While these self-regulatory approaches vary significantly in scope and enforcement, they represent growing recognition of neural rights concerns within the industry.

5.5.12 Neural Rights in an AI-Enhanced World

The convergence of artificial intelligence with neurotechnology creates both opportunities and challenges for neural rights protection:

5.5.12.1 Human-AI Cognitive Systems and Distributed Agency

As AI systems become increasingly integrated with human cognition through brain-computer interfaces and other neurotechnologies, questions arise about agency, control, and responsibility within these hybrid systems. Neural rights frameworks must address how to maintain meaningful human control and appropriate attribution of responsibility when cognition spans biological and technological components.

The concept of "distributed agency" acknowledges that actions may result from complex interactions between neural processes, device algorithms, and external systems rather than from any single component alone. This distribution complicates traditional notions of individual autonomy and responsibility, requiring new conceptual frameworks that accommodate these hybrid cognitive systems while preserving core human values.

5.5.13 Neural Security in an Era of Advanced AI

The security of neural technologies takes on particular importance in the context of sophisticated AI capabilities. Potential vulnerabilities include adversarial attacks on neural decoding algorithms, manipulation of brain-computer interface function, or unauthorized access to neural data for malicious purposes.

As AI systems develop increasing capabilities to analyze patterns across large datasets, the inferential vulnerability of neural data grows. Advanced machine learning approaches might identify subtle patterns in seemingly innocuous neural data that reveal sensitive information not explicitly recorded or shared. These inferential capabilities necessitate security approaches that address not only direct access but also secondary inferences from collected data.

5.5.14 AI as Both Threat and Protector of Neural Rights

AI systems play dual roles in neural rights contexts—potentially threatening these rights through surveillance capabilities while also offering powerful tools for their protection. This duality requires nuanced approaches that harness beneficial applications while establishing appropriate safeguards against potential harms.

On the one hand, advanced AI systems can analyze neural data with unprecedented sophistication, potentially enabling detailed inference about thoughts, emotions, or intentions without explicit consent. These capabilities could enable novel forms of privacy violation, manipulation, or coercion if deployed without appropriate constraints. The combination of neural monitoring with sophisticated predictive algorithms raises particularly significant concerns about preserving mental privacy and cognitive liberty.

On the other hand, AI approaches can enhance neural rights protection through privacy-preserving analytics, anomaly detection for identifying potential security breaches, and personalized interfaces that support informed decision-making about neurotechnology use. Techniques like federated learning, differential privacy, and homomorphic encryption leverage AI capabilities to enable beneficial applications while minimizing privacy risks [28]. These protective applications highlight the importance of directing AI development toward rights-enhancing rather than rights-compromising implementations.

5.5.15 Challenges in Neural Rights Implementation

Despite growing recognition of neural rights' importance, significant challenges remain in implementing effective protections:

5.5.15.1 Balancing Competing Values and Interests

Neural rights implementation requires balancing multiple legitimate values that sometimes conflict. For example, mental privacy protections might constrain beneficial research or medical applications that require detailed neural data, while cognitive liberty might tension with security or public health considerations in some contexts.

Different stakeholders, including researchers, medical professionals, industry, security agencies, and diverse user communities, may legitimately prioritize different values in neurotechnology development and deployment. Establishing frameworks that appropriately balance these perspectives without unduly privileging any single viewpoint represents an ongoing challenge for neural rights governance [29].

5.5.16 Definition and Operationalization Challenges

Translating abstract neural rights concepts into specific, implementable protections poses significant definitional challenges. What precisely constitutes a violation of mental privacy or cognitive liberty in different contexts? How should these boundaries be defined across diverse applications from medical treatment to consumer entertainment to workplace monitoring?

Technical complexities further complicate operationalization, as determining what information can be legitimately derived from neural data requires a sophisticated understanding of both current capabilities and potential future developments. Establishing appropriate limitations requires ongoing dialogue between technical experts, ethicists, legal scholars, and affected communities to develop workable definitions and boundaries.

5.5.17 Access and Equity Considerations

As with many emerging technologies, neurotechnology and associated protections may be unevenly distributed across populations and contexts. Ensuring that neural rights frameworks address the needs and perspectives of marginalized communities represents a crucial challenge for equitable implementation.

Economic barriers might limit access to both beneficial neurotechnologies and effective protections against potential harms. If privacy-preserving technologies remain expensive or complicated to implement, privacy could become a luxury rather than a universal right. Similarly, if neural enhancement technologies are available only to privileged populations, existing inequalities might be exacerbated.

5.5.18 Future Directions for Neural Rights

As neurotechnology and AI continue advancing, several emerging considerations will shape neural rights development:

5.5.18.1 Integration with Broader Digital Rights Frameworks

Neural rights connect to and extend existing digital rights frameworks addressing privacy, security, and self-determination in technological contexts. Rather than developing entirely separate systems, effective governance may involve integrating neural-specific protections into evolving digital rights approaches while recognizing the distinctive aspects of neural information.

5.5.19 Anticipatory Governance for Emerging Capabilities

Given the rapid pace of both neurotechnology and AI development, forward-looking governance approaches become essential for effective neural rights protection. Anticipatory frameworks consider not only current capabilities but also potential future developments and their implications for human rights and dignity [34].

5.5.20 Participatory Approaches and Collective Governance

Meaningful involvement of diverse stakeholders, particularly those most affected by neurotechnology applications, represents an essential element of effective neural rights protection. Participatory approaches engage potential users and affected communities throughout the development of both technologies and governance frameworks, ensuring that protections address actual rather than presumed concerns.

The concept of neural rights represents an emerging framework for protecting human dignity, autonomy, and identity in contexts where AI and neurotechnology create unprecedented interfaces between computational systems and human brains. Building on philosophical foundations including human dignity, autonomy, and privacy, neural rights address the unique implications of technologies that can monitor, decode, and potentially influence neural activity.

Specific protections including mental privacy, cognitive liberty, mental integrity, psychological continuity, and mental agency have begun receiving recognition in various legal, policy, and institutional contexts. From constitutional amendments to international declarations to industry standards, these emerging frameworks acknowledge the distinctive considerations raised by neural technologies while connecting them to broader human rights traditions.

The continued convergence of AI with neurotechnology creates both opportunities and challenges for neural rights protection. As human-AI cognitive systems become increasingly sophisticated and integrated, questions about agency, identity, security, and equity take on new dimensions requiring thoughtful anticipatory governance. Balancing competing values, operationalizing abstract concepts, and ensuring equitable access represent significant implementation challenges requiring ongoing interdisciplinary collaboration.

The future of neural rights will involve integration with broader digital rights frameworks, anticipatory governance approaches for emerging capabilities, and participatory models that engage diverse stakeholders in collective governance. Through these approaches, we can work toward technological futures that enhance human flourishing while protecting fundamental values of mental privacy, cognitive liberty, and authentic identity in an era of unprecedented brain-machine convergence.

The ethical dimensions of neurotechnology and artificial intelligence converge in particularly significant ways as these fields enable unprecedented interfaces between computational systems and human cognition. From neural decoding capabilities that might reveal private mental states to brain-computer interfaces that establish direct communication between neural tissue and external devices, these technologies raise profound questions about mental privacy, cognitive liberty, agency, and identity that extend beyond traditional ethical frameworks.

Understanding the current technical reality of "mind reading" technologies provides an essential foundation for ethical analysis. While significant advances in neural decoding have enabled reconstruction of visual experiences, semantic processing, and emotional states from brain activity, important limitations remain regarding resolution, generalizability, and access to many aspects of thought. Bridging the gap between technical reality and public perception—often influenced by science fiction or sensationalized media coverage—represents a crucial challenge for developing appropriate ethical and governance approaches.

Brain-computer interfaces demonstrate both the transformative potential and ethical complexity of direct brain-machine interaction. These technologies offer remarkable possibilities for treating neurological conditions, enhancing human capabilities, and enabling novel forms of interaction while simultaneously raising questions about autonomy, privacy, identity, and equity that traditional bioethical principles only partially address. Emerging ethical frameworks specifically designed for neurotechnology emphasize considerations like cognitive liberty, neural privacy, and mental integrity that recognize the unique implications of technologies interfacing directly with neural function.

Privacy considerations take on distinctive dimensions when applied to neural data, which represents an exceptionally intimate and sensitive information category. The unique characteristics of neural information—including its connection to thoughts and consciousness, the challenges of true anonymization, and its significant inferential potential when analyzed with advanced algorithms—necessitate specialized protection approaches combining technical measures, legal frameworks, and ethical guidelines. Balancing innovation with protection requires nuanced, contextual approaches that enable beneficial applications while safeguarding mental privacy.

The concept of "neural rights" has emerged to address these novel considerations, extending traditional human rights frameworks to specifically protect mental privacy, cognitive liberty, mental integrity, psychological continuity, and agency in contexts where technologies can monitor or influence neural function. These rights have begun receiving recognition in various legal, policy, and industry contexts, from constitutional amendments to international declarations to professional guidelines. As AI and neurotechnology continue converging, neural rights frameworks must address increasingly complex questions about human-AI cognitive systems, neural security, identity, and equity.

Across these domains, several common themes emerge regarding ethical approaches to neurotechnology and AI. First, the importance of accurate technical understanding that neither exaggerates current capabilities nor dismisses potential future developments. Second, the need for anticipatory governance that prepares for emerging possibilities rather than simply reacting to established technologies. Third, the value of inclusive, participatory approaches that engage diverse stakeholders—particularly those most affected by these technologies—in developing both technical systems and governance frameworks. Fourth, the significance of balancing multiple legitimate values, including innovation, protection, autonomy, and equity, without privileging any single consideration exclusively.

As neurotechnology and AI continue advancing, ethical frameworks must evolve alongside technical capabilities, addressing novel challenges while remaining grounded in fundamental values of human dignity, autonomy, and well-being. By developing comprehensive, forward-looking approaches to the ethical dimensions of brain-machine interaction, we can work toward technological futures that enhance human flourishing while protecting the mental privacy, cognitive liberty, and authentic identity that remain essential to human experience.

5.6 Summary

This chapter delves into the profound ethical frontier where neuroscience meets artificial intelligence, exploring technologies that can peer into the mind's neural activity and potentially decode thoughts, emotions, and intentions. As machines learn to read brain signals with increasing sophistication, fundamental questions emerge about the sanctity of mental privacy and human autonomy.

The Mind Reading Reality
Current neural decoding technologies can reconstruct visual experiences from brain scans, translate imagined speech into text, and identify emotional states from neural patterns. Yet a dangerous gap persists between these laboratory achievements and public perception—media hype alternately exaggerates current capabilities while underestimating future potential, creating confusion about what's possible today versus tomorrow.

Emerging Ethical Landscapes
Traditional medical ethics prove insufficient for technologies that interface directly with consciousness. New concepts emerge:

- **Cognitive liberty**: The fundamental right to control one's own mental processes
- **Neural privacy**: Protecting the mind's most intimate information from unauthorized access
- **Mental integrity**: Safeguarding against technological manipulation of thoughts and emotions

Neural Data Dilemma
Brain activity generates uniquely sensitive information—more intimate than fingerprints, harder to anonymize than DNA, and capable of revealing thoughts we ourselves may not consciously recognize. As AI algorithms grow more sophisticated, neural data collected today may unveil secrets decades from now that we cannot yet imagine.

Toward Neural Rights

Countries like Chile are pioneering "neural rights"—constitutional protections for mental privacy and cognitive freedom in an age of brain-computer interfaces. These frameworks represent humanity's first attempts to legally safeguard the mind itself as technology increasingly blurs the boundary between thought and machine.

This chapter ultimately argues that protecting human dignity in an AI-enhanced world requires unprecedented ethical innovation—frameworks that evolve alongside technology while preserving the essence of what makes us human.

References

1. Naselaris T, Allen E, Kay K. Extensive sampling for complete models of individual brains. Curr Opin Behav Sci. 2020;32:65–71.
2. Moses DA, Metzger SL, Liu JR, Anumanchipalli GK, Makin JG, Sun PF, et al. Neuroprosthesis for decoding speech in a paralyzed person with anarthria. N Engl J Med. 2021;385(3):217–27.
3. Kragel PA, LaBar KS. Decoding the nature of emotion in the brain. Trends Cogn Sci. 2016;20(6):444–55.
4. Musk E, Neuralink. An integrated brain-machine interface platform with thousands of channels. J Med Internet Res. 2019;21(10):e16194.
5. Ezzyat Y, Wanda PA, Levy DF, Kadel A, Aka A, Pedisich I, et al. Closed-loop stimulation of temporal cortex rescues functional networks and improves memory. Nat Commun. 2018;9(1):1–8.
6. Clewett D, Huang R, Velasco R, Lee TH, Mather M. Locus coeruleus activity strengthens prioritized memories under arousal. J Neurosci. 2018;38(6):1558–74.
7. Yuste R, Goering S, Agüera y Arcas B, Bi G, Carmena JM, Carter A, et al. Four ethical priorities for neurotechnologies and AI. Nature. 2017;551(7679):159–63.
8. Farahany NA. The costs of changing our minds. Houston Law Rev. 2019;57(5):1057.
9. Nishimoto S, Vu AT, Naselaris T, Benjamini Y, Yu B, Gallant JL. Reconstructing visual experiences from brain activity evoked by natural movies. Curr Biol. 2011;21(19):1641–6.
10. Shen G, Horikawa T, Majima K, Kamitani Y. Deep image reconstruction from human brain activity. PLoS Comput Biol. 2019;15(1):e1006633.
11. Horikawa T, Kamitani Y. Generic decoding of seen and imagined objects using hierarchical visual features. Nat Commun. 2017;8(1):1–15.
12. Mitchell TM, Shinkareva SV, Carlson A, Chang KM, Malave VL, Mason RA, Just MA. Predicting human brain activity associated with the meanings of nouns. Science. 2008;320(5880):1191–5.
13. Willett FR, Avansino DT, Hochberg LR, Henderson JM, Shenoy KV. High-performance brain-to-text communication via attempted handwriting. Nature. 2023;593(7858):249–54.
14. Haynes JD, Sakai K, Rees G, Gilbert S, Frith C, Passingham RE. Reading hidden intentions in the human brain. Curr Biol. 2007;17(4):323–8.
15. Poldrack RA. The physics of representation. Synthese. 2020;199(4):2157–73.
16. Gordon EM, Laumann TO, Gilmore AW, Newbold DJ, Greene DJ, Berg JJ, et al. Precision functional mapping of individual human brains. Neuron. 2017;95(4):791–807.
17. Johnson BD. Science fiction prototyping: designing the future with science fiction. Synthesis Lectures on Computer Science. 2011;3(1):1–190.
18. Illes J, Weiss S, Bains J, Chandler JA, Conrod P, De Koninck Y, Sokolowski MB. A neuroethics backbone for the evolving Canadian brain research strategy. Neuron. 2020;107(3):591–5.
19. Owen R, Macnaghten P, Stilgoe J. Responsible research and innovation: from science in society to science for society, with society. Sci Public Policy. 2012;39(6):751–60.

20. Hochberg LR, Bacher D, Jarosiewicz B, Masse NY, Simeral JD, Vogel J, Donoghue JP. Reach and grasp by people with tetraplegia using a neurally controlled robotic arm. Nature. 2012;485(7398):372–5.
21. Rao RP. Towards neural co-processors for the brain: combining decoding and encoding in brain-computer interfaces. Curr Opin Neurobiol. 2019;55:142–51.
22. Ienca M, Ignatiadis K, Andorno R, Barazzetti G, Biller-Andorno N, Büchler A, et al. Towards a governance framework for brain data. Neuroethics. 2022;15(1):1–14.
23. Pycroft L, Boccard SG, Owen SL, Stein JF, Fitzgerald JJ, Green AL, Aziz TZ. Brainjacking: implant security issues in invasive neurotechnology. World Neurosurg. 2016;92:454–62.
24. Bublitz JC. The concept of cognitive liberty. In: Philosophical foundations of neuroscience and neuroethics. Singapore: Springer; 2020. p. 169–93.
25. Ienca M, Andorno R. Towards new human rights in the age of neuroscience and neurotechnology. Life Sci Soc Policy. 2017;13(1):1–27.
26. Friedman B, Kahn PH, Borning A. Value sensitive design and information systems. In: The handbook of information and computer ethics. Wiley; 2013. p. 55–95.
27. Finn ES, Shen X, Scheinost D, Rosenberg MD, Huang J, Chun MM, Constable RT. Functional connectome fingerprinting: identifying individuals using patterns of brain connectivity. Nat Neurosci. 2015;18(11):1664–71.
28. Dwork C, Feldman V, Hardt M, Pitassi T, Reingold O, Roth A. Preserving statistical validity in adaptive data analysis. In: Proceedings of the forty-seventh annual ACM symposium on Theory of computing; 2014. p. 117–26.
29. Yuste R, Genser J, Herrmann S. It's time for neuro-rights. Horizons J Int Relat Sustain Dev. 2021;18:154–64.
30. Budin-Ljøsne I, Teare HJ, Kaye J, Beck S, Bentzen HB, Caenazzo L, et al. Dynamic consent: a potential solution to some of the challenges of modern biomedical research. BMC Med Ethics. 2017;18(1):1–10.
31. Bostrom N. Dignity and enhancement. Human dignity and bioethics: essays commissioned by the president's council on bioethics. Washington, DC: President's Council on Bioethics; 2008. p. 173–206.
32. OECD. Recommendation of the Council on Responsible Innovation in Neurotechnology, OECD/LEGAL/0457. Organisation for Economic Co-operation and Development; 2019.
33. IEEE. IEEE 2799-2022: IEEE standard for neuroethically driven artificial intelligence and autonomous systems design and development. IEEE Standards Association; 2022.
34. Nordmann A, Rip A. Mind the gap revisited. Nat Nanotechnol. 2009;4(5):273–4.

6 Consent and Autonomy in the Age of Predictive Neuroscience

The mind is its own place, and in itself can make a heaven of hell, a hell of heaven.

—John Milton

6.1 Introduction

The unprecedented ability of artificial intelligence to analyze neuroimaging data has ushered in an era of predictive neuroscience. In this new era, algorithms can forecast neurological outcomes, predict disease progression, and identify neural markers of conditions years before clinical symptoms manifest. These capabilities offer tremendous potential for early intervention, personalized treatment planning, and improved patient outcomes across a spectrum of neurological and psychiatric conditions. From predicting conversion to Alzheimer's disease in individuals with mild cognitive impairment to forecasting treatment response in depression, these technologies are reshaping clinical neuroscience practice and research.

However, this predictive power raises profound ethical questions. These concern informed consent, patient autonomy, and the right to know—or not know—information about one's future brain health. When an algorithm predicts a high likelihood of developing a neurodegenerative condition decades in the future, particularly one with limited treatment options, how should this information be communicated to patients? What constitutes meaningful consent when the implications of neurological predictions may not be fully understood even by clinicians? How do we preserve patient autonomy when algorithmic predictions may influence treatment decisions, insurance coverage, or life planning in ways that could become self-fulfilling prophecies?

These questions take on additional complexity in the context of incidental findings—unexpected abnormalities or predictions discovered during neuroimaging that were not the primary target of investigation. While incidental findings have

T. Rudroff, *AI for the Healthy Brain and Neurological Disorders*, Neurocultural Health and Wellbeing, https://doi.org/10.1007/978-3-032-11596-6_6

always been a consideration in neuroimaging, AI-enhanced analysis dramatically increases both their frequency and the subtlety of patterns that can be detected. This shift from occasional, visually apparent abnormalities to routine, algorithm-detected patterns fundamentally changes the ethical landscape surrounding neuroimaging research and clinical applications.

Cultural and cross-cultural perspectives further complicate these considerations, as attitudes toward medical disclosure, privacy, autonomy, and brain-related information vary significantly across societies and communities. What constitutes appropriate consent in one cultural context may be inadequate or inappropriate in another, creating challenges for global research collaborations, multinational clinical trials, and the development of universal ethical frameworks for predictive neurotechnology.

This chapter explores these ethical dimensions of predictive neuroscience, examining how traditional principles of informed consent and patient autonomy must evolve to address the unique challenges posed by AI-enhanced neurological predictions. We begin by analyzing the specific consent challenges raised by predictive neurotechnology, considering both theoretical frameworks and practical implementation approaches. We then explore the particularly complex issue of incidental findings in AI-enhanced neuroimaging, examining potential management protocols and disclosure considerations. Next, we address how predictive algorithms may influence patient autonomy and decision-making, considering approaches for preserving meaningful choice in an era of algorithmic forecasting. We then examine cultural and cross-cultural dimensions of neural data privacy and predictive disclosure, highlighting the importance of culturally responsive approaches. Finally, we present case studies from clinical research experience that illustrate these ethical dilemmas in practice and suggest potential resolution approaches.

Throughout this chapter, we emphasize that the ethical challenges of predictive neuroscience cannot be resolved through one-size-fits-all solutions. Rather, they require nuanced, context-sensitive approaches that balance scientific innovation with respect for patient values, cultural perspectives, and the fundamental human right to cognitive liberty and self-determination. By developing frameworks that address these complexities proactively, we can ensure that the transformative potential of predictive neuroscience advances human flourishing while respecting individual autonomy and dignity.

6.2 Informed Consent Challenges in Predictive Neuroscience

The traditional model of informed consent faces significant challenges when applied to AI-enabled predictive neuroscience. This model involves individuals receiving information about procedures, risks, benefits, and alternatives before making a decision. These challenges stem from both the nature of the predictions themselves and the technological complexity of the systems generating them.

6.2.1 The Nature of Neurological Predictions

Neurological predictions differ from traditional diagnostic or prognostic information in several ways that complicate informed consent processes:

Probabilistic and Statistical Nature
Predictions generated by AI algorithms are inherently probabilistic rather than deterministic, expressing likelihoods rather than certainties. For example, an algorithm might predict a 70% probability of developing Alzheimer's disease within 5 years. This prediction would be based on patterns in neuroimaging data. Communicating such probabilistic information effectively is challenging, as many individuals struggle to interpret statistical information and may either overinterpret predictions as definitive or dismiss them as meaningless [1].

Research on risk communication indicates that presentation format significantly affects understanding. Visual aids like icon arrays improve comprehension of statistical information compared to numerical presentations alone [2]. However, even with optimal presentation, fundamental challenges remain in helping nonspecialists understand concepts like confidence intervals, false positive rates, and the difference between population-level statistics and individual predictions [3].

The statistical literacy required to fully understand neurological predictions exceeds what can reasonably be expected from most patients, creating tensions between complete disclosure and effective communication.

Figure 6.1 illustrates this fundamental shift in medical communication complexity. Traditional neuroimaging results provided clear binary classifications that patients could easily understand and act upon. In contrast, AI-generated predictions

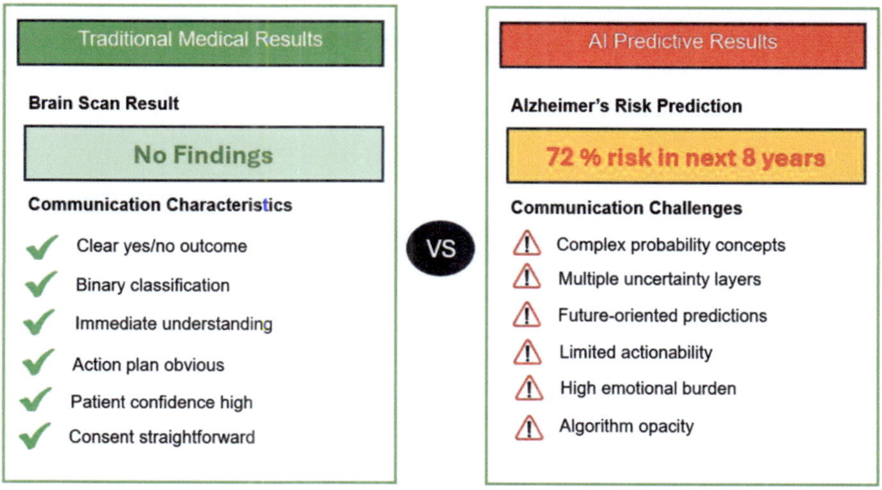

Fig. 6.1 Traditional versus AI medical communication. The fundamental shift from binary results to probabilistic predictions

introduce multiple layers of probabilistic information, confidence intervals, and temporal uncertainty that challenge conventional consent processes. This comparison highlights why established informed consent models, designed for straightforward yes/no medical decisions, prove inadequate for the complex statistical forecasts generated by predictive neuroscience algorithms.

As Peters et al. [4] note, "Statistical information that cannot be effectively processed by decision makers may actually impair rather than improve informed choice." This creates ethical challenges for consent processes that must balance comprehensive disclosure with comprehensible information.

The challenge of probabilistic risk communication is further complicated by framing effects, where identical statistical information can drive dramatically different patient responses depending on presentation. Figure 6.2 demonstrates how the same algorithmic prediction—a 72% risk of developing Alzheimer's disease—can be framed to emphasize either the likelihood of disease (creating anxiety and urgency) or the probability of remaining healthy (providing reassurance). Research indicates that patients shown risk-focused framing are three times more likely to pursue aggressive interventions, even when treatments offer limited effectiveness. This sensitivity to presentation format represents a critical challenge for informed consent, as the manner of disclosure may have a greater influence on patient decisions than the underlying statistical reality.

Temporal Distance and Uncertainty

Many neurological predictions involve conditions that may develop for years or decades in the future, creating significant temporal distance between prediction and potential manifestation. This temporal element introduces additional uncertainty and complexity for consent processes, as both predictive algorithms and available treatments may evolve significantly during the intervening period [5].

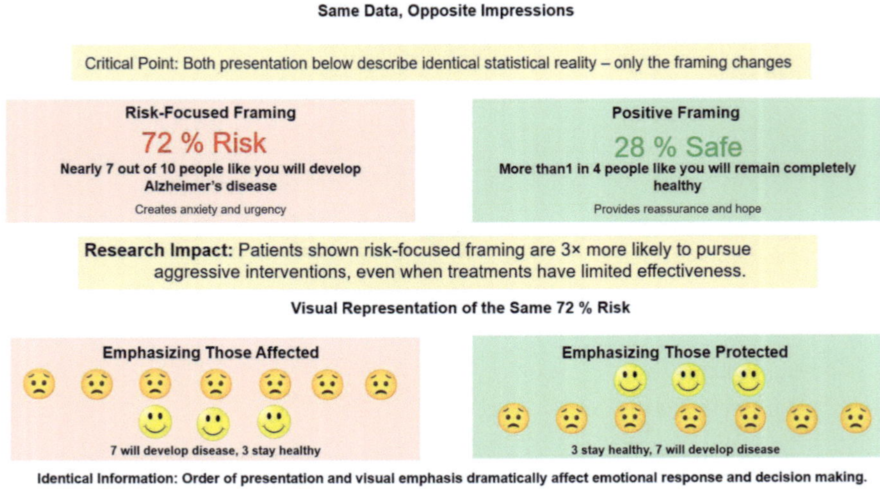

Fig. 6.2 Framing effects in risk communication. How identical statistical information creates different patient responses

For example, when an algorithm identifies neuroimaging markers suggesting elevated risk for Parkinson's disease decades before symptomatic onset, the clinical significance of this prediction depends not only on its statistical accuracy but also on how treatment options might evolve over that timeframe. Individuals consenting to receiving such predictions cannot be fully informed about these future developments, creating what the predictive testing literature has identified as fundamental temporal challenges in consent processes, where complete information is inherently unavailable at the time of decision-making [6]. This temporal dimension of consent has been extensively documented in Huntington's disease genetic testing, where similar gaps exist between prediction and potential manifestation.

Variable Actionability of Predictions

The actionability of neurological predictions—whether they can inform meaningful intervention or prevention strategies—varies significantly across conditions and predictions, further complicating consent processes. For conditions with established preventive measures or treatments, predictive information offers clearer benefits. However, for conditions where treatments remain limited or nonexistent, predictive knowledge may offer limited medical benefit while potentially imposing significant psychological burdens [7].

This variable actionability creates challenges for weighing risks and benefits during consent, particularly given evidence that clinicians and researchers tend to overestimate the actionability of their findings compared to how patients perceive them [8]. As Roberts et al. [9] note, "The value of predictive information exists in a complex relationship with available interventions, making standard risk-benefit calculations inadequate for many neurological predictions."

These communication challenges create six critical problems that undermine traditional informed consent processes in predictive neuroscience contexts (Fig. 6.3). From cognitive overload caused by complex probability processing to algorithm opacity that prevents truly "informed" decision-making, these interconnected challenges demonstrate why current consent frameworks prove inadequate for AI-generated neurological predictions. Unlike traditional medical decisions, where

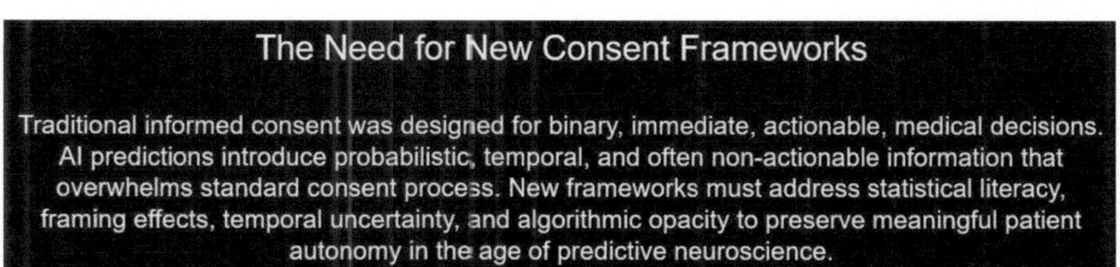

Fig. 6.3 Impact on informed consent. Why probabilistic AI predictions break traditional consent models

patients can meaningfully weigh clear risks and benefits, predictive neuroscience introduces multiple uncertainty layers, temporal disconnects, and potential self-fulfilling prophecies that exceed the processing capabilities assumed by standard consent models.

6.2.2 Technological Complexity and Transparency

Beyond the nature of predictions themselves, the technological systems generating them introduce additional consent challenges:

Algorithmic Complexity and Explainability
Many high-performing predictive algorithms in neuroscience, particularly deep learning approaches, function as "black boxes" where the relationship between inputs and outputs cannot be easily explained in human-understandable terms. This opacity complicates informed consent, as patients cannot fully understand how predictions about their brain health are being generated [10].

The explainability challenge extends beyond technical understanding to include interpretation of what algorithmic outputs mean for individual patients. Even when algorithms can generate post hoc explanations identifying features that influence a prediction, translating these into clinically meaningful information accessible to patients remains difficult [11]. Without such translation, patients may be asked to consent to receiving information generated through processes they cannot meaningfully understand.

Various technical approaches to explainable AI in neuroscience have emerged, including attention maps for neuroimaging, feature importance rankings, and case-based explanations [12]. However, these approaches often trade off comprehensiveness for comprehensibility, raising questions about what level of explanation is necessary for valid consent. As Amann et al. [13] argue, "The threshold of explainability required for ethical consent remains contested, with different stakeholders emphasizing different aspects of understanding."

Evolving Performance and Validation
Unlike established diagnostic tests with well-characterized performance metrics, many predictive neuroscience algorithms continue evolving through additional training, architecture modifications, or application to new populations. This evolution creates challenges for consent processes that must communicate not only current performance characteristics but also how these might change over time [14].

Patients consenting to predictive analysis may not realize that algorithm performance varies across different populations and contexts, potentially leading to misplaced confidence in predictions. Mittelstadt et al. [15] term this the "validation gap" in AI-based healthcare, noting that "performance metrics established in development environments may not translate to real-world clinical contexts, creating scenarios where patients consent based on idealized rather than realized capabilities."

This challenge intensifies for federated or continuous learning systems that update based on new data, potentially changing performance characteristics after initial consent was obtained. As these systems become more common in neuroscience applications, traditional one-time consent models become increasingly inadequate [16].

6.2.3 Evolving Consent Models for Predictive Neuroscience

Addressing these challenges requires rethinking traditional informed consent approaches to develop models better suited to the realities of predictive neuroscience:

Layered and Dynamic Consent

Layered consent models provide information at multiple levels of detail, allowing individuals to access basic information necessary for decision-making while making more comprehensive information available for those who desire it. This approach has shown promise for complex genomic testing and may offer advantages for predictive neuroscience as well [17].

A typical layered consent process might include an initial layer covering essential information about the predictive analysis, its general purpose, and key risks and benefits; intermediate layers providing more detailed information about specific conditions being assessed and their implications; and detailed layers explaining technical aspects, performance characteristics, and data handling practices [18].

Dynamic consent extends this approach by enabling ongoing communication and decision-making rather than one-time authorization. Through digital platforms, individuals can update their preferences over time, receive new information about evolving capabilities or findings, and maintain greater control over how their data and predictions are handled [19]. This approach seems particularly well-suited to predictive neuroscience, where both technological capabilities and individual circumstances may change significantly over time.

Implementation of these models requires both technological infrastructure for secure, accessible information management and cultural shifts in how consent is conceptualized. As Williams et al. [20] note, "Moving from document-centric to relationship-centric models of consent represents not merely a procedural change but a fundamental reconceptualization of the consent relationship."

Process Consent and Longitudinal Engagement

Process consent approaches emphasize consent as an ongoing dialogue rather than a discrete event, recognizing that understanding and decision-making evolve over time through continued engagement. This conceptual shift aligns with the longitudinal nature of many neurological predictions and the need for sustained relationships between patients, clinicians, and research teams [21].

In practice, process consent involves regular check-ins, opportunities to revisit and revise decisions, and recognition that capacity and preferences may change over time. This approach has been successfully implemented in dementia research and may offer valuable insights for predictive neuroscience, particularly for conditions where cognitive changes might affect decision-making capacity over time [22].

Process consent requires greater resource investment than traditional models but may ultimately better protect autonomy while facilitating more meaningful engagement with predictive information. As Goering et al. [23] argue, "The additional time and resources required for process consent should be understood not as optional extras but as essential components of ethical predictive neuroscience."

Comprehension-Focused Approaches

Recognizing the challenges in understanding complex probabilistic information, some researchers have proposed consent models that prioritize comprehension assessment and improvement rather than simply providing information. These approaches focus on developing materials and processes that effectively communicate key concepts, verifying understanding before proceeding, and adapting approaches based on individual needs and capacities [24].

Comprehension-focused approaches might include interactive decision aids, scenario-based explanations, personalized risk communication, and teach-back methods where individuals explain information in their own words to verify understanding. Research suggests that these approaches can significantly improve the comprehension of complex medical information compared to standard consent processes [25].

For predictive neuroscience specifically, comprehension-focused approaches might emphasize understanding of key concepts like probability, actionability, limitations, and the distinction between prediction and diagnosis rather than technical details of algorithmic function. As Biller-Andorno et al. [1] note, "Meaningful consent requires not comprehensive technical knowledge but sufficient understanding of implications for one's life and choices."

6.2.4 Implementation Considerations and Practical Challenges

Translating these conceptual approaches into practical implementations requires addressing several challenges:

Time and Resource Constraints

More comprehensive consent models generally require greater time and resource investments from both healthcare providers and patients. In resource-constrained healthcare environments with limited appointment times and high patient volumes, implementing ideal consent processes may prove challenging [26].

Digital tools and decision aids offer potential solutions by enabling asynchronous information review, adaptive content presentation, and remote engagement with consent materials. However, these approaches introduce their own challenges regarding digital access, technical literacy, and the potential loss of human connection in the consent process [27].

Finding an appropriate balance between thoroughness and practicality remains an ongoing challenge, particularly as predictive applications become more integrated into routine clinical care rather than specialized research contexts with dedicated resources for consent processes [28].

Professional Training and Preparedness

Clinicians and researchers obtaining consent for predictive neuroscience applications require specialized training in both technical aspects of the technologies and effective communication strategies for complex, probabilistic information. Current evidence suggests significant gaps in professional preparedness for these responsibilities, with many clinicians reporting discomfort discussing algorithmic predictions with patients [29].

Addressing these gaps requires curriculum development, continuing education, and potentially new specialized roles in healthcare settings. As Mittelstadt et al. [15] suggest, "The emergence of AI-driven prediction may necessitate new professional roles combining technical expertise with patient communication skills to bridge gaps between algorithm developers and clinical teams."

Regulatory Frameworks and Standards

Current regulatory frameworks for informed consent were largely developed before the emergence of AI-enabled prediction and may inadequately address its unique challenges. While general principles of autonomy and understanding remain applicable, specific requirements regarding disclosure, documentation, and oversight may require updating [30].

Various organizations have begun developing guidance specific to AI in healthcare, including the World Health Organization's guidance on ethics in AI for health and the American Medical Association's policy on augmented intelligence. However, specific standards for predictive neuroscience applications remain limited, creating uncertainty for implementation [28].

This regulatory uncertainty affects not only clinical applications but also research and development, potentially creating barriers to innovation if requirements remain unclear or inconsistent across jurisdictions. As Cheshire et al. [31] note, "Balancing adequate oversight with innovation-enabling flexibility represents a central tension in governance of predictive neurotechnology."

The informed consent challenges in predictive neuroscience reflect broader tensions between technological capabilities and human comprehension, between standardization and personalization, and between advancing knowledge and respecting individual autonomy. Addressing these challenges requires not only procedural innovations but also deeper engagement with the values and purposes that informed consent is meant to serve in healthcare and research contexts.

6.3 The Problem of Incidental Findings in AI-Enhanced Neuroimaging

The increasing sensitivity and comprehensiveness of AI-enhanced neuroimaging analysis have transformed the landscape of incidental findings. These are unexpected abnormalities or predictions unrelated to the primary purpose of the imaging study. While incidental findings have always been a consideration in neuroimaging, AI approaches dramatically expand both the frequency and types of findings that may be detected, creating new ethical challenges for researchers, clinicians, and patients.

6.3.1 The Changing Nature of Incidental Findings

Traditional incidental findings in neuroimaging typically involved visually apparent anatomical abnormalities noticed by radiologists during image review, such as unexpected tumors, vascular malformations, or evidence of prior stroke. AI-enhanced analysis fundamentally changes this paradigm in several ways:

Expanded Detection Capabilities
AI algorithms can detect subtle patterns invisible to human observers. These include statistical deviations from normative data, microstructural abnormalities, functional connectivity alterations, and correlations between imaging features and disease risk factors. This expanded detection capability means that virtually every brain scan contains potential "findings" when subjected to comprehensive AI analysis [32].

Studies examining incidental findings in large-scale neuroimaging research cohorts illustrate this expansion. For example, in the Rotterdam Scan Study, conventional radiological review identified clinically significant incidental findings in approximately 2.5% of participants, whereas automated analysis of the same images identified deviations from population norms in over 30% when applying multiple assessment algorithms [33].

This quantitative increase creates new challenges for managing findings and determining appropriate disclosure thresholds. As O'Sullivan et al. [34] note, "The shift from occasional, binary findings to routine, continuous metrics fundamentally changes the ethical calculus surrounding incidental findings management."

Novel Finding Categories

Beyond detecting known abnormalities more sensitively, AI analysis can identify entirely new categories of findings without established clinical significance. These might include unusual connectivity patterns, subtle volumetric changes, or statistical associations between imaging features and future disease risk [35].

These novel findings often exist in an ambiguous space between normal variation and pathology, with unclear implications for individual health and well-being. Traditional clinical guidelines for managing incidental findings, developed for established radiological abnormalities, provide limited guidance for these algorithmically detected patterns [36].

The emergence of predictive biomarkers represents a particularly significant category of novel findings. When algorithms identify patterns associated with future disease risk—such as imaging signatures suggesting an elevated likelihood of developing Alzheimer's disease or depression—these constitute a fundamentally different type of finding than acute abnormalities, requiring distinct ethical frameworks and management approaches [37].

Changing Associations and Responsibilities

Traditional incidental findings management relied on clear professional roles and responsibilities, with radiologists or other imaging specialists serving as designated reviewers responsible for identifying and reporting clinically significant abnormalities. AI-enhanced analysis disrupts these established roles, raising questions about who bears responsibility for reviewing algorithmic outputs, determining clinical significance, and communicating findings [38].

When images are analyzed by multiple algorithms for different purposes, potentially generating dozens of statistical outputs and predictions, traditional review models become impractical. This shift necessitates new frameworks for determining which findings merit human review, how review responsibilities should be allocated, and what disclosure thresholds should apply across different finding types [39].

6.3.2 Ethical Frameworks for Managing AI-Enhanced Incidental Findings

Several ethical frameworks have been proposed for addressing the challenges of incidental findings in AI-enhanced neuroimaging:

Risk-Stratified Approaches

Risk-stratified frameworks categorize potential findings based on their clinical actionability, severity, and certainty, applying different management protocols to each category. This approach acknowledges that not all algorithm-detected anomalies warrant the same level of review or disclosure [40].

A typical risk-stratified framework might include categories such as:

- Critical findings requiring immediate clinical attention (e.g., large tumors and acute hemorrhage)
- Clinically significant findings requiring follow-up (e.g., small aneurysms and significant atrophy)
- Findings of uncertain significance requiring case-by-case evaluation (e.g., unusual connectivity patterns)
- Expected variations not requiring disclosure (e.g., minor asymmetries and common age-related changes)

For each category, specific protocols would define review requirements, disclosure processes, and follow-up responsibilities. This structured approach helps manage the potentially overwhelming volume of AI-detected findings while prioritizing those most likely to impact patient welfare [36].

Critics of risk-stratified approaches note that categorization itself involves value judgments about what constitutes "actionability" or "significance"—judgments that may vary across different contexts or conflict with individual patient values. As Mackenzie et al. [41] argue, "Seemingly technical decisions about categorization embed substantive ethical choices about what information patients should receive about their own brains."

Autonomy-Centered Frameworks

Autonomy-centered frameworks prioritize individual choice in determining what incidental findings should be disclosed, allowing research participants or patients to specify their disclosure preferences in advance rather than relying solely on expert determinations of clinical significance [42].

In practice, these approaches might involve tiered consent options where individuals can choose different disclosure thresholds—from receiving all identified findings to receiving only immediately life-threatening abnormalities. Some implementations allow for more granular choices regarding specific finding types or categories, particularly for predictive findings related to specific conditions [17].

Digital platforms can facilitate these approaches by enabling dynamic preference-setting that can be updated over time as individual circumstances or finding categories evolve. For example, a participant might initially decline receiving predictive information about dementia risk but later opt in after a relative's diagnosis changes their perspective [22].

While autonomy-centered approaches respect individual values, they face practical challenges regarding informed preference-setting. Individuals may struggle to anticipate how they would want to handle hypothetical future findings, particularly for novel finding types with complex implications. As Berkman et al. [8] note, "Meaningful autonomy requires not just offering choices but ensuring adequate understanding of options and implications."

Contextual and Relational Approaches
Contextual frameworks emphasize that appropriate handling of incidental findings depends not only on finding characteristics but also on the context in which imaging occurs, the relationship between parties involved, and broader social and cultural factors [43].

These approaches reject one-size-fits-all protocols in favor of context-sensitive decision-making that considers factors like:

- The purpose and setting of the imaging (clinical care vs. research vs. commercial applications)
- The relationship and ongoing communication channels between parties
- Available resources for follow-up and support
- Individual circumstances and vulnerabilities
- Cultural and community norms regarding medical information

Contextual approaches align with broader movements in bioethics toward more relational, case-based reasoning rather than principle-based rule application. As Samuel et al. [43] argue, "The complexity of AI-enhanced incidental findings requires moving beyond algorithmic ethics toward a more nuanced, contextual approach to ethical decision-making."

6.3.3 Implementation Challenges and Practical Approaches

Translating ethical frameworks into practical protocols for managing AI-enhanced incidental findings involves addressing several implementation challenges:

Technical Infrastructure and Workflow Integration
Effective management requires technical infrastructure for flagging, reviewing, and communicating findings within existing clinical or research workflows. This infrastructure must balance automation for efficiency with human oversight for quality assurance and ethical judgment [36].

Pioneering implementations have developed tiered review systems where algorithms initially flag potential findings, trained technicians conduct first-level reviews, and specialists evaluate findings that meet certain threshold criteria. These systems include structured documentation, communication protocols, and quality assurance mechanisms to ensure appropriate handling across diverse finding types [44].

The integration of these systems into existing workflows remains challenging, particularly in resource-constrained settings or when imaging occurs outside traditional medical environments. As neuroimaging applications expand beyond specialized centers to community settings, primary care, or even consumer applications, ensuring appropriate infrastructure for incidental findings management becomes increasingly complex [27].

Expertise and Training Requirements

AI-enhanced findings often require specialized expertise for interpretation that may exceed what is available in many clinical or research settings. Determining the clinical significance of novel biomarkers, connectivity patterns, or statistical anomalies requires not only radiological knowledge but also understanding of the specific algorithms, their validation status, and the literature on similar findings [38].

Addressing this expertise gap requires both expanded training for imaging professionals and new collaborative models involving algorithm developers, imaging specialists, and clinical experts. Some institutions have established dedicated AI findings review committees that combine multiple expertise domains for evaluating complex or novel findings [36].

The shortage of professionals with appropriate training represents a significant barrier to ethical management of AI-enhanced findings, particularly as applications expand faster than training programs can adapt. As Grande et al. [29] observe, "The expertise gap represents perhaps the most immediate practical constraint on ethical implementation of advanced neuroimaging analysis."

Resource Implications and Equity Considerations

Comprehensive management of AI-enhanced incidental findings requires significant resources for review, disclosure, counseling, and follow-up—resources that are unevenly distributed across healthcare systems and research contexts. This resource requirement raises important equity considerations, as underserved populations may have less access to appropriate management infrastructure [45].

The resource intensity of proper management could potentially exacerbate existing healthcare disparities if AI-enhanced analysis becomes standard in well-resourced settings while remaining unavailable in resource-constrained environments. Alternatively, if AI analysis is implemented without adequate resources for findings management, it could create new ethical challenges in settings unable to provide appropriate follow-up [10].

Addressing these equity concerns requires both resource allocation decisions and innovation in scalable, cost-effective approaches to findings management. Potential solutions include telehealth models for specialized review, standardized educational materials and decision aids, and tiered follow-up protocols that target resources according to finding severity and urgency [38].

6.3.4 Predictive Incidental Findings: A Special Case

Predictive findings—algorithm-detected patterns suggesting future disease risk rather than current pathology—represent a particularly challenging subcategory of incidental findings in AI-enhanced neuroimaging. These findings raise distinct ethical questions beyond those associated with traditional anatomical abnormalities:

Temporal and Probabilistic Complexity

Unlike findings indicating current pathology, predictive findings involve temporal projection and probabilistic uncertainty. An algorithm might identify patterns suggesting elevated risk for Alzheimer's disease decades before potential symptom onset, with significant uncertainty about whether the condition will actually develop given potential changes in both brain trajectories and environmental factors [9].

This temporal and probabilistic complexity creates challenges for determining disclosure thresholds, as the "clinical significance" of such findings depends not only on immediate implications but also on long-term trajectories, intervention possibilities, and individual life circumstances. As Johansson-Kark et al. [7] note, "The delayed and uncertain nature of predictive findings fundamentally changes their ethical calculus compared to acute abnormalities."

Variable Actionability and Prevention Potential

The appropriate management of predictive findings depends significantly on their actionability—whether effective preventive measures or risk-reducing interventions exist. This actionability varies dramatically across conditions and continues evolving as new interventions emerge [17].

For conditions with established preventive approaches, like discovery of modifiable stroke risk factors, disclosure may offer clear benefits by enabling risk-reducing interventions. For conditions with limited prevention options, like many neurodegenerative diseases, the benefit-risk balance of disclosure becomes more uncertain, potentially imposing psychological burdens without corresponding medical benefits [9].

This variable actionability creates challenges for developing consistent disclosure protocols across different prediction types. As Berkman et al. [8] argue, "The appropriate handling of predictive incidental findings requires not categorical rules but condition-specific evaluation of potential benefits, harms, and uncertainties."

Psychological Impact and Right Not to Know

Predictive findings raise particular concerns regarding psychological impact and the "right not to know" about potential future conditions. Learning about elevated risk for serious, progressive neurological conditions may cause significant anxiety, altered self-perception, or changes in life planning—impacts that some individuals might reasonably prefer to avoid, particularly for conditions with limited prevention options [7].

Research on the disclosure of genetic risk information provides some insights, suggesting significant individual variation in psychological responses, with most individuals showing initial anxiety that diminishes over time without long-term psychological harm. However, substantial minorities experience more persistent distress, particularly those with preexisting anxiety or depression [46].

The limited research specifically examining the disclosure of neuroimaging-based predictions suggests similar patterns, although with important differences from genetic risk disclosure due to the visual and apparently concrete nature of brain images. As Biller-Andorno et al. [1] note, "Neuroimaging findings may carry particular psychological weight due to their perceived scientific authority and apparent direct visualization of the self."

6.3.5 Case Examples and Emerging Approaches

Several research programs and clinical implementations have developed pioneering approaches to managing AI-enhanced incidental findings, offering instructive examples for broader application:

The Human Connectome Project Approach
The Human Connectome Project (HCP), a large-scale effort to map human brain connectivity, developed a structured, multitier approach to incidental findings management that has influenced many subsequent protocols [47].

The HCP protocol includes the following:

- Automated quality control algorithms to identify potential abnormalities
- Initial review by trained technicians
- Referral of potential findings to neuroradiologists
- Categorization into actionability tiers determining disclosure approach
- Standardized disclosure processes with appropriate clinical referral
- Follow-up to ensure appropriate clinical evaluation

This approach balances efficiency with thoroughness by using a staged review process while maintaining clear responsibility allocation and documentation. However, it was designed primarily for structural abnormalities rather than the full spectrum of AI-detected patterns and predictions emerging in contemporary applications [48].

The UK Biobank Imaging Enhancement
The UK Biobank, a large-scale biomedical database including brain imaging from over 100,000 participants, has developed one of the most comprehensive approaches to AI-enhanced incidental findings management in a research context [49].

Their protocol incorporates the following:

- Clear participant information about the possibility of incidental findings
- Tiered disclosure options during consent
- Automated analysis for both established abnormalities and novel biomarkers
- Structured review processes with specialist input
- Standardized communication materials with appropriate context and support
- Research on participant experiences and outcomes following disclosure

This approach explicitly addresses both traditional abnormalities and predictive findings, with specialized protocols for different finding categories. It also incorporates ongoing research on the impact of different disclosure approaches, enabling evidence-based refinement over time [49].

The All of Us Research Program

The NIH's All of Us Research Program, aiming to gather data from one million participants in the United States, has developed a particularly expansive approach to incidental findings that emphasizes participant preferences and cultural responsiveness [50].

Their approach includes the following:

- Comprehensive education about potential findings before enrollment
- Granular choice regarding disclosure preferences across finding categories
- Culturally adapted communication materials in multiple languages
- Navigation support for participants receiving significant findings
- Community engagement in protocol development and refinement
- Ongoing monitoring of disclosure impacts across diverse populations

This model emphasizes individual autonomy while providing substantial support resources to enable informed decision-making and appropriate follow-up. Its attention to cultural factors and health literacy makes it particularly relevant for diverse implementation contexts [50].

The management of incidental findings in AI-enhanced neuroimaging remains an evolving field, with approaches continuing to develop as both technological capabilities and ethical frameworks mature. The most promising approaches combine clear protocols with flexibility for context-specific adaptation, emphasize ongoing consent and communication rather than one-time disclosure, and allocate resources according to finding significance and individual needs. As these approaches continue developing, they will require ongoing evaluation regarding their impacts on participants, researchers, healthcare systems, and broader society.

6.4 Patient Autonomy When Facing Algorithmic Predictions

The intersection of predictive algorithms and patient autonomy represents one of the most profound ethical challenges in contemporary neuroscience. AI systems increasingly generate probabilistic forecasts about neurological conditions, cognitive trajectories, and treatment responses. This development means traditional conceptions of medical decision-making and patient self-determination require reconsideration. This section examines how algorithmic predictions may influence autonomy in neurological contexts and explores approaches for preserving meaningful patient choice in an era of predictive neuroscience.

Autonomy in Medical Contexts: Traditional Frameworks and New Challenges
Patient autonomy has become a central value in contemporary healthcare ethics. This concept refers to the capacity for self-determination regarding medical decisions, although its conceptualization and implementation vary across different theoretical frameworks.

Traditional Conceptions of Autonomy
In traditional bioethical frameworks, autonomy is often conceptualized as an independent choice by competent individuals based on adequate information and free from undue influence. This concept emphasizes rational deliberation, voluntary choice, and protection from paternalism or coercion [51].

The principle-based approach popularized by Beauchamp and Childress positions respect for autonomy as one of four key bioethical principles alongside beneficence, nonmaleficence, and justice. In this framework, respecting autonomy requires both negative obligations (avoiding interference with autonomous choices) and positive obligations (enabling informed decision-making through information provision and capacity support) [51].

This traditional conception has significantly influenced medical practice, research ethics, and healthcare policy, shaping informed consent requirements, patient rights frameworks, and shared decision-making approaches. However, it has also faced criticism for overemphasizing individual independence while underemphasizing relational aspects of autonomy and social determinants of choice [52].

Relational and Narrative Perspectives
More recent conceptualizations emphasize autonomy as inherently relational, recognizing that self-determination occurs within social contexts and relationships rather than in isolation. These approaches highlight how autonomy develops through interpersonal connections and depends on supportive social environments rather than merely the absence of interference [52].

Relational autonomy frameworks particularly emphasize how autonomy can be enhanced or diminished by social conditions, including power dynamics, access to resources, and recognition of decision-making authority. These perspectives have particular relevance for neurological contexts, where cognitive conditions and social perceptions may significantly influence how patients' decision-making capacity is understood and supported [52].

Narrative approaches to autonomy emphasize how medical decisions fit within individuals' ongoing life stories and identity development rather than representing isolated choices. These approaches consider how medical information and choices align with or disrupt patients' sense of self and future projections, recognizing that autonomy involves maintaining narrative coherence as well as making specific decisions [53].

6.4.1 Algorithmic Predictions and Autonomy Challenges

Algorithmic predictions introduce several novel challenges to patient autonomy that extend beyond those associated with traditional prognostic information:

Epistemic Opacity and Authority

Many advanced predictive algorithms function as "black boxes" whose internal operations cannot be fully explained even by their developers. This epistemic opacity creates challenges for autonomous decision-making, as patients cannot fully understand the basis for predictions about their brain health that may significantly influence their medical choices and life planning [54].

The technical complexity and perceived scientific authority of algorithmic predictions may create what [55] terms "epistemic asymmetry"—a knowledge gap that potentially shifts decision-making authority from patients to technical systems or those who interpret them. When predictions are presented as outputs from advanced computing systems with apparent mathematical precision, patients may feel unable to question or evaluate them independently, potentially diminishing their decision-making agency [55].

Research in other domains suggests that algorithmic recommendations often carry particular persuasive weight, with individuals frequently deferring to automated systems despite theoretical options to override them—a phenomenon termed "automation bias" [56]. In medical contexts, this deference effect may be amplified by the already significant epistemic authority accorded to healthcare providers and scientific information [54].

Predictive Uncertainty and Risk Framing

The inherently probabilistic nature of algorithmic predictions introduces challenges regarding uncertainty communication and risk framing. Research consistently demonstrates that how uncertainty and risk information are presented significantly influences decision-making, even when the underlying information remains identical [2].

For example, presenting a 30% risk of developing a neurological condition might be framed as "70% likelihood of remaining condition-free" or as "nearly one-third chance of developing the condition"—framings that may evoke different emotional responses and decision paths despite mathematical equivalence. Algorithm developers and clinicians make numerous framing choices, both in algorithm design and result communication, that shape how patients perceive predictions [2].

The selection of reference classes and comparison groups also significantly influences interpretation. A prediction that an individual has "twice the average risk" for cognitive decline carries different implications depending on whether the reference group is the general population, age-matched controls, or individuals with similar risk factors. These reference choices often embed value judgments that may not be transparent to patients [15].

Temporal Disruption and Identity Impacts
Algorithmic predictions about future neurological states can significantly impact individuals' sense of identity and future projections, potentially creating what Schermer [57] terms "temporal disruption" in one's self-narrative. When an algorithm predicts elevated risk for future cognitive impairment or neuropsychiatric conditions, this information may fundamentally alter how individuals perceive their future selves and life possibilities.

This disruption may be particularly significant for neurological predictions given the close association between brain function, cognition, and personal identity. Unlike predictions about conditions affecting other organ systems, neurological predictions may be perceived as directly implicating the self rather than merely bodily function. Neurological predictions are uniquely challenging because they appear to forecast not just what conditions one might have but who one might become [41].

Research on the psychological impact of predictive testing for Huntington's disease and other neurogenetic conditions suggests that such predictions can significantly influence major life decisions including career choices, relationship formation, reproductive planning, and financial management, illustrating their profound impact on autonomy beyond immediate medical decision-making [6].

Algorithmic Determinism and Self-Fulfilling Prophecies
Algorithmic predictions risk creating what Bjerring et al. [10] term "algorithmic determinism"—a sense of inevitability about predicted outcomes that may itself influence whether predictions materialize. When individuals receive predictions about their neurological future, these forecasts may influence their behaviors, self-perception, and life choices in ways that either increase or decrease the likelihood of the predicted outcome.

For example, an individual predicted to have elevated risk for depression might respond with increased vigilance and preventive behaviors—or alternatively, might develop a fatalistic attitude that actually increases vulnerability. Similarly, a predicted high risk for cognitive decline might motivate cognitive engagement and lifestyle modifications or might create resignation that reduces protective behaviors [6].

This potential for self-fulfilling or self-negating prophecies creates complex feedback loops between prediction and outcome that traditional autonomy frameworks struggle to address. As Mittelstadt et al. [15] note, "Predictive algorithms do not merely forecast futures but may help create them through their influence on perception and behavior."

6.4.2 Preserving Autonomy in the Era of Algorithmic Prediction

Several approaches have been proposed for preserving and enhancing patient autonomy in contexts where algorithmic predictions influence neurological care:

Transparency and Explainability Enhancements
Efforts to make predictive algorithms more transparent and explainable represent one approach to enhancing autonomy by reducing the epistemic asymmetry between algorithm developers and patients. These efforts operate at multiple levels, from technical explainability mechanisms to enhanced communication practices [54].

Technical approaches include developing inherently interpretable algorithms rather than black-box systems, creating post hoc explanation methods that identify features influencing predictions, and designing user interfaces that visualize algorithmic processes and uncertainties. For example, some neuroimaging algorithms now generate attention maps highlighting brain regions that most influenced a particular prediction, making the basis for forecasts more transparent [58].

Beyond technical transparency, Turkle [59] emphasizes the importance of "translation transparency"—making algorithmic operations understandable to nonspecialists through appropriate metaphors, visualizations, and explanations tailored to different knowledge levels. This approach recognizes that technical explanations alone may be insufficient for supporting autonomous decision-making without appropriate contextual interpretation.

Critics note that transparency mechanisms face both technical and practical limitations. Some high-performing algorithms use computational approaches that resist straightforward explanation, creating tensions between performance and explainability. Additionally, explanations themselves require interpretation, potentially creating new dependencies on expert mediators rather than directly enhancing patient autonomy [11].

Shared Decision-Making Adaptations

Shared decision-making (SDM) approaches—which emphasize collaboration between patients and clinicians in medical decisions—have been adapted to address the challenges of algorithmic predictions. These adaptations focus on integrating algorithmic forecasts into deliberative processes that respect patient values while providing appropriate contextual support [60].

Adapted SDM models typically include the following:

- Providing both algorithm-generated predictions and their limitations
- Contextualizing predictions within individual patient circumstances and values
- Discussing multiple potential courses of action rather than single recommendations
- Explicitly addressing how predictions align with or challenge patient self-understanding
- Creating space for emotional processing and narrative integration of predictions
- Recognizing the ongoing nature of decision-making rather than one-time choices

Research evaluating these adapted approaches suggests they can enhance patients' sense of agency when facing algorithmic predictions while mitigating tendencies toward uncritical algorithmic differences. For example, Hargraves et al. [61] found that structured SDM interventions surrounding predictive analytics for depression management improved patients' decision satisfaction and perceived involvement compared to standard disclosure approaches.

Agency-Enhancing Design and Implementation

Some researchers advocate focusing on how predictive systems are designed and implemented rather than merely how their outputs are communicated. This approach emphasizes creating systems that actively enhance rather than diminish patient agency throughout their design, deployment, and use [62].

Agency-enhancing design principles include the following:

- Including diverse patient perspectives in algorithm development and implementation
- Designing systems that provide options and contingent recommendations rather than singular predictions
- Implementing interfaces that emphasize patient control over how predictions are used
- Creating mechanisms for patients to contextualize predictions with their own knowledge
- Enabling dynamic interaction with predictive systems rather than static outputs
- Establishing feedback mechanisms that incorporate patient experiences and outcomes

McDougall [55] argues that such approaches move beyond merely mitigating autonomy challenges to actively supporting what she terms "computational autonomy"—the capacity to engage meaningfully with algorithmic systems as part of one's decision-making processes rather than being positioned merely as their subject.

Temporal and Narrative Approaches
Recognizing the potential for algorithmic predictions to disrupt patients' temporal self-understanding and narrative coherence, some approaches emphasize supporting autonomy by helping individuals integrate predictions into their ongoing life stories and identity development [53].

Narrative medicine approaches adapted for predictive contexts focus on creating space for patients to explore how predictions affect their self-understanding and future projections. Rather than treating predictions as simple forecasts requiring specific decisions, these approaches frame them as elements to be incorporated into evolving self-narratives that patients continue to author [53].

Practical applications include the following:

- Exploring how predictions fit within patients' existing life stories and identity
- Discussing multiple possible futures rather than single trajectories
- Supporting patients in maintaining narrative agency despite predictive uncertainty
- Acknowledging emotional and existential dimensions of neurological predictions
- Creating ongoing conversations rather than one-time disclosures
- Recognizing cultural and social contexts that shape narrative responses to predictions

Research evaluating narrative approaches in genetic counseling contexts suggests they may be particularly valuable for predictions with significant identity implications, helping individuals maintain coherent self-understanding while incorporating new prognostic information [53].

6.4.3 Special Considerations for Neurological Contexts

Autonomy considerations take on particular dimensions in neurological contexts, where the conditions being predicted may themselves affect decision-making capacity and where brain-related predictions may carry distinctive psychological and social implications:

Autonomy and Cognitive Change
When algorithms predict conditions that may eventually affect decision-making capacity, special considerations arise regarding autonomy across time. Predictions of future cognitive decline raise questions about the association between current and future selves and how present decisions should account for potential future capacity changes [63].

Advance care planning takes on expanded significance in these contexts, potentially including not only traditional advance directives but also "predictive directives" specifying how individuals wish predictive information to be handled as their condition evolves. For example, patients might specify whether they wish to receive updated predictions as algorithms improve or whether certain threshold probabilities should trigger specific interventions [63].

Beyond formal planning, supporting autonomy in these contexts may involve what Jaworska [63] terms "capacity preservation"—focusing not only on respecting current decisions but also on maintaining decision-making capacity as long as possible. This approach emphasizes autonomy as an ongoing capability to be sustained rather than merely a principle to be respected at discrete decision points.

Stigma, Discrimination, and Social Dimensions
Autonomy in neurological prediction contexts cannot be considered solely at the individual level but must address social dimensions including potential stigma, discrimination, and altered social perceptions that may follow from predictive labels [64].

Algorithmic predictions about future neurological or psychiatric conditions may trigger what Corrigan and Watson [65] term "label avoidance"—individuals declining potentially beneficial assessment or care to avoid stigmatizing diagnostic categories or predictions. This avoidance represents an autonomy constraint arising not from the prediction itself but from its anticipated social consequences.

Legal protections against neurological discrimination remain inconsistent and incomplete in many jurisdictions, creating legitimate concerns about how predictive information might influence employment, insurance, education, or other opportunities. These concerns may significantly constrain autonomous choice regarding whether to obtain predictive testing or how to respond to incidental predictive findings [66].

Misaligned Incentives and Structural Constraints
Broader healthcare system factors significantly influence how algorithmic predictions affect autonomy, including potential misalignment between different stakeholders' interests and structural constraints on available options [62].

Commercial interests in predictive algorithms may not align perfectly with patient autonomy interests, particularly when business models involve ongoing monitoring, data collection, or subscription services. These misalignments may subtly influence how predictions are generated, framed, and updated in ways that prioritize commercial rather than autonomy-enhancing goals [62].

Healthcare payment systems and resource constraints may limit available responses to algorithmic predictions, potentially creating situations where patients receive forecasts without access to corresponding preventive measures or support services. These structural constraints represent significant autonomy limitations beyond individual decision-making capacities or preferences [45].

Addressing these broader constraints requires moving beyond individual-level interventions to consider system-level changes that better align incentives, expand available options, and create supportive environments for autonomous decision-making in predictive contexts.

6.4.4 Empirical Research and Future Directions

Empirical research on how algorithmic predictions influence patient autonomy in neurological contexts remains limited, with most evidence extrapolated from adjacent fields like genetic testing, general medical decision-making, or nonmedical algorithmic decision support. Several promising research directions would enhance understanding of these issues:

Longitudinal Studies of Prediction Impact
Longitudinal research following individuals who receive algorithmic predictions about neurological conditions could provide crucial insights into how these predictions influence decision-making, identity, and life trajectories over time. Such research could help identify factors that support or undermine autonomy in different contexts and at different time points postdisclosure [46].

Comparative studies examining different disclosure and decision support approaches could inform best practices for autonomy-enhancing implementation. By systematically varying factors like prediction framing, explanation methods, or follow-up support, researchers could identify which approaches best facilitate autonomous decision-making across diverse patient populations and prediction types [2].

Participatory Research and Codesign
Participatory research involving individuals with lived experience of neurological conditions represents another promising direction for understanding autonomy in predictive contexts. By engaging those directly affected by various conditions in research design, implementation, and interpretation, such approaches can better capture experiential dimensions that may be overlooked in more clinically oriented studies [23].

Codesign methodologies that involve patients in developing and evaluating predictive systems and communication approaches may help create more autonomy-enhancing implementations. These approaches recognize that preserving autonomy requires not merely studying affected populations but actively involving them in shaping how prediction technologies are designed and deployed [67].

Implementation Science and Health Systems Research
Research examining how predictive algorithms function within actual healthcare systems and clinical workflows could provide important insights about structural

factors influencing autonomy. Implementation science approaches can identify barriers and facilitators to autonomy-enhancing practices at individual, organizational, and system levels [14].

Health systems research comparing different organizational models and incentive structures could help identify approaches that better align system factors with patient autonomy in predictive contexts. By examining how different payment models, decision support tools, workforce configurations, and regulatory frameworks influence autonomy outcomes, such research could inform broader system reforms beyond individual-level interventions [45].

The challenge of preserving patient autonomy when facing algorithmic predictions about neurological conditions requires addressing multiple dimensions—from technical transparency and explanation methods to narrative integration and structural constraints. By developing comprehensive approaches that address these various dimensions, we can work toward implementations that harness the potential benefits of predictive neuroscience while respecting and enhancing individual self-determination about brain health.

6.5 Cultural and Cross-Cultural Perspectives on Neural Data Privacy

Cultural contexts and values deeply influence the ethical considerations surrounding neural data privacy and predictive disclosure. As neurotechnology expands globally, understanding diverse cultural perspectives becomes essential. This understanding is needed for developing appropriate consent processes, privacy protections, and disclosure approaches that respect varying conceptions of personhood, privacy, and medical information. This section explores how cultural frameworks shape attitudes toward neural data across different societies and examines approaches for developing culturally responsive neurotechnology practices.

6.5.1 Cultural Variations in Privacy Conceptions

Privacy represents a culturally shaped value rather than a universal concept with identical meaning across all societies. Several key dimensions of cultural variation significantly influence how neural data privacy is understood and prioritized:

Individualism Versus Collectivism

One of the most extensively studied cultural dimensions—the continuum between individualism and collectivism—significantly shapes privacy conceptions and practices. Societies with more individualistic orientations typically emphasize personal privacy as a fundamental right protecting individual autonomy and self-determination. These orientations are common in Western Europe and North America. In contrast, more collectivist societies (prevalent in many East Asian, African, and Latin American contexts) often emphasize relational harmony and shared decision-making over individual privacy claims [68].

These different orientations influence attitudes toward neural data sharing, familial access to predictive information, and appropriate decision-making processes surrounding neural technologies. The individualism-collectivism cultural dimension [68] suggests that societies with more collectivist orientations would likely prioritize potential societal benefits and community welfare over individual privacy concerns when considering neural data sharing, while more individualistic societies would emphasize personal privacy rights and individual control over such information. These cultural frameworks shape expectations about appropriate decision-making processes, with collectivist contexts potentially favoring family or community involvement in neural technology decisions, whereas individualistic contexts typically emphasize individual autonomy and personal choice.

The implications extend to consent processes as well, with research suggesting that participants from more collectivist contexts often prefer family involvement in neurotechnology decisions rather than the strict individual consent typical of Western medical practice. The Western bioethical emphasis on individual autonomy may inadequately capture how many societies conceptualize appropriate decision-making about brain information, where family units rather than individuals may be seen as the proper locus of authority [9].

High Versus Low Context Communication

The distinction between high-context and low-context communication cultures significantly influences expectations surrounding information disclosure and appropriate communication practices. Low-context cultures (common in Germanic and Anglo societies) typically value explicit, comprehensive information provision and direct discussion of medical possibilities. High-context cultures (prevalent in many East Asian, Middle Eastern, and Mediterranean societies) often communicate more implicitly, with greater emphasis on contextual understanding and sometimes deliberate nondisclosure of potentially distressing information [69].

These communication patterns shape expectations regarding how neural predictions should be disclosed, what information should be explicitly stated versus implied, and appropriate levels of directness in discussing potential neurological outcomes. In some high-context cultures, practices like "benevolent deception" or selective nondisclosure may be considered appropriate expressions of care rather than violations of autonomy [70].

Research by Hunt et al. [71] examining cross-cultural attitudes toward the disclosure of dementia predictions found significant variations in preferred disclosure approaches. While North American and Northern European participants generally preferred direct, comprehensive disclosure regardless of prognosis severity, participants from several East Asian and Southern European contexts often preferred gradual, filtered disclosure mediated by family members, particularly for severe prognoses.

Power Distance and Authority Perceptions
Cultural variations in power distance—the extent to which hierarchical differences are expected and accepted—influence perceptions of medical and technological authority, appropriate questioning of expert opinions, and the balance between professional judgment and patient choice. Societies with higher power distance orientations (common in many Asian, African, and Latin American contexts) typically accord greater deference to medical authorities and may expect paternalistic guidance rather than shared decision-making [71].

These differences shape how algorithmic predictions are received and questioned across cultural contexts. Research by Linzer et al. [72] found that patients from higher power distance cultures were significantly less likely to question or seek explanations for AI-generated medical recommendations compared to those from lower power distance backgrounds. This tendency may be particularly significant for neural predictions, given their technical complexity and brain-related authority.

Professional expectations also vary accordingly, with clinicians in higher power distance contexts often expected to provide definitive guidance rather than presenting multiple options for patient selection. As Nie et al. [73] observe, "The Western bioethical emphasis on patient choice may conflict with expectations in many cultural contexts where authoritative guidance is considered a proper expression of medical responsibility rather than inappropriate paternalism."

Uncertainty Avoidance and Risk Perception
Cultural variations in uncertainty avoidance—comfort with ambiguity and unpredictability—significantly influence attitudes toward predictive information, particularly probabilistic forecasts with inherent uncertainty. Societies with higher uncertainty avoidance (common in East Asian, Germanic, and Latin American contexts) typically seek greater predictability and clear guidelines, while those with lower uncertainty avoidance (prevalent in Anglo and Nordic societies) often demonstrate greater comfort with ambiguous or probabilistic information [68].

These differences shape preferences regarding whether to obtain predictive neural information, how such information should be framed, and appropriate responses to uncertain predictions. Research by Kwan et al. [74] examining attitudes toward receiving probabilistic information about future dementia risk found that participants from higher uncertainty avoidance contexts generally preferred binary over probabilistic information and often expressed greater distress when presented with ambiguous risk assessments compared to those from lower uncertainty avoidance backgrounds.

This dimension may be particularly significant for neural predictions given their inherently probabilistic nature and the substantial uncertainties surrounding many neurological conditions. The discomfort with ambiguity characteristic of high uncertainty avoidance contexts may create particular challenges for implementing predictive neurotechnology in ways that respect cultural preferences while maintaining scientific accuracy.

6.5.2 Cultural Dimensions of Brain-Related Information

Beyond general privacy and medical information attitudes, several cultural factors specifically influence how brain-related information is understood and valued across different societies:

Concepts of Selfhood and Brain-Identity Relationships
Cultural variations in concepts of selfhood and the relationship between brain and identity significantly shape attitudes toward neural data privacy. Western philosophical and scientific traditions often closely associate brain function with personal identity, potentially elevating concerns about neural data privacy due to its perceived intimate connection with the self. Other cultural traditions may emphasize different loci of selfhood, including the heart, social relationships, or spiritual dimensions beyond physical substrates [64].

Research by Racine et al. [64] examining neuroimaging perceptions across cultural contexts found significant variations in how brain images were interpreted relative to personal identity. While North American participants typically viewed brain scans as direct representations of "who they are," participants from several East Asian contexts were more likely to view them as representing just one aspect of selfhood among many, potentially influencing the perceived sensitivity of such information.

These varying conceptions influence how invasive neural monitoring or prediction is perceived to be relative to other forms of medical surveillance or forecasting.

Stigma and Mental Health Perceptions
Cultural attitudes toward mental health conditions, cognitive impairment, and neurological differences significantly influence the perceived risks of neural data disclosure and predictive information. Societies vary considerably in levels of stigma associated with various brain-related conditions, from relatively destigmatized framings focused on biological causation to highly stigmatized perceptions associating mental illness with moral failure, spiritual possession, or family dishonor [75].

These variations shape the potential social consequences of neural predictions and thus the perceived risks of insufficient privacy protections. In contexts where neurological or psychiatric conditions carry significant stigma, predictions suggesting elevated risk for such conditions may have more severe social implications, potentially affecting marriage prospects, employment opportunities, or family status [76].

Fate, Control, and Prediction Ethics

Cultural frameworks regarding fate, personal control, and the ethics of prediction itself vary significantly across societies, influencing attitudes toward predictive neurotechnology. Some cultural and religious traditions emphasize the acceptance of divine will or predetermined fate, potentially viewing attempts to predict future health states as inappropriate interference with proper acceptance. Others emphasize personal agency and responsibility for health outcomes, potentially viewing predictive information as enabling appropriate preventive action [68].

These frameworks shape the perceived appropriateness of seeking predictive neural information and preferred responses to such forecasts. Research examining attitudes toward genetic testing across cultural contexts found that participants from more fatalistic cultural backgrounds often expressed discomfort with predictive testing, sometimes considering it presumptuous to seek knowledge properly left to divine determination [77].

Religious perspectives add additional dimensions, with some traditions expressing concern about technological encroachment on domains properly left to spiritual authorities or divine mystery. As Ilkilic [77] notes, "In many religious frameworks, uncertainty about future health is understood not merely as a practical limitation but as a designed aspect of human existence with important spiritual purposes."

6.5.3 Cross-Cultural Challenges in Global Neurotechnology

As neurotechnology development and implementation expand globally, several specific cross-cultural challenges emerge regarding consent, privacy, and data governance:

Language and Translation Challenges

Effective communication about neurotechnology across cultural and linguistic contexts faces significant challenges related to terminology, concept translation, and metaphor appropriateness. Technical terms like "brain-computer interface," "neural decoding," or "predictive biomarker" may lack direct equivalents in many languages, creating risks of miscommunication or concept distortion during translation [78].

Even seemingly straightforward terms like "privacy," "consent," or "prediction" carry different connotations and cultural baggage across languages. Research by Yang et al. [79] analyzing consent documents translated across multiple languages found significant meaning shifts regarding core concepts like voluntariness, risk, and confidentiality, potentially undermining informed decision-making despite technical translation accuracy.

The metaphors commonly used to explain neurotechnology may also function differently across cultural contexts. Western explanations often employ computer metaphors for brain function that may seem intuitive in digitally saturated societies but potentially misleading or inaccessible in contexts with different technological exposure or conceptual frameworks [80].

Power Asymmetries in Global Research

Cross-cultural neurotechnology research and implementation often occur within contexts of significant power asymmetries between technology developers (predominantly from wealthy Western nations) and research participants or users from less resourced contexts. These asymmetries create risks of technological colonialism, where Western ethical frameworks and priorities are imposed without appropriate adaptation or respect for local values and practices [81].

Historical patterns of exploitation in cross-cultural medical research amplify concerns about neural data collection from vulnerable populations, particularly when data flow predominantly from Global South to Global North contexts for analysis and commercialization. As Vayena, & Tasioulas [82] note, "The extraction of valuable neural data from less resourced contexts without corresponding benefit sharing represents a concerning continuation of exploitative research patterns into the neurotechnology domain."

These power dynamics influence consent processes as well, with individuals in more vulnerable contexts potentially agreeing to neural monitoring or data sharing out of deference to perceived authority, hope for healthcare access, or limited understanding of long-term implications rather than genuinely autonomous choice [81].

Regulatory Fragmentation and Jurisdiction Questions

The global landscape of neural data protection features significant regulatory fragmentation, with different jurisdictions adopting varying approaches to data ownership, privacy requirements, and subject rights. This fragmentation creates particular challenges for international research collaborations, cross-border data sharing, and global neurotechnology deployment [67].

While some regions have developed comprehensive frameworks addressing neural data specifically (like Chile's constitutional "neurorights" protections) or general data protection regulations with relevance to neural information (like the European Union's GDPR), many jurisdictions lack specific provisions for this emerging data category. This uneven protection creates risks of "regulatory arbitrage," where neural data collection might shift to jurisdictions with fewer restrictions or protections [67].

The jurisdictional questions become particularly complex for cloud-based analysis of neural data collected across multiple countries or for international research collaborations with participants and researchers subject to different regulatory regimes. These complexities may disadvantage participants from less regulated contexts, whose neural data may receive fewer protections than those from more stringently regulated jurisdictions [81].

6.5.4 Toward Culturally Responsive Approaches

Developing approaches to neural data privacy and consent that respect cultural diversity while maintaining core ethical protections requires several key elements:

Cultural Humility and Localized Adaptation
Rather than imposing standardized approaches developed in Western contexts, culturally responsive neurotechnology requires what Tervalon et al. [82] term "cultural humility"—an orientation of openness to different frameworks and recognition of the validity of diverse ethical approaches. This orientation involves acknowledging the culturally contingent nature of many privacy assumptions rather than treating Western models as universal defaults.

Practical implementation involves localizing consent processes, privacy protections, and disclosure approaches to align with cultural context while maintaining core protections against exploitation or harm. This localization might include adapting decision-making processes to include family involvement where culturally appropriate, modifying communication approaches to respect high-context preferences, or adjusting disclosure timing to align with cultural expectations [71].

Research by Campbell et al. [83] evaluating culturally adapted consent approaches for neuroimaging research found that protocols developed through community consultation and incorporating local values and communication practices achieved significantly better comprehension and satisfaction compared to translated versions of Western-developed protocols. This finding suggests the value of substantive rather than merely linguistic adaptation.

Participatory Design and Community Engagement
Meaningful involvement of diverse cultural communities in neurotechnology design, implementation, and governance represents a crucial approach for developing appropriate privacy and consent models. Participatory methodologies engage community members as active collaborators rather than passive subjects or recipients of predetermined approaches [23].

In practice, this engagement might include community advisory boards providing ongoing input on neural data collection and sharing practices, collaborative protocol development involving local stakeholders, and iterative refinement based on community feedback and experiences. The emphasis shifts from seeking approval for predetermined approaches to cocreating appropriate practices through genuine dialogue [23].

The Global Neuroethics Working Group exemplifies this approach, bringing together scholars and community representatives from diverse cultural contexts to develop frameworks for responsible neurotechnology that respect cultural variation while protecting core human rights [84]. Their work emphasizes moving beyond token consultation toward meaningful power-sharing in how neural technologies are developed and governed.

Identifying Cross-Cultural Ethical Convergence
While respecting cultural diversity, efforts to identify areas of ethical convergence across different cultural frameworks can help establish core protections with broad cross-cultural legitimacy. This approach seeks principles or concerns that resonate across diverse cultural contexts despite different philosophical or religious foundations [85].

Research by the International Neuroethics Society examining cross-cultural perspectives on neural privacy identified several convergent concerns despite significant variation in their specific articulation, including protection from exploitation by powerful entities, maintenance of dignity in vulnerable states, respect for individual or family choice regarding sensitive information, and equitable distribution of benefits from neural data use [84].

These convergent concerns can form the basis for broadly legitimate governance frameworks while allowing flexibility in specific implementation approaches. The goal is not uniform global standards that override cultural differences but rather identification of core protections with cross-cultural resonance that can be implemented through culturally appropriate mechanisms.

Tiered Consent and Flexible Disclosure Models
Recognizing that preferences regarding information disclosure and decision-making processes vary across cultural contexts, tiered consent and flexible disclosure models offer promising approaches for respecting diversity while maintaining essential protections [17].

Tiered consent allows individuals to select different levels of involvement in decision-making, information disclosure, and family participation according to their personal and cultural preferences. Rather than imposing either complete individual autonomy or fixed family authority, this approach enables culturally congruent variability in how decisions about neural data and predictions are navigated [17].

Flexible disclosure models similarly allow for variation in how predictive information is communicated, potentially including options for gradual disclosure, family-mediated communication, or even designated nondisclosure in contexts where such approaches align with cultural values and preferences. This flexibility respects cultural diversity in communication norms while still ensuring appropriate authorization for information management [71].

Research by Bunnik et al. [17] evaluating flexible disclosure protocols for neurogenetic testing found that approaches allowing participant choice regarding disclosure process, timing, and family involvement achieved higher satisfaction across diverse cultural groups compared to standardized protocols, suggesting the value of process flexibility for cross-cultural implementation.

6.5.5 Case Studies in Cross-Cultural Neuroethics

Several pioneering projects have developed approaches to neural data privacy and consent that explicitly address cross-cultural considerations, offering instructive examples for broader application:

The Human Brain Project's Cultural Analysis
The European Human Brain Project incorporated cross-cultural ethical analysis explicitly within its governance structure, recognizing that its global impact necessitated engagement with diverse cultural perspectives rather than purely European ethical frameworks [86].

Their approach includes the following:

- Cultural impact assessment as a required component of technology development
- Engagement with scholars and stakeholders from multiple cultural traditions
- Analysis of how core neuroscience concepts translate across different cultural and linguistic contexts
- Consideration of how data sharing practices might impact differently across global contexts
- Development of adaptable governance frameworks allowing for cultural context variation

This systematic attention to cultural dimensions within a major neuroscience initiative represents an important model for incorporating cross-cultural considerations from early development stages rather than as afterthoughts in implementation [86].

The NeuroRights Initiative's Global Engagement
The NeuroRights Initiative, which successfully advocated for the constitutional protection of neural privacy in Chile, has employed an explicitly cross-cultural approach in its ongoing global advocacy work, seeking to develop frameworks with broad transcultural legitimacy rather than simply exporting Western concepts internationally [87].

Their methodology includes the following:

- Regional consultations with diverse stakeholders across different cultural contexts
- Systematic analysis of how neural privacy concerns manifest across different societies
- Identification of both universal and culturally variable aspects of neural rights
- Flexible implementation guidelines that maintain core protections while allowing cultural adaptation
- Translation of key concepts and principles into diverse linguistic and conceptual frameworks

This approach has facilitated engagement across significantly different cultural and political contexts while maintaining coherent core principles, demonstrating the possibility of combining cross-cultural responsiveness with substantive ethical protection [87].

Indigenous Data Sovereignty in Neuroscience
Indigenous communities globally have developed particularly thoughtful approaches to neuroscience participation that explicitly address cultural dimensions of data governance and decision-making, often termed "Indigenous data sovereignty" frameworks [88].

These approaches typically involve the following:

- Community-level decision-making about participation in neural data collection
- Collectively determined data sharing and privacy protocols
- Recognition of communal interests in individual neural data
- Incorporation of traditional knowledge systems alongside scientific frameworks
- Ongoing involvement in data interpretation and application

The Native BioData Consortium exemplifies this approach, developing community-governed protocols for neurogenetic and neural data that incorporate traditional values while engaging with contemporary neuroscience. Their work demonstrates how cultural frameworks can substantively shape neurotechnology governance rather than merely receiving superficial accommodation [88].

Cultural and cross-cultural perspectives on neural data privacy highlight both the diversity of frameworks through which neurotechnology may be understood and the importance of developing approaches that respect this diversity while maintaining core ethical protections. By moving beyond simplistic universalism toward thoughtful engagement with different cultural frameworks, we can develop more robust, widely legitimate approaches to neural privacy in our increasingly interconnected global context.

6.6 Case Studies: Ethical Dilemmas from Clinical Research Experience

Translating ethical principles into practice often reveals tensions and complexities not fully captured by theoretical frameworks alone. This section presents several case studies drawn from clinical research experience that illustrate key ethical dilemmas in predictive neuroscience and explores how these challenges might be navigated. These cases highlight the contextual, relational, and often ambiguous nature of ethical decision-making in this emerging field.

6.6.1 Case 1: Unexpected Predictive Findings in Multiple Sclerosis Research

6.6.1.1 Clinical Scenario

In a longitudinal study examining structural and functional brain changes in multiple sclerosis (MS), researchers implemented an experimental machine learning algorithm to analyze multimodal imaging data. While the primary research focus was on identifying biomarkers of disease progression, the algorithm unexpectedly demonstrated high accuracy in predicting future cognitive decline 5–7 years before clinical manifestation, based on subtle patterns in functional connectivity and volumetric measurements not previously recognized as predictive.

These predictions were particularly significant because they identified participants who showed minimal current symptoms but had neural signatures suggesting a high likelihood of developing substantial cognitive impairment within the next decade. The original consent process had not specifically addressed the possibility of such predictive findings, stating only that participants would be informed of "clinically significant abnormalities" identified during imaging.

The research team now faced several interconnected ethical dilemmas:

1. Whether to disclose these predictive findings to participants, given the limited validation of the algorithm in independent samples
2. How to handle the lack of specific consent for receiving such predictive information
3. What clinical follow-up would be appropriate given the limited preventive options currently available for cognitive decline in MS
4. How to balance the potential benefits of early warning against psychological harms from this knowledge

6.6.1.2 Ethical Analysis

This case illustrates several core challenges in predictive neuroscience ethics:

Incidental Predictive Findings: Unlike traditional incidental findings showing current abnormalities, these algorithmic predictions represented a novel category of incidental findings—future-oriented predictions not specifically sought but nonetheless generated through research analysis. The unexpectedness of these capabilities highlights how rapidly advancing algorithms may generate information beyond what was anticipated during study design and consent processes [39].

Consent Limitations: The original consent language regarding "clinically significant abnormalities" was developed with traditional radiological findings in mind, not algorithmic predictions of future conditions. This language reflects the common gap between consent processes and emerging technological capabilities, raising questions about whether participants' general consent to be informed of significant findings extends to this novel predictive category [17].

Scientific Uncertainty: While the algorithm showed impressive predictive accuracy within the study dataset, it had not been validated in independent samples or clinical settings, creating uncertainty about its real-world reliability. This illustrates the tension between generating potentially valuable predictive information and establishing sufficient validation to justify clinical disclosure and action [15].

Actionability Considerations: Current preventive options for cognitive decline in MS remain limited, although some evidence suggests cognitive reserve-building activities and aggressive disease-modifying therapies might offer modest protection. This limited actionability complicates risk-benefit analysis for disclosure, illustrating how the value of predictive information depends significantly on available interventions [8].

6.6.1.3 Resolution Approach

After extensive consultation with neuroethicists, MS specialists, participant representatives, and the institutional review board, the research team developed a staged approach to address these dilemmas:

Reconsent Process: They implemented a specific reconsent process for all study participants, explaining the new predictive capabilities that had emerged and offering options regarding whether to receive individual predictive information. This process included detailed educational materials about the preliminary nature of the predictions, potential benefits and risks of receiving this information, and available (although limited) preventive approaches [17].

Tiered Disclosure Options: Rather than an all-or-nothing approach to disclosure, participants were offered multiple options, including receiving complete predictive information with uncertainty estimates, receiving only binary high/low risk categorization, having information shared with their clinical provider but not themselves, or declining to receive any predictive information. This tiered approach respected different preferences regarding the detail and directness of predictive information [22].

Collaborative Clinical Integration: For participants electing to receive predictive information, disclosure occurred through a collaborative process involving both research team members familiar with the algorithm and participants' clinical providers who could integrate this information into ongoing care plans. This approach helped bridge the gap between research findings and clinical management [39].

Longitudinal Support: Recognizing the potential psychological impact of receiving predictions about future cognitive decline, the team established a longitudinal support protocol including follow-up at 1 week, 1 month, and 6 months postdisclosure, with psychological support resources available as needed. This approach acknowledged that adapting to predictive information occurs over time rather than at a single disclosure moment [46].

Research Integration: With appropriate consent, the team systematically studied participants' experiences receiving or declining predictive information, their psychological responses, and subsequent behavioral changes. This research component transformed an ethical challenge into an opportunity to generate evidence about real-world impacts of neurological predictions, potentially informing future protocols [23].

This case illustrates the complex intersection of consent practices, scientific uncertainty, and clinical integration in predictive neuroscience. The resolution approach emphasizes respect for participant autonomy through reconsent and choice options while acknowledging obligations for appropriate support and follow-up when generating potentially impactful predictive information.

6.6.2 Case 2: Algorithmic Predictions and Therapeutic Decision-Making in Depression

6.6.2.1 Clinical Scenario

In a clinical trial evaluating personalized approaches to treatment-resistant depression, researchers developed an algorithm that analyzed resting-state functional connectivity patterns to predict individual response to transcranial direct current stimulation (tDCS). The algorithm showed 78% accuracy in identifying likely responders versus nonresponders before treatment initiation, potentially allowing more targeted therapy application.

During the trial, the algorithm classified participants into three categories: likely responders (high probability of significant improvement), uncertain cases (insufficient confidence to make prediction), and likely nonresponders (high probability of minimal or no improvement). The protocol initially called for using these predictions to stratify analysis but not to determine treatment assignment, as all participants would receive the intervention regardless of predicted response.

However, as the trial progressed, several ethical questions emerged:

1. Whether continuing to provide the full treatment course to predicted nonresponders represented an appropriate use of participant time and research resources
2. How algorithmic predictions should be integrated with clinician judgment when they conflicted
3. Whether and how to disclose prediction results to participants themselves

4. How to handle cases where participants explicitly asked about their predicted response category

The situation became particularly challenging when several participants classified as likely nonresponders reported feeling subjectively worse after initial sessions, raising questions about whether continuing treatment aligned with their best interests despite the protocol requirements.

6.6.2.2 Ethical Analysis

This case highlights several critical ethical dimensions of integrating algorithmic predictions into therapeutic decision-making:

Prediction Disclosure: The original protocol had not explicitly addressed whether participants would be informed of their algorithmically predicted response category, creating ambiguity when participants directly asked about their likelihood of benefit. This situation illustrates the need for prospective consideration of disclosure plans when implementing predictive algorithms in clinical research [1].

Algorithmic Authority: When algorithmic predictions and clinician assessments diverged, questions arose about appropriate weighting of these different information sources. This tension illustrates broader questions about the proper relationship between algorithmic forecasts and human clinical judgment in therapeutic decision-making [55].

Adaptive Protocol Ethics: The emergence of apparent nonresponders who reported symptom worsening created tension between protocol adherence (continuing treatment regardless of predicted response) and participant welfare considerations. This situation highlights how predictive information can create ethical challenges for fixed research protocols that were designed without anticipating such precise response predictions [11].

Therapeutic Misconception: Some participants appeared to believe that their inclusion in the trial indicated they were personally likely to benefit, despite standard consent language about uncertain benefits. This misconception became more problematic once algorithmic predictions suggested that some would likely not improve, illustrating how predictive technologies can complicate the already challenging issue of therapeutic misconception in clinical research [89].

6.6.2.3 Resolution Approach

After consultation with the data safety monitoring board, ethics committee, and participant representatives, the research team implemented several adaptations to address these emerging ethical challenges:

Protocol Amendment: They amended the protocol to include a formal early stopping rule based on combined algorithmic prediction and observed initial response. Participants with both predicted nonresponse and actual symptom worsening after initial sessions could discontinue the intervention phase while remaining in follow-up assessments. This change balanced protocol integrity with participant welfare concerns [11].

Expanded Consent Process: The team developed a specific consent addendum explaining the predictive algorithm, its accuracy limitations, and how predictions might be used in the study. This addendum was presented to all current and future participants, clarifying the role of algorithmic forecasts in the research context [54].

Shared Decision Framework: For participants expressing concerns about continued participation, researchers implemented a structured shared decision process incorporating both algorithmic predictions and clinical assessments while emphasizing participant values and preferences. This approach acknowledged both the value of predictive information and its appropriate contextualization within broader clinical and personal considerations [60].

Transparent Communication Guidelines: The team developed explicit guidelines for discussing algorithmic predictions with participants who requested this information, including standardized language about uncertainty, limitations, and the relationship between predictions and actual outcomes. These guidelines helped ensure consistent, appropriate communication about complex predictive information [15].

Alternative Intervention Access: For participants who discontinued tDCS due to predicted nonresponse and symptom worsening, the team established a pathway to access alternative interventions outside the study, ensuring that protocol changes did not simply abandon participants without treatment options. This approach recognized ongoing obligations to participants even when specific research interventions proved inappropriate [23].

This case demonstrates the dynamic ethical challenges that emerge when predictive algorithms influence therapeutic decision-making in neuropsychiatric research. The resolution emphasizes the importance of protocol flexibility, transparent communication, and continued commitment to participant welfare when navigating the complex intersection of algorithmic prediction and clinical care.

6.6.3 Case 3: Cross-Cultural Implementation of Predictive Alzheimer's Research

6.6.3.1 Clinical Scenario

An international research collaboration initiated a study using machine learning analysis of multimodal data (neuroimaging, fluid biomarkers, cognitive measures, and digital biomarkers) to develop predictive models for Alzheimer's disease progression. The study included sites in the United States, Germany, Japan, Brazil, and Nigeria, aiming to develop algorithms with cross-cultural applicability and validity.

While the core scientific protocol was standardized across sites, significant challenges emerged regarding consent processes, disclosure protocols, and data governance practices due to cultural variations in attitudes toward:

- Family involvement in medical decision-making
- Appropriate disclosure of dementia risk information
- Concepts of data ownership and sharing
- Trust in algorithmic versus human assessment
- Stigma associated with cognitive impairment
- Religious or cultural interpretations of predictive information

These differences created tensions between implementing a scientifically consistent protocol across sites while respecting diverse cultural contexts and values. Particularly challenging situations arose when:

1. Family members in some cultural contexts requested access to predictive information without participant knowledge.
2. Some sites reported participant distress following direct disclosure practices developed in Western contexts.
3. Community leaders in certain regions raised concerns about group stigmatization from algorithmic predictions.
4. Digital biomarker collection faced varying levels of acceptance and understanding across cultural contexts.

6.6.3.2 Ethical Analysis

This case highlights several key dimensions of cross-cultural implementation challenges in predictive neuroscience:

Cultural Variation in Autonomy Concepts: The significant differences in how individual versus family-based medical decision-making was understood across sites illustrated the culturally contingent nature of autonomy conceptions. Western assumptions about individual informed consent did not translate seamlessly to contexts where family or community involvement in health decisions was normative [73].

Disclosure Practice Variations: The apparent distress following direct disclosure in some cultural contexts highlighted how communication practices appropriate in one setting may be harmful in others. This situation illustrates the need for disclosure approaches sensitive to cultural communication norms and expectations regarding medical information [71].

Collective Stigmatization Concerns: The concerns raised about group stigmatization reflected important dimensions of predictive ethics not fully captured in individualistic ethical frameworks. These community-level concerns highlighted how predictive technologies may have impacts beyond individual participants that require dedicated ethical attention [88].

Technology Access Disparities: The varying levels of digital literacy and technology access across sites created potential inequities in how digital biomarker data collection was experienced and understood by participants. These disparities illustrate how technological implementation without adequate contextual adaptation may exacerbate rather than mitigate existing inequalities [81].

6.6.3.3 Resolution Approach

After extensive consultation with ethics committees, community advisory boards, and cultural consultants at each site, the research team developed a culturally responsive framework that maintained scientific integrity while respecting cultural diversity:

Localized Consent Pathways: Rather than imposing a single consent model across all sites, they developed core required elements alongside culturally adapted implementation pathways for each location. These pathways included options for family-based consent processes where culturally appropriate while maintaining core protections against coercion [83].

Flexible Disclosure Protocols: The team created adaptable disclosure guidelines that allowed for variation in how predictive information was communicated, by whom, and with what level of family involvement. These guidelines maintained respect for participant values while acknowledging cultural variation in appropriate information-sharing practices [71].

Community Engagement Structures: At each site, dedicated community engagement processes were established to address group-level concerns, develop appropriate safeguards against stigmatization, and ensure research benefits reached affected communities. These structures acknowledged ethical obligations beyond individual participants to broader communities [88].

Collaborative Governance Model: The team implemented a multilevel governance structure incorporating representation from all sites in key decisions about data management, algorithm development, and result communication. This approach avoided imposing dominant cultural perspectives while maintaining necessary coordination across sites [84].

Technological Adaptation: Digital biomarker collection was modified across sites to accommodate different technology access levels and cultural contexts, including options for research assistant-supported data collection where direct digital interaction was unfamiliar or inaccessible. These adaptations maintained scientific consistency while acknowledging varying technological contexts [27].

Ethical Learning Framework: Throughout the project, the team documented ethical challenges and resolution approaches, creating a learning framework that explicitly acknowledged the evolving nature of cross-cultural neuroethics rather than assuming complete solutions from the outset. This approach recognized the ongoing, developmental nature of ethical practice in this emerging field [85].

This case illustrates the complex challenges of implementing predictive neuroscience protocols across diverse cultural contexts. The resolution approach emphasizes the importance of substantive rather than merely superficial cultural adaptation, meaningful power-sharing in governance, and ongoing ethical learning rather than presumed ethical expertise.

6.6.4 Case 4: Predictive Monitoring in Children at Risk for Neurodevelopmental Disorders

6.6.4.1 Clinical Scenario

A research program developed a predictive monitoring approach for infants at elevated genetic risk for autism spectrum disorder (ASD), combining eye-tracking measures, electroencephalography (EEG), and machine learning analysis to generate individualized risk predictions as early as 6–9 months of age—well before behavioral symptoms typically emerge. Early validation studies suggested that the approach could identify approximately 80% of children who would later meet diagnostic criteria, potentially enabling much earlier intervention.

Parents of at-risk infants were enthusiastic about the technology's potential benefits, often specifically seeking enrollment to gain access to early detection capabilities not available in standard clinical care. However, as the program expanded, several challenging situations emerged:

1. The algorithm identified an infant as high-risk, but the parents expressed that they did not want to know this information despite having initially consented to receive results.
2. A parent of a predicted high-risk child shared the prediction extensively on social media, potentially creating a "digital diagnosis" that might follow the child for years.
3. An insurance company requested documentation of all predictive assessments when one family applied for supplemental health coverage.
4. Childcare providers for one identified high-risk infant began treating the child differently after learning of the prediction, despite researchers' emphasis that this remained a probability rather than diagnosis.

These situations highlighted the complex social dimensions of early predictive information and raised questions about appropriate boundaries and protections surrounding neurodevelopmental predictions.

6.6.4.2 Ethical Analysis

This case highlights several distinctive ethical dimensions of predictive neuroscience applications in pediatric contexts:

Parental Authority and Child Future Autonomy: The situation with parents withdrawing interest in receiving results after initially consenting highlights tensions between respecting current parental authority and protecting the child's future autonomy. This tension illustrates broader questions about who should control access to predictive information with lifelong implications [90].

Digital Permanence of Predictions: The social media sharing situation illustrates how predictive information, once disclosed, may persist in ways that create long-term impacts not anticipated during initial consent processes. This digital permanence represents a novel ethical challenge requiring attention to the long-term social life of predictive information beyond immediate clinical or research contexts [27].

Third-Party Access and Discrimination Risks: The insurance request situation highlights how predictive information, once generated, may be subject to access requests from entities with interests potentially conflicting with the individual's welfare. This situation illustrates broader concerns about how predictive neural information might influence insurance coverage, educational opportunities, or other third-party decisions [45].

Prediction-Based Stigmatization: The childcare provider's situation demonstrates how predictive information may influence others' perceptions and behaviors toward the child, potentially creating self-fulfilling prophecies or unjustified limitations. This risk of prediction-based stigmatization requires attention to the social contexts surrounding predictive technologies, not merely their technical accuracy [76].

6.6.4.3 Resolution Approach

After consultation with bioethicists, advocacy organizations, parents, and regulatory experts, the research team implemented several approaches to address these emerging challenges:

Dynamic Consent Platform: They developed a digital platform enabling ongoing consent management, allowing parents to update their disclosure preferences throughout the research process rather than making a single decision at enrollment. This approach acknowledged that preferences about predictive information might evolve as parents processed implications or as family circumstances changed [19].

Result Communication Framework: They established a structured framework for communicating predictions that emphasized their probabilistic nature, included appropriate contextualizing information about interventions and supports, and specifically addressed appropriate and inappropriate sharing of this information with third parties. This framework helped set expectations about responsible information management beyond the immediate disclosure context [6].

Legal and Policy Protections: The team worked with legal experts to develop clear documentation for families regarding their rights, existing legal protections against discrimination, and appropriate responses to third-party information requests. They also engaged with policy advocates to promote regulatory protections specifically addressing predictive neurodevelopmental information [45].

Educational Interventions: They developed targeted educational materials for key stakeholders in children's lives—including extended family, educators, and childcare providers—explaining the meaning of predictive information, appropriate interpretations, and the importance of avoiding limitation-imposing responses based on probabilistic forecasts. These materials helped construct a supportive social environment around children with identified risk [76].

Longitudinal Impact Assessment: The team implemented a systematic assessment of how predictive information influenced family experiences, third-party responses, and child developmental trajectories over time. This research component created an evidence base regarding real-world impacts of early prediction that could inform future ethical guidelines and protective measures [23].

This case illustrates the complex social dimensions of predictive neurodevelopmental information and the importance of considering impacts beyond immediate clinical or research settings. The resolution approach emphasizes both technological solutions like dynamic consent platforms and broader social interventions addressing how predictive information functions within community contexts and institutional systems.

6.6.4.4 Synthesis and Integration

These case studies illustrate several recurring themes and lessons for ethical practice in predictive neuroscience:

Beyond Procedural Ethics to Contextual Responsiveness

Across all cases, effective ethical approaches required moving beyond standardized procedures to responsive engagement with specific contexts, relationships, and emerging challenges. While clear protocols provided necessary foundations, the most successful approaches maintained flexibility for adaptation as new ethical dimensions emerged during implementation [43].

This contextual responsiveness required ongoing attention to ethical dimensions throughout research and clinical processes rather than treating ethics as a procedural checkpoint addressed solely during initial planning or approval stages. The cases demonstrate how ethical practice in predictive neuroscience represents a continuous, developmental process rather than a fixed set of requirements to be satisfied at discrete timepoints [53].

Balancing Scientific Standardization with Ethical Adaptation

A recurring tension across cases involved balancing scientific requirements for standardization with ethical needs for contextual adaptation. Successfully navigating this tension required identifying core scientific and ethical elements that required consistency while allowing appropriate flexibility in implementation approaches [85].

This balance was particularly evident in the cross-cultural implementation case, where maintaining scientific validity across sites required consistent methodological approaches while respecting cultural diversity necessitated adaptive ethical practices. The most successful resolution approaches distinguished between essential scientific requirements and contextually variable implementation methods, allowing adaptation without compromising core scientific integrity [83].

Anticipatory Ethics and Adaptive Governance

The cases collectively demonstrate the importance of anticipatory ethics—considering potential ethical challenges before they emerge—while maintaining adaptive governance mechanisms to address unanticipated issues. The most successful approaches combined thoughtful advance planning with flexibility to respond to unexpected ethical dimensions as they developed [91].

This combination was evident in the depression treatment case, where the original protocol had not fully anticipated how algorithmic predictions might create ethical tensions during implementation. The successful resolution involved both protocol amendments addressing the immediate situation and governance adaptations to better identify and respond to similar challenges in the future [11].

Beyond Individual Autonomy to Relational and Collective Dimensions
All cases illustrated ethical dimensions extending beyond individual autonomy to include relational and collective considerations. From family involvement in decision-making to community-level stigmatization concerns to social media sharing impacts, the ethical landscape of predictive neuroscience consistently involved interconnected relationships rather than isolated individual choices [52].

Successfully addressing these dimensions required frameworks that acknowledged both individual rights and broader relational contexts, avoiding false dichotomies between individual and collective perspectives. The most effective approaches recognized how individual decisions about predictive information occurred within and affected wider social contexts, requiring attention to these broader dimensions alongside individual autonomy considerations [23].

The Longitudinal Nature of Predictive Ethics
The cases highlight how predictive technologies create ethical considerations extending over time rather than confined to discrete decision points. From changing disclosure preferences to evolving social impacts of predictions to long-term discrimination concerns, the ethical dimensions consistently involved temporal extension rather than momentary choices [63].

Addressing this longitudinal nature required approaches that maintained engagement over time rather than treating consent or disclosure as one-time events. The most successful resolution strategies incorporated ongoing communication, periodic reassessment of preferences, and longitudinal support structures that acknowledged the extended temporal dimension of predictive information impacts [22].

These recurring themes suggest that effective ethical practice in predictive neuroscience requires moving beyond standardized, procedure-focused approaches toward more adaptive, contextual, relational, and temporally extended frameworks. While such approaches demand greater resources and ongoing engagement than one-time procedural compliance, they better address the complex ethical landscape of technologies that generate predictions with far-reaching implications for individuals, families, and communities.

6.7 Conclusion

Consent and autonomy in predictive neuroscience present multifaceted ethical challenges that extend beyond traditional bioethical frameworks. As artificial intelligence increasingly enables forecasting neurological trajectories, treatment responses, and disease risks from neural data, established approaches to informed consent and patient self-determination require significant reconsideration and expansion. The complex nature of probabilistic predictions, the black-box character of many algorithms, the temporal distance between prediction and manifestation, and the deeply personal implications of brain-related forecasts all contribute to a novel ethical landscape requiring thoughtful navigation.

The informed consent challenges in this domain stem from both the nature of neurological predictions themselves and the technological systems generating them. Traditional consent models struggle to address the probabilistic, temporally distant, and variably actionable nature of many neural predictions. Similarly, the algorithmic complexity, evolving performance, and opacity of many predictive systems create significant barriers to the understanding typically considered necessary for meaningful consent. Addressing these challenges requires developing innovative approaches such as layered consent models, process consent frameworks, and comprehension-focused methods that better align with the realities of predictive neurotechnology.

Incidental findings take on new dimensions in AI-enhanced neuroimaging, as advanced algorithms can detect subtle patterns invisible to human observers and identify statistical associations with future disease risks. These expanded detection capabilities fundamentally change the ethical calculus surrounding findings management, necessitating new frameworks for determining review responsibilities, disclosure thresholds, and appropriate follow-up. Predictive incidental findings—algorithm-detected patterns suggesting future rather than current conditions—represent a particularly challenging subcategory requiring careful attention to temporal and probabilistic complexity, variable actionability, and potential psychological impacts.

Patient autonomy faces distinctive challenges when confronting algorithmic predictions about neurological futures. The epistemic opacity of many advanced algorithms, the influential framing of risk information, the potential for predictions to disrupt temporal self-understanding, and the risk of self-fulfilling prophecies all complicate autonomous decision-making in this context. Preserving meaningful autonomy requires approaches that enhance transparency and explainability, adapt shared decision-making models to algorithmic contexts, implement agency-enhancing design principles, and support narrative integration of predictive information into ongoing life stories.

Cultural dimensions add further complexity to these ethical considerations, as attitudes toward privacy, disclosure, autonomy, and brain-related information vary significantly across societies and communities. Cultural frameworks regarding individualism versus collectivism, high versus low context communication, power distance, and uncertainty avoidance all influence how neural data privacy and predictive disclosure are understood across different contexts. Cross-cultural implementation of neurotechnology faces significant challenges related to language and translation, power asymmetries in global research, and regulatory fragmentation across jurisdictions. Developing culturally responsive approaches requires moving beyond imposing standardized Western models toward cultural humility, participatory design, identification of cross-cultural ethical convergence, and flexible implementation frameworks.

The case studies presented illustrate how these theoretical considerations manifest in real-world research and clinical contexts, highlighting the contextual, relational, and often ambiguous nature of ethical decision-making in predictive neuroscience. From managing unexpected algorithm-generated predictions to integrating algorithmic forecasts into treatment decisions to implementing predictive protocols across diverse cultural contexts to addressing the social dimensions of early neurodevelopmental predictions, these cases demonstrate the complex ethical landscape requiring thoughtful navigation. The most successful approaches move beyond standardized procedures toward contextual responsiveness, balance scientific standardization with ethical adaptation, combine anticipatory ethics with adaptive governance, address relational and collective dimensions beyond individual autonomy, and acknowledge the longitudinal nature of predictive ethics.

As predictive neuroscience continues advancing, ethical frameworks must evolve alongside technological capabilities, maintaining core commitments to human dignity, welfare, and self-determination while addressing the novel considerations introduced by neural predictions. This evolution requires ongoing collaboration among neuroscientists, ethicists, clinicians, patients, and diverse cultural communities to develop approaches that harness the promising benefits of predictive technologies while respecting the profound personal and social implications of forecasting neurological futures. By engaging thoughtfully with these ethical dimensions, we can work toward implementations that enhance human flourishing through improved neurological understanding while preserving the values and protections essential to respectful care.

6.8 Summary

This chapter examines the profound ethical challenges arising from AI's ability to predict neurological conditions and outcomes from brain imaging data. As algorithms increasingly forecast disease progression, treatment responses, and cognitive trajectories years before symptoms appear, traditional frameworks for informed consent and patient autonomy require fundamental reconsideration.

6.8.1 Key Challenges Identified

Informed Consent Limitations Traditional consent models struggle with the probabilistic, temporally distant nature of neural predictions. Patients face difficulties understanding complex statistical forecasts, while algorithmic "black boxes" create barriers to truly informed decision-making.

Expanded Incidental Findings AI-enhanced neuroimaging dramatically increases both the frequency and types of unexpected discoveries, detecting subtle patterns invisible to human observers and identifying novel categories of findings with unclear clinical significance.

Threatened Patient Autonomy Algorithmic predictions can disrupt individuals' sense of identity and future planning, while the authoritative nature of AI-generated forecasts may undermine patients' ability to make truly autonomous decisions about their neurological care.

Cultural Variations Attitudes toward privacy, disclosure, and medical decision-making vary significantly across cultures, complicating the global implementation of neurotechnology and requiring culturally responsive approaches rather than one-size-fits-all solutions.

Proposed Solutions
This chapter advocates for evolving consent models including layered consent (providing information at multiple detail levels), dynamic consent (allowing ongoing preference updates), and process consent (emphasizing continued dialogue). It recommends risk-stratified approaches to managing incidental findings, enhanced transparency in algorithmic systems, and culturally adaptive implementation frameworks.

Through detailed case studies from clinical research, this chapter demonstrates that effective ethical practice requires moving beyond standardized procedures toward contextual responsiveness, balancing scientific rigor with respect for diverse values and maintaining ongoing engagement rather than one-time procedural compliance.

The overarching message emphasizes that while predictive neuroscience offers tremendous potential benefits, realizing these benefits requires the proactive development of ethical frameworks that preserve human dignity, autonomy, and cultural diversity in an age of algorithmic brain analysis.

References

1. Biller-Andorno N, Jaeger S. Predictive medicine and artificial intelligence: a clinical perspective. In: Machine medical ethics. Springer; 2016. p. 21–37.
2. Fagerlin A, Zikmund-Fisher BJ, Ubel PA. Helping patients decide: ten steps to better risk communication. J Natl Cancer Inst. 2011;103(19):1436–43.
3. Christin A, Rosenblat A, Boyd D. Courts and predictive algorithms. In: Data & civil rights: a new era of policing and justice. New America; 2016. p. 1–15.

4. Peters E, Tompkins MK, Knoll MA, Ardoin SP, Shoots-Reinhard B, Meara AS. Despite high objective numeracy, lower numeric confidence relates to worse financial and medical outcomes. Proc Natl Acad Sci. 2017;116(39):19386–91.
5. Fins JJ. Neuroethics and disorders of consciousness: discerning brain states in clinical practice and research. AMA J Ethics. 2016;18(12):1182–91.
6. Tibben A. Predictive testing for Huntington's disease. Brain Res Bull. 2007;72(2–3):165–71.
7. Fresson M, Meulemans T, Dardenne B, Geurten M. Overdiagnosis of ADHD in boys: stereotype impact on neuropsychological assessment. Appl Neuropsychol Child. 2019;8(3):231–45.
8. Berkman BE, Hull SC, Eckstein L. The unintended implications of blurring the line between research and clinical care in a genomic age. Pers Med. 2014;11(3):285–95.
9. Roberts JS, Uhlmann WR. Genetic susceptibility testing for neurodegenerative diseases: ethical and practice issues. Prog Neurobiol. 2013;110:89–101.
10. Bjerring JC, Busch J. Artificial intelligence and patient-centered decision-making. Philos Technol. 2021;34(2):349–71.
11. London AJ. Artificial intelligence and black-box medical decisions: accuracy versus explainability. Hastings Cent Rep. 2019;49(1):15–21.
12. Ribeiro MT, Singh S, Guestrin C. "Why should I trust you?": explaining the predictions of any classifier. In: Proceedings of the 22nd ACM SIGKDD international conference on knowledge discovery and data mining. 2016. (pp 1135–1144).
13. Amann J, Blasimme A, Vayena E, Frey D, Madai VI. Explainability for artificial intelligence in healthcare: a multidisciplinary perspective. BMC Med Inform Decis Mak. 2020;20(1):1–9.
14. Wiens J, Saria S, Sendak M, Ghassemi M, Liu VX, Doshi-Velez F, et al. Do no harm: a roadmap for responsible machine learning for health care. Nat Med. 2019;25(9):1337–40.
15. Mittelstadt BD, Allo P, Taddeo M, Wachter S, Floridi L. The ethics of algorithms: mapping the debate. Big Data Soc. 2016;3(2) https://doi.org/10.1177/2053951716679679.
16. Goodman B, Flaxman S. European Union regulations on algorithmic decision-making and a "right to explanation". AI Mag. 2017;38(3):50–7.
17. Bunnik EM, Janssens ACJW, Schermer MH. A tiered-layered-staged model for informed consent in personal genome testing. Eur J Hum Genet. 2013;21(6):596–601.
18. Kaye J, Whitley EA, Lund D, Morrison M, Teare H, Melham K. Dynamic consent: a patient interface for twenty-first century research networks. Eur J Hum Genet. 2015;23(2):141–6.
19. Budin-Ljøsne I, Teare HJ, Kaye J, Beck S, Bentzen HB, Caenazzo L, et al. Dynamic consent: A potential solution to some of the challenges of modern biomedical research. BMC Med Ethics. 2017;18(1):1–10.
20. Williams H, Spencer K, Sanders C, Lund D, Whitley EA, Kaye J, Dixon WG. Dynamic consent: a possible solution to improve patient confidence and trust in how electronic patient records are used in medical research. JMIR Med Inform. 2021;3(1):e3.
21. Dewing J. Participatory research: A method for process consent with persons who have dementia. Dementia. 2007;6(1):11–25.
22. Thorogood A, Dalpé G, Knoppers BM. Return of individual genomic research results: are laws and policies keeping step? Eur J Hum Genet. 2018;27(4):535–46.
23. Goering S, Klein E, Dougherty DD, Widge AS. Staying in the loop: relational agency and identity in next-generation DBS for psychiatry. AJOB Neurosci. 2017;8(2):59–70.
24. Kraft SA, Cho MK, Gillespie K, Halley M, Varsava N, Ormond KE, et al. Beyond consent: building trusting relationships with diverse populations in precision medicine research. Am J Bioeth. 2017;18(4):3–20.
25. Nishimura A, Carey J, Erwin PJ, Tilburt JC, Murad MH, McCormick JB. Improving understanding in the research informed consent process: A systematic review of 54 interventions tested in randomized control trials. BMC Med Ethics. 2013;14(1):1–15.
26. Grady C. Enduring and emerging challenges of informed consent. N Engl J Med. 2015;372(9):855–62.
27. Torous J, Roberts LW. Needed innovation in digital health and smartphone applications for mental health: transparency and trust. JAMA Psychiatry. 2017;74(5):437–8.

28. Vayena E, Blasimme A, Cohen IG. Machine learning in medicine: addressing ethical challenges. PLoS Med. 2018;15(11):e1002689.
29. Grande D, Asch DA, Wan F, Bradbury AR, Jagsi R, Mitra N. Are patients with cancer less willing to share their health information? Privacy, sensitivity, and social purpose. J Oncol Pract. 2020;11(5):378–83.
30. Rainey S, Martin S, Christen A, Mégevand P, Fourneret E. Brain recording, mind-reading, and neurotechnology: ethical issues from consumer devices to brain-based speech decoding. Sci Eng Ethics. 2019;25(6):1721–31.
31. Cheshire WP. Multidimensional ethical decision making in the age of neurotechnology. Ethics Med. 2020;36(1):19–36.
32. Vernooij MW, Ikram MA, Tanghe HL, Vincent AJ, Hofman A, Krestin GP, et al. Incidental findings on brain MRI in the general population. N Engl J Med. 2007;357(18):1821–8.
33. Bos D, Poels MM, Adams HH, Akoudad S, Cremers LG, Zonneveld HI, et al. Prevalence, clinical management, and natural course of incidental findings on brain MR images: the population-based Rotterdam Scan Study. Radiology. 2016;281(2):507–15.
34. O'Sullivan JW, Muntinga T, Grigg S, Ioannidis JP. Prevalence and outcomes of incidental imaging findings: umbrella review. BMJ. 2018;361:k2387.
35. Cole JH, Poudel RP, Tsagkrasoulis D, Caan MW, Steves C, Spector TD, Montana G. Predicting brain age with deep learning from raw imaging data results in a reliable and heritable biomarker. NeuroImage. 2020;163:115–24.
36. Wardlaw JM, Davies H, Booth TC, Laurie G, Compston A, Freeman C, et al. Acting on incidental findings in research imaging. BMJ. 2015;351:h5190.
37. Illes J, Chin VN. Bridging philosophical and practical implications of incidental findings in brain research. J Law Med Ethics. 2008;36(2):298–304.
38. Mittelstadt BD, Floridi L. The ethics of big data: current and foreseeable issues in biomedical contexts. Sci Eng Ethics. 2016;22(2):303–41.
39. Illes J, Kirschen MP, Edwards E, Stanford LR, Bandettini P, Cho MK, et al. Incidental findings in brain imaging research: what should happen when a researcher sees a potential health problem in a brain scan from a research subject? Science. 2010;311(5762):783–4.
40. Wolf SM, Lawrenz FP, Nelson CA, Kahn JP, Cho MK, Clayton EW, et al. Managing incidental findings in human subjects research: analysis and recommendations. J Law Med Ethics. 2008;36(2):219–48.
41. Mackenzie IS, Wei L, Rutherford D, Findlay EA, Saywood W, Campbell MK, MacDonald TM. Promoting public awareness of randomized clinical trials using the media: the 'get randomized' campaign. Br J Clin Pharmacol. 2017;69(2):128–35.
42. Ormond KE, Wheeler MT, Hudgins L, Klein TE, Butte AJ, Altman RB, et al. Challenges in the clinical application of whole-genome sequencing. Lancet. 2016;375(9727):1749–51.
43. Samuel G, Cribb A, Owens J, Williams C. Are brain imaging technologies transforming our understanding of moral experiences? AJOB Neurosci. 2020;11(4):155–6.
44. Vernooij MW. Incidental findings on brain MRI. N Engl J Med. 2019;381(20):1880–1.
45. Hoffman S, Podgurski A. Artificial intelligence and the healthcare workforce: the role of markets, professions, and the state. Yale J Health Policy Law Ethics. 2017;18(3):1–32.
46. Christensen KD, Roberts JS, Whitehouse PJ, Royal CD, Obisesan TO, Cupples LA, et al. Disclosing pleiotropic effects during genetic risk assessment for Alzheimer disease: A randomized trial. Ann Intern Med. 2016;164(3):155–63.
47. Van Essen DC, Smith SM, Barch DM, Behrens TE, Yacoub E, Ugurbil K, Human Connectome Project Consortium. The WU-Minn human connectome project: an overview. Neuroimage. 2013;80:62–79.
48. Boes AD, Prasad S, Liu H, Liu Q, Pascual-Leone A, Caviness VS, Fox MD. Network localization of neurological symptoms from focal brain lesions. Brain. 2020;143(4):1080–99.
49. Miller KL, Alfaro-Almagro F, Bangerter NK, Thomas DL, Yacoub E, Xu J, et al. Biobank-scale neuroimaging: UK biobank brain imaging and early outcomes. Neuroimage. 2022;259:119432.

50. Ormond KE, Laurino MY, Barlow-Stewart K, Wessels TM, Macaulay S, Austin J, Middleton A. Genetic counseling globally: where are we now? Am J Med Genet C: Semin Med Genet. 2018;178(1):98–107.
51. Beauchamp TL, Childress JF. Principles of biomedical ethics. 8th ed. Oxford University Press; 2019.
52. Mackenzie C, Stoljar N, editors. Relational autonomy: feminist perspectives on autonomy, agency, and the social self. Oxford University Press; 2000.
53. Walker P. Narrative therapy in the neuropsychology practice: addressing the impact of disease narrative on patients' mental health. J Clin Neuropsychol. 2018;25(3):1–12.
54. Burrell J. How the machine 'thinks': understanding opacity in machine learning algorithms. Big Data Soc. 2016;3(1):1–12. https://doi.org/10.1177/2053951715622512.
55. McDougall RW. Computer knows best? The need for value-flexibility in medical AI. J Med Ethics. 2019;45(3):156–60.
56. Goddard K, Roudsari A, Wyatt JC. Automation bias: empirical results assessing influencing factors. Int J Med Inform. 2014;83(5):368–75.
57. Schermer M. The mind and the machine. On the conceptual and moral implications of brain-machine interaction. NanoEthics. 2009;3(3):217–30.
58. Holzinger A, Langs G, Denk H, Zatloukal K, Müller H. Causability and explainability of artificial intelligence in medicine. Wiley Interdiscip Rev Data Min Knowl Discov. 2019;9(4):e1312.
59. Turkle S. Reclaiming conversation: the power of talk in a digital age. Penguin Books; 2016.
60. Elwyn G, Frosch DL, Kobrin S. Implementing shared decision-making: consider all the consequences. Implement Sci. 2016;11(1):1–10.
61. Hargraves I, LeBlanc A, Shah ND, Montori VM. Shared decision making: the need for patient-clinician conversation, not just information. Health Aff. 2019;35(4):627–9.
62. Ibrahim H, Liu X, Rivera SC, Moher D, Chan AW, Sydes MR, et al. Reporting guidelines for clinical trials of artificial intelligence interventions: the SPIRIT-AI and CONSORT-AI guidelines. Trials. 2021;22(1):1–5.
63. Jaworska A. Respecting the margins of agency: Alzheimer's patients and the capacity to value. Philos Public Aff. 1999;28(2):105–38.
64. Racine E, Aspler J, Forlini C, Chandler JA. Contextualized autonomy and liberalism: broadening the lenses on complementary and alternative medicines in preclinical Alzheimer's disease. Kennedy Inst Ethics J. 2017;27(1):1–41.
65. Corrigan PW, Watson AC. The paradox of self-stigma and mental illness. Clin Psychol Sci Pract. 2009;9(1):35–53.
66. Yuste R, Goering S, Agüera y Arcas B, Bi G, Carmena JM, Carter A, et al. Four ethical priorities for neurotechnologies and AI. Nature. 2017;551(7679):159–63.
67. Ienca M, Ignatiadis K, Andorno R, Barazzetti G, Biller-Andorno N, Büchler A, et al. Towards a governance framework for brain data. Neuroethics. 2022;15(1):1–14.
68. Hofstede G. Dimensionalizing cultures: the Hofstede model in context. Online Read Psychol Cult. 2011;2(1):8.
69. Hall ET. Beyond culture. Anchor Books; 1989.
70. Beyene Y. Medical disclosure and refugees: telling bad news to Ethiopian patients. West J Med. 1992;157(3):328–32.
71. Hunt LM, de Voogd KB. Are good intentions good enough? Informed consent without trained interpreters. J Gen Intern Med. 2007;22(5):598–605.
72. Linzer M, Bitton A, Tu SP, Plews-Ogan M, Horowitz KR, Schwartz MD. The end of the 15–20 minute primary care visit. J Gen Intern Med. 2018;30(11):1584–6.
73. Nie JB, Cheng Y, Zou X, Gong N, Tucker JD, Wong B, Kleinman A. The vicious circle of patient-physician mistrust in China: health professionals' perspectives, institutional conflict of interest, and building trust through medical professionalism. Dev World Bioeth. 2018;18(1):26–36.
74. Kwan VS, Wojcik SP, Miron-Shatz T, Votruba AM, Olivola CY. Effects of symptom presentation order on perceived disease risk. Psychol Sci. 2021;23(4):381–5.

75. Abdullah T, Brown TL. Mental illness stigma and ethnocultural beliefs, values, and norms: an integrative review. Clin Psychol Rev. 2011;31(6):934–48.
76. Hinshaw SP, Stier A. Stigma as related to mental disorders. Annu Rev Clin Psychol. 2008;4:367–93.
77. Ilkilic I. Bioethical conflicts between Muslim patients and German physicians and the principles of biomedical ethics. Med Law. 2009;28(4):595–610.
78. Yang LH, Purdie-Vaughns V, Kotabe H, Link BG, Saw A, Wong G, Phelan JC. Culture, threat, and mental illness stigma: identifying culture-specific threat among Chinese-American groups. Soc Sci Med. 2017;88:56–67.
79. Parens E. Shaping our selves: on technology, flourishing, and a habit of thinking. Oxford University Press; 2014.
80. Daes EIA. Study on the protection of the cultural and intellectual property of indigenous peoples. United Nations Economic and Social Council; 1993.
81. Vayena E, Tasioulas J. The dynamics of big data and human rights: the case of scientific research. Philos Trans R Soc A Math Phys Eng Sci. 2016;374(2083):20160129.
82. Tervalon M, Murray-García J. Cultural humility versus cultural competence: a critical distinction in defining physician training outcomes in multicultural education. J Health Care Poor Underserved. 1998;9(2):117–25.
83. Campbell MM, Susser E, de Vries J, Baldinger A, Sibeko G, Mndini MM, et al. Exploring researchers' experiences of working with a researcher-driven, population-specific community advisory board in a south African schizophrenia genomics study. BMC Med Ethics. 2018;19(1):1–12.
84. Global Neuroethics Summit Delegates, Rommelfanger KS, Jeong SJ, Ema A, Fukushi T, Kasai K, et al. Neuroethics questions to guide ethical research in the international brain initiatives. Neuron. 2018;100(1):19–36.
85. Evers K, Salles A, Farisco M. Theoretical framing of neuroethics: the need for a conceptual approach. In: Debates about neuroethics. Springer; 2017. p. 89–107.
86. Salles A, Bjaalie JG, Evers K, Farisco M, Fothergill BT, Guerrero M, et al. The human brain project: responsible brain research for the benefit of society. Neuron. 2019;101(3):380–4.
87. Yuste R, Genser J, Herrmann S. It's time for neuro-rights. Horizons. 2021;18:154–64.
88. Garrison NA, Hudson M, Ballantyne LL, Garba I, Martinez A, Taualii M, et al. Genomic research through an indigenous lens: understanding the expectations. Annu Rev Genomics Hum Genet. 2019;20:495–517.
89. Appelbaum PS, Lidz CW, Grisso T. Therapeutic misconception in clinical research: frequency and risk factors. IRB: Ethic Hum Res. 2004;26(2):1–8.
90. Feinberg J. The child's right to an open future. In: Aiken W, LaFollette H, editors. Whose child? Children's rights, parental authority, and state power. Rowman and Littlefield; 1980. p. 124–53.
91. Stahl BC, Wright D. Ethics and privacy in AI and big data: implementing responsible research and innovation. IEEE Secur Priv. 2018;16(3):26–33.

Bias and Justice in Neurological AI

7

We must take sides. Neutrality helps the oppressor, never the victim. Silence encourages the tormentor, never the tormented.

—Elie Wiesel

7.1 Introduction

The intersection of artificial intelligence and neuroscience represents one of the most promising frontiers in modern healthcare, offering unprecedented opportunities to unravel the mysteries of the human brain and revolutionize patient care. From sophisticated neuroimaging analysis that can detect the earliest signs of neurological disorders to brain-computer interfaces that restore communication for paralyzed patients, neurological AI technologies are transforming our understanding of neural function and opening new therapeutic pathways.

However, beneath this veneer of technological progress lies a troubling reality. These powerful systems often perpetuate and amplify the very inequities they promise to address. Like a mirror reflecting society's deepest prejudices, neurological AI can inadvertently encode historical biases, systematically disadvantage marginalized populations, and create new forms of digital discrimination. The stakes could not be higher. When AI systems influence decisions about brain health, cognitive assessment, and neurological treatment, biased algorithms do more than misclassify data points. They can profoundly impact human lives, shaping who receives accurate diagnoses, effective treatments, and access to cutting-edge interventions.

This chapter embarks on a comprehensive examination of how bias infiltrates neurological AI systems, from the datasets that train these algorithms to the clinical settings where they make consequential decisions. We explore the multifaceted sources of inequity, analyze their real-world manifestations, and chart a path toward more just and inclusive neurological technologies that truly serve all of humanity's diverse neurological landscape.

T. Rudroff, *AI for the Healthy Brain and Neurological Disorders*, Neurocultural Health and Wellbeing, https://doi.org/10.1007/978-3-032-11596-6_7

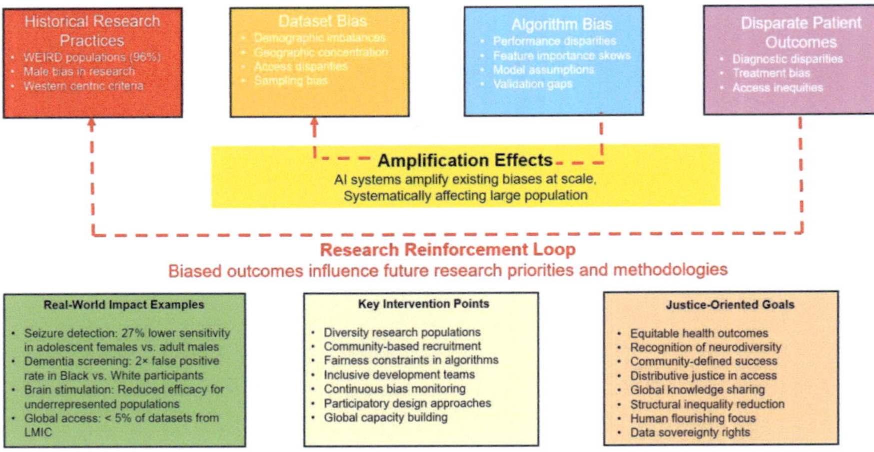

Fig. 7.1 The bias propagation pipeline in neurological AI

Understanding how bias infiltrates neurological AI systems requires examining the systematic pathways through which inequities enter, amplify, and perpetuate throughout the AI development lifecycle. Figure 7.1 illustrates this **bias propagation pipeline**, revealing how seemingly independent decisions at each stage compound to create profound disparities in AI system performance and outcomes.

The pipeline begins with **historical research practices** that have systematically favored WEIRD populations and Western-centric approaches to neurological research. These foundational biases directly shape the composition of **datasets**, creating profound demographic imbalances where participants in major neuroimaging studies predominantly identify as White despite representing only a small fraction of the global population. **Sampling methods** that rely on convenience recruitment from elite institutions and impose language or access barriers further exacerbate these representation gaps. These methods systematically exclude children, older adults, racial minorities, and neurodivergent individuals from the very research intended to serve them.

As these biased datasets flow into the **technical development** phase, AI systems learn patterns that reflect the limited experiences captured in their training data. Without explicit fairness constraints or adequate validation across demographic subgroups, these systems develop **algorithmic biases** that manifest as performance disparities and differential error rates across populations. The final stage—**deployment decisions**—often fails to account for these disparities, leading to **outcome inequities** documented across multiple neurological AI applications.

Perhaps most troubling are the **feedback loops** depicted at the bottom of the pipeline. Deployed biased systems generate new biased data that trains future models while normalizing disparate outcomes and eroding trust among underrepresented communities. This creates a self-reinforcing cycle where each iteration of AI development becomes more, rather than less, exclusionary, transforming individual biases into systematic patterns of digital discrimination that compound over time.

This pipeline framework guides our examination throughout this chapter, with Sects. 7.2, 7.3 and 7.4 providing detailed evidence for each stage and Sect. 7.5 outlining technical approaches to interrupt these harmful cascades.

7.2 Sources of Bias in Neurological Datasets

The foundation of any AI system rests upon the data used to train it. In neurological applications, these datasets often carry the accumulated biases of decades of research practices, clinical traditions, and societal inequities. Understanding these sources of bias is crucial for recognizing how they propagate through AI systems and impact patient outcomes.

The extent of demographic bias in neurological datasets becomes starkly apparent when systematic representation data are compared against global population statistics. While researchers often acknowledge the need for diverse samples, the actual composition of major neuroimaging datasets reveals profound and systematic exclusions that fundamentally limit the generalizability of neurological AI systems trained on these data.

To quantify these representation gaps, we examined demographic characteristics across five major neuroimaging datasets. These datasets serve as training foundations for numerous neurological AI applications. They include the Human Connectome Project (HCP), the Alzheimer's Disease Neuroimaging Initiative (ADNI), the UK Biobank neuroimaging subset, the Adolescent Brain Cognitive Development Study (ABCD), and aggregated datasets from OpenNeuro. These datasets collectively represent tens of thousands of participants and serve as the backbone for hundreds of published AI studies. They demonstrate consistent patterns of demographic limitations that reflect broader inequities in neurological research.

Table 7.1 presents a comprehensive comparison of demographic representation in these datasets, revealing systematic patterns of underrepresentation across multiple dimensions including race/ethnicity, geographic distribution, age ranges, and socioeconomic status. The contrasts revealed in this analysis are not merely statistical artifacts but represent fundamental misalignments between the populations used to train neurological AI systems and the global populations these systems are intended to serve.

The data presented in Table 7.1 reveal representation gaps of concerning magnitude across virtually every demographic dimension examined. Most striking is the persistent pattern of racial and ethnic underrepresentation, with datasets like ADNI showing less than 5% participation from Black/African American individuals despite this population's disproportionately higher risk for Alzheimer's disease. The UK Biobank's 94.6% White ethnicity representation, while reflecting the UK's demographic composition at the time of recruitment, nonetheless limits the generalizability of findings to more diverse global populations. These patterns reflect not random sampling variation but systematic exclusions that fundamentally shape the knowledge base upon which neurological AI systems are built.

Table 7.1 Demographic representation in major neuroimaging datasets

Dataset	Sample size	Age range	Sex distribution	Race/ethnicity	Geographic distribution	Key limitations
HCP	1206 participants	22–35 years	51% female, 49% male	Recruited to reflect ethnic diversity of US; White, Hispanic, Asian, and African American families included	Missouri, USA	Convenience sampling from Missouri families with twins
ADNI	~2366 participants (868 cognitively unimpaired, 1090 MCI, 408 dementia)	Elderly (mean ~75 years)	Approximately equal M/F in normal and dementia groups; 64.6% male in MCI group	88.9% White/87.4% Non-Hispanic White, 7.3% Black/African American, 3.4% Hispanic/Latino, minimal Asian representation	59 sites across North America	<5% Black/African American participants despite higher dementia risk
UK Biobank	36,184 participants (neuroimaging subset)	40–70 years	47% male, 53% female	94.6% White ethnicity	22 assessment centers in England, Wales, Scotland	Overrepresents health conscious, higher SES individuals
ABCD	11,880 participants	9–10 years at baseline	Approximately balanced	Designed to approximate the US population demographics by race/ethnicity	21 sites across the United States	Overrepresents families with higher income and education
OpenNeuro	>20,000 participants across 600+ datasets	Variable by study	Variable by study	Variable by study	Predominantly US and European studies	Aggregated repository; demographics vary significantly by individual dataset

Footnotes:

HCP Human Connectome Project

ADNI Alzheimer's Disease Neuroimaging Initiative

ABCD Adolescent Brain Cognitive Development Study

Data compiled from peer-reviewed publications and official dataset documentation

Key Findings:

Geographic Concentration: Most datasets primarily represent North American and European populations

Racial Underrepresentation: Consistent underrepresentation of racial and ethnic minorities across datasets, particularly concerning given higher disease burden in some conditions

Age Gaps: Limited representation of very young children and very elderly populations

Socioeconomic Bias: Multiple datasets overrepresent higher-income, more educated populations

Reporting Inconsistencies: Demographic reporting practices vary widely, with some studies providing minimal detail

The geographic concentration documented in Table 7.1 is equally problematic, with datasets predominantly representing North American and European populations despite these regions comprising a minority of the global population. This geographic bias is compounded by socioeconomic disparities, with multiple datasets overrepresenting higher-income, more educated participants. Even the ABCD study, explicitly designed to approximate US population demographics, still overrepresents families with higher socioeconomic status. These disparities have direct implications for algorithmic performance, clinical utility, and health equity outcomes, as AI systems trained on such demographically skewed datasets may systematically underperform or fail entirely when applied to underrepresented populations.

7.2.1 Limited Demographic Representation

Neuroimaging datasets, the backbone of many neurological AI applications, suffer from profound demographic imbalances that reflect broader patterns of exclusion in medical research. Comprehensive analyses have revealed a stark reality: participant demographic information is frequently unreported in publicly available neuroimaging datasets and, when reported, shows substantial underrepresentation of non-White populations [1]. This pattern of demographic invisibility creates a cascade of problems, as AI models trained on these datasets effectively learn to "see" the brain through a narrow demographic lens.

Major neuroimaging initiatives, despite being among the most influential and well funded in history, exemplify these representation challenges. Critics have highlighted substantial overrepresentation of White participants in foundational datasets, creating resources that, while scientifically valuable, fundamentally misrepresent the global population's neurological diversity [2]. This overrepresentation becomes particularly problematic when considering that neurological conditions often manifest differently across racial and ethnic groups, with varying genetic predispositions, environmental influences, and symptom presentations.

The geographic concentration of neurological data collection further exacerbates these representation gaps. Systematic reviews have revealed striking imbalances: less than 5% of public neuroimaging datasets include significant participation from low- and middle-income countries, despite these regions comprising over 80% of the global population affected by neurological disorders [3]. This geographic bias creates a profound disconnect between where neurological AI is developed and where neurological diseases exact their greatest toll, potentially rendering these technologies less effective or entirely inappropriate for the populations that need them most.

7.2.2 Sampling Biases in Clinical Settings

Clinical neurological datasets, while often more diverse than research-focused collections, carry their own set of systematic biases that reflect the inequities embedded in healthcare systems themselves. These biases operate through multiple interconnected mechanisms that systematically exclude or underrepresent certain populations.

Access disparities represent perhaps the most fundamental source of bias in clinical datasets. Patients with better healthcare access—those with comprehensive insurance, proximity to specialized medical centers, and the resources to navigate complex healthcare systems—are dramatically overrepresented in hospital-based datasets [4]. This creates a skewed picture of neurological conditions, as the patients captured in these datasets may represent only those with the most severe symptoms or the greatest resources to seek care, entirely missing the experiences of those who suffer in silence or seek care in underresourced settings.

Socioeconomic factors create additional layers of bias that extend far beyond simple access issues. Individuals from lower socioeconomic backgrounds face multiple barriers to participation in neurological research studies or clinical trials. Time constraints from multiple jobs, lack of reliable transportation, inability to take time off work without financial hardship, and the absence of childcare options all contribute to systematic underrepresentation of economically disadvantaged populations [5]. These barriers are not merely inconveniences; they represent structural inequities that fundamentally shape which voices and experiences are captured in the data that trains neurological AI systems.

Language barriers present another significant source of bias, often excluding nonnative speakers from studies due to communication challenges or explicit language requirements [6]. This exclusion is particularly problematic in neurological research, where subtle changes in language and communication can be early indicators of cognitive decline or neurological dysfunction. When AI systems are trained on data that systematically excludes multilingual populations, they may fail to recognize culturally specific patterns of symptom presentation or misinterpret language differences as pathological signs.

7.2.3 Instrumentation and Measurement Bias

The technical aspects of neurological data collection introduce their own forms of systematic bias, often invisible to researchers but profoundly influential in shaping AI model behavior. These instrumentation biases operate at the intersection of technology and human biology, creating subtle but consistent differences that can mislead AI systems about the nature of neurological variation.

Device variability represents a significant challenge in neurological AI, as different neuroimaging equipment produces systematically different data patterns. MRI machines from different manufacturers, operating at different field strengths, or using different software versions can create consistent differences in image quality, resolution, and signal characteristics [7]. When AI models trained on data from one type of equipment are applied to data from another, these technical differences can be misinterpreted as biological variations, leading to systematic errors in diagnosis or treatment recommendations.

Signal quality disparities add another layer of complexity to neurological data collection. Neuroimaging techniques like functional MRI (fMRI) are exquisitely sensitive to participant movement, and head motion during scanning can significantly degrade image quality. However, head motion is not randomly distributed across populations—it correlates with factors including age, neurological condition, anxiety levels, and socioeconomic status [8]. Younger children, older adults with movement disorders, and individuals experiencing stress or unfamiliarity with medical procedures may be more likely to move during scanning, systematically affecting the quality of their data and potentially leading AI systems to perform differently across these groups.

Protocol inconsistencies across research sites introduce additional sources of bias that can confound AI model development. Variations in data collection protocols—from the specific sequences used in neuroimaging to the instructions given to participants—contribute to systematic differences in datasets [9]. These protocol variations often reflect resource constraints, local expertise, or institutional preferences, but they can create artificial patterns in data that AI systems may learn to exploit, leading to models that work well within specific institutional contexts but fail to generalize to broader populations or settings.

7.2.4 Historical and Cultural Biases in Diagnostic Criteria

The diagnostic frameworks that provide ground-truth labels for neurological AI systems carry significant problems. They contain the accumulated biases of medical history and cultural perspectives on neurological conditions. These biases, often invisible to contemporary researchers, profoundly shape how AI systems learn to categorize and understand neurological phenomena.

Western-centric definitions dominate most standardized neurological assessments, having been developed and validated primarily in Western populations with specific cultural, educational, and linguistic backgrounds [10]. These assessments may not adequately account for cultural variations in symptom presentation, interpretation of experiences, or communication styles. For example, descriptions of pain, cognitive experiences, or emotional states can vary dramatically across cultures, potentially leading to systematic misclassification when Western-developed diagnostic criteria are applied to diverse populations.

The evolution of diagnostic categories over time creates additional challenges for AI development. Changes in diagnostic criteria—such as those seen with successive versions of the Diagnostic and Statistical Manual of Mental Disorders (DSM) or the International Classification of Diseases (ICD)—create inconsistencies in historical datasets [11]. Conditions that were once considered distinct may be merged, or single conditions may be subdivided into multiple categories as scientific understanding evolves. These temporal shifts in diagnostic frameworks can create confusion for AI systems that attempt to learn patterns from historical data, potentially leading to models that reflect outdated medical understanding rather than current best practices.

Reporting biases introduce another layer of complexity, as cultural differences in symptom reporting and help-seeking behaviors can systematically affect which conditions are diagnosed and recorded [12]. In some cultures, psychological or cognitive symptoms may be expressed through somatic complaints, while in others, certain types of neurological experiences may be normalized or attributed to non-medical causes. These cultural variations in symptom expression and medical help-seeking create systematic biases in the types of cases that appear in medical datasets, potentially leading AI systems to miss or misinterpret culturally specific patterns of neurological conditions.

7.3 Disparities in Representation Across Demographic Groups

The consequences of uneven representation in neurological datasets manifest across multiple demographic dimensions. These create intersecting patterns of exclusion that compound the challenges faced by marginalized populations. These disparities are not merely statistical artifacts; they represent systematic gaps in knowledge that directly impact the effectiveness and equity of neurological AI applications.

7.3.1 Age-Related Disparities

The representation of different age groups in neurological datasets reveals striking imbalances that reflect both practical challenges and implicit biases in research priorities. Children and older adults face significant underrepresentation in the datasets that train AI systems designed to understand and treat their conditions. This is ironic, as these groups are among the populations most affected by neurological conditions. Pediatric neurological data remain severely limited due to the complex ethical considerations surrounding research with minors [13]. The additional protections required for child participants, while necessary and appropriate, create practical barriers that many research studies choose to avoid rather than navigate. This results in a profound knowledge gap where AI systems designed to detect early signs of developmental disorders, childhood epilepsy, or pediatric brain tumors may lack adequate training data to recognize the unique patterns of these conditions in developing brains.

The exclusion of older adults, particularly those with cognitive impairments, creates equally problematic gaps in representation. Despite being among the populations most affected by neurological conditions such as dementia, Parkinson's disease, and stroke, older adults are often excluded from studies due to concerns about their ability to provide informed consent, higher rates of comorbidities that may confound results, or simply implicit ageism that views their conditions as "normal" aspects of aging rather than treatable medical conditions [14]. This systematic exclusion means that AI systems may be poorly equipped to distinguish between normal age-related changes and pathological processes, potentially leading to both overdiagnosis and underdiagnosis in older populations.

7.3.2 Sex and Gender Underrepresentation

The history of sex and gender bias in neurological research reflects broader patterns of discrimination in biomedical science, with consequences that extend far beyond academic knowledge to impact real-world health outcomes. These biases operate at multiple levels, from basic animal research to human clinical studies, creating compounding effects that limit the applicability of neurological AI to diverse populations.

Animal models in neuroscience have historically displayed a profound male bias, with research predominantly using male specimens under the misguided assumption that female hormonal cycles introduce unwanted variability [15]. This male-centric approach to basic neuroscience research has created foundational knowledge gaps about how neurological processes differ between sexes, gaps that inevitably propagate into AI systems trained on this biased knowledge base.

Human neuroimaging datasets have perpetuated these biases, often skewing toward male participants, particularly in studies of conditions that affect all sexes [16]. This male overrepresentation is especially problematic given mounting evidence that many neurological conditions manifest differently in men and women, from the symptom patterns of stroke and migraine to the progression rates of neurodegenerative diseases.

The binary conceptualization of sex that dominates most neurological research fails to account for the full spectrum of biological sex characteristics, systematically excluding intersex individuals whose neurological patterns may provide crucial insights into the relationship between hormones, genetics, and brain function [17]. This binary framework not only excludes important populations but also oversimplifies the complex interactions between biological and social factors that influence neurological health.

Gender diversity receives even less attention in neurological datasets, with extremely limited data on transgender and nonbinary individuals [18]. This exclusion is particularly problematic given evidence that gender-affirming hormone treatments may influence neurological function and disease risk, creating important knowledge gaps that could impact the care of transgender patients with neurological conditions.

7.3.3 Racial and Ethnic Representation Issues

The underrepresentation of racial and ethnic minorities in neurological research reflects broader patterns of medical racism and creates profound challenges for developing equitable AI systems. These representation issues operate through multiple mechanisms and have far-reaching consequences for health equity.

In the United States, Black, Hispanic, and Native American populations are consistently underrepresented in neurological research relative to their prevalence in the general population and their burden of neurological disease [19]. This underrepresentation is particularly problematic given evidence that many neurological conditions show different patterns of prevalence, progression, and treatment response across racial and ethnic groups.

When racial and ethnic data are collected, they are often recorded using broad categories that fail to capture the genetic and cultural diversity within these groups [20]. For example, lumping together all Asian populations ignores significant genetic and cultural differences between East Asian, South Asian, and Southeast Asian groups, each of which may have distinct patterns of neurological disease risk and symptom presentation.

The practice of using race as a variable in neurological models without accounting for its social construction can reinforce harmful stereotypes and obscure the true causes of health disparities [21]. When AI systems identify racial differences in neurological patterns without considering the social determinants of health that drive these differences—such as exposure to environmental toxins, chronic stress from discrimination, or differences in healthcare access—they risk perpetuating racist assumptions about biological differences between racial groups.

7.3.4 Disability and Neurodiversity Considerations

Perhaps most paradoxically, individuals with disabilities and neurodivergent individuals often face exclusion from the very research studies designed to understand their conditions. This exclusion creates a troubling circularity where neurological AI systems are developed without adequate input from the populations they are intended to serve.

Individuals with certain neurological conditions or disabilities are frequently excluded from studies due to practical challenges in data collection or concerns about data quality [22]. For example, individuals with movement disorders may be excluded from neuroimaging studies due to motion artifacts, despite the fact that understanding movement-related brain patterns could provide crucial insights into their conditions. Similarly, individuals with communication difficulties may be excluded from cognitive assessments, missing opportunities to develop more inclusive assessment methods.

The perspectives and experiences of neurodivergent individuals are rarely incorporated into the design and evaluation of neurological AI systems intended to study or treat their conditions [23]. This exclusion reflects broader patterns of discrimination against disabled individuals in research and technology development, but it also represents a significant missed opportunity to develop more effective and acceptable interventions.

7.4 How Algorithmic Bias Manifests in Neurological Applications

The biases embedded in neurological datasets do not remain dormant. They actively shape the behavior of AI systems in ways that can profoundly impact patient care and health outcomes. Understanding how these biases manifest in real-world applications is crucial for recognizing their potential harms and developing effective mitigation strategies.

While the sources of bias in neurological datasets are concerning from a theoretical perspective, the true measure of their impact lies elsewhere. The real impact appears in how these biases translate into differential performance when AI systems encounter real patients. The abstract notion of "algorithmic bias" becomes a concrete patient safety issue when detection algorithms miss critical events, diagnostic systems generate false alarms, or treatment protocols fail to achieve optimal outcomes for certain populations.

Documented examples of algorithmic bias in healthcare AI reveal the scope and severity of these disparities. In facial recognition technology, research by Buolamwini and Gebru demonstrated dramatic performance gaps across demographic groups, with error rates for darker-skinned women reaching 20-34% compared to less than 1% for lighter-skinned men—a disparity of over 40-fold [24]. While facial recognition may seem distant from neurological applications, similar biases appear throughout medical AI systems that rely on visual analysis or demographic-correlated data patterns.

Healthcare risk assessment algorithms show equally troubling disparities. Obermeyer and colleagues' landmark study of a widely used algorithm revealed that Black patients required 26% more chronic conditions than White patients to receive the same risk score, effectively creating a higher threshold for accessing care [25]. This systematic bias meant that black patients had to be demonstrably sicker before being flagged for additional medical attention, potentially delaying crucial interventions.

In dermatology, where AI systems analyze images similar to neuroimaging applications, performance disparities persist across skin tones. Research has documented reduced sensitivity for detecting malignant lesions on darker skin, with accuracy gaps that could lead to delayed cancer diagnoses in underrepresented populations [26]. These biases raise particular concerns for neurological AI systems that rely on brain imaging, as subtle systematic differences in image processing could similarly affect diagnostic accuracy.

Studies of seizure detection technology have revealed that patient demographics significantly impact system performance, with age-related variations in detection accuracy that could leave certain populations more vulnerable to missed seizures [27]. Data quality issues also vary systematically across demographic groups, creating compounding effects where certain populations receive both lower-quality data collection and reduced algorithmic performance.

The systematic nature of these disparities across different healthcare AI applications is particularly concerning. The consistent pattern of reduced performance for underrepresented populations suggests that these are not random variations or isolated technical failures. Rather, they represent predictable consequences of the biased training processes documented in earlier sections. This systematic nature implies that as neurological AI systems proliferate, so will the disparate impacts on vulnerable populations, potentially exacerbating existing health inequities rather than addressing them.

Research has also documented how these biases compound over time through feedback loops. When biased AI systems are deployed in clinical settings, they generate new data that reflects their biased decision-making patterns. These biased data then influence future model training, creating a cycle where algorithmic discrimination becomes increasingly entrenched in healthcare systems. For neurological conditions that disproportionately affect certain populations—such as stroke risk in specific demographic groups or neurodevelopmental differences across communities—these feedback loops could systematically worsen care disparities.

7.4.1 Diagnostic Disparities

Perhaps nowhere is the impact of algorithmic bias more concerning than in diagnostic applications, where AI systems increasingly influence decisions about patient care. These diagnostic disparities manifest in multiple ways, each with potentially serious consequences for patient health and well-being.

Performance variations across demographic groups represent one of the most direct manifestations of algorithmic bias in neurological AI. Models trained predominantly on data from certain demographic groups often show systematically decreased accuracy when applied to underrepresented populations. Research has documented instances where algorithms trained primarily on data from specific demographics showed reduced sensitivity when applied to different populations [25]. These performance gaps mean that patients from underrepresented groups would be more likely to have critical conditions missed by AI systems, potentially leading to delayed diagnosis and inadequate treatment.

False positive and false negative rates often vary systematically across demographic groups, creating patterns of bias that can have profound clinical consequences. Studies have found that AI systems designed to detect neurological conditions can have substantially different error rates across different patient populations [26]. These disparities mean that certain patients would be much more likely to receive false alarms about their conditions, potentially leading to unnecessary anxiety, additional testing, and premature interventions while also undermining trust in medical AI systems.

These diagnostic disparities are particularly concerning because they often operate in ways that are difficult to detect in routine clinical practice. Unlike human biases, which may be more obvious and subject to conscious correction, algorithmic biases operate consistently and systematically, potentially affecting large numbers of patients before being recognized and addressed.

7.4.2 Treatment Recommendation Biases

When AI systems move beyond diagnosis to influence treatment decisions, the stakes become even higher. Biased treatment recommendations can directly impact patient outcomes, potentially leading to suboptimal care for underrepresented populations.

Personalized treatment protocols, an increasingly important modality for conditions such as depression and Parkinson's disease, may be less effective for underrepresented groups when developed using biased datasets [27]. These protocols rely on a detailed understanding of individual variations in anatomy and physiological patterns, but if the training data primarily represent certain demographic groups, the resulting algorithms may not adequately account for variations in other populations.

Medication recommendations from AI systems present another area of concern, particularly given known population-specific differences in drug metabolism and response. When AI systems fail to account for pharmacogenomic differences that vary across populations—differences that may be underrepresented in training data—they risk recommending inappropriate treatments that could be either ineffective or harmful [28].

Treatment planning algorithms may optimize for outcomes based on limited definitions of success that do not align with the priorities of diverse patient populations [29]. If training data primarily represent patients from certain cultural backgrounds with specific values about independence, mobility, and quality of life, the resulting algorithms may recommend interventions that are not culturally appropriate or aligned with individual patient goals.

7.4.3 Research Direction Skews

The influence of biased AI systems extends beyond direct clinical care to shape the trajectory of neurological research itself. These research direction skews can create self-perpetuating cycles where bias in AI systems leads to misdirected research efforts, which in turn generate more biased data and reinforce existing inequities.

Feature importance metrics from biased models may mislead researchers about which brain regions or neural connections are most relevant to particular conditions [30]. If these metrics are derived from models trained on unrepresentative data, they may direct research attention away from neural mechanisms that are more relevant to underrepresented populations, leading to a narrowing rather than broadening of scientific understanding.

Publication patterns in AI studies can create a distorted picture of AI capabilities, particularly when negative results disproportionately involve applications to underrepresented populations [31]. This bias can create false confidence in AI systems' general applicability while obscuring systematic performance gaps that affect marginalized groups.

7.4.4 Accessibility and User Experience Disparities

The design and implementation of neurological AI systems can create additional barriers for certain groups, manifesting biases not just in algorithmic performance but also in accessibility and usability.

Brain-computer interfaces, one of the most promising applications of neurological AI, may be optimized for specific neural signal patterns, creating usability challenges for users with different neurological presentations [32]. These interfaces often assume certain patterns of neural activity and cognitive processing that may not apply to individuals with autism, ADHD, or other forms of neurodivergence, potentially excluding them from accessing these transformative technologies.

Language barriers in AI-based cognitive assessments can lead to systematic misclassification of nonnative speakers [33]. When AI systems are trained primarily on monolingual populations, they may interpret the cognitive patterns associated with multilingualism—such as increased cognitive flexibility and executive control—as signs of impairment rather than assets.

The high cost and complexity of advanced neurological AI technologies can exacerbate existing healthcare disparities, with cutting-edge tools remaining available primarily to privileged populations [34]. This accessibility gap means that the benefits of neurological AI may primarily accrue to those who are already advantaged in the healthcare system, potentially widening rather than narrowing health disparities.

7.5 Technical Approaches to Fair and Equitable AI in Neuroscience

Addressing bias in neurological AI requires a comprehensive approach. This approach must intervene at every stage of the AI development lifecycle. Technical solutions must be deployed systematically to create more equitable AI systems. These range from improving the diversity and quality of training datasets to implementing sophisticated fairness constraints during model development.

7.5.1 Data Collection and Preprocessing Strategies

The foundation of fair neurological AI lies in improving the quality and representativeness of the datasets used to train these systems. This requires moving beyond passive data collection approaches to actively seek out and include underrepresented populations.

Targeted recruitment strategies represent a crucial first step in addressing dataset bias. Rather than relying on traditional recruitment methods that may systematically exclude certain populations, researchers are increasingly adopting community-based participatory research approaches that partner with community organizations to reach underrepresented groups [35]. These approaches not only improve demographic representation but also ensure that research priorities align with community needs and values.

International data sharing initiatives offer another pathway to improving dataset diversity, but these efforts must be designed with equity as a central goal [36]. Rather than simply aggregating existing datasets, which may perpetuate existing biases, effective data sharing frameworks prioritize the inclusion of underrepresented populations and regions, potentially through targeted funding mechanisms or capacity-building initiatives.

Standardized reporting requirements for demographic characteristics in neurological datasets can help make bias visible and actionable [37]. When researchers are required to report detailed demographic information and justify any exclusions, it becomes easier to identify gaps in representation and develop strategies to address them.

Synthetic data augmentation techniques, such as generative adversarial networks, offer a promising but complex approach to addressing representation gaps [38]. These techniques can generate artificial data points that help balance demographic representation, but they must be carefully validated to ensure that they do not introduce new biases or unrealistic patterns that could mislead AI systems.

7.5.2 Algorithmic Fairness Techniques

Technical approaches to embedding fairness directly into AI algorithms represent a rapidly evolving field with particular relevance to neurological applications. These techniques move beyond simply hoping for fair outcomes to explicitly constraining algorithms to behave equitably across different groups.

Fairness constraints can be implemented during model training to ensure similar performance across demographic groups [39]. Techniques such as equalized odds ensure that AI systems have similar true positive and false positive rates across different groups, while demographic parity approaches ensure that positive predictions are made at similar rates across groups. The choice of a fairness metric depends on the specific application and the types of disparities that are most concerning.

Adversarial debiasing techniques use sophisticated machine learning approaches to remove information that could lead to biased predictions based on protected attributes [40]. These methods train AI systems to make accurate predictions while simultaneously making it difficult for an adversarial system to determine demographic characteristics from the model's internal representations, effectively forcing the model to ignore irrelevant demographic information.

Multitask learning approaches attempt to optimize for both prediction performance and fairness metrics simultaneously [41]. Rather than treating fairness as a constraint that limits performance, these approaches recognize that truly effective AI systems must be fair across diverse populations and optimize for both goals jointly.

Uncertainty quantification techniques, particularly Bayesian approaches, provide clinicians with an important context about model reliability [42]. When AI systems can communicate their confidence in predictions and identify cases where they are uncertain, clinicians can exercise appropriate caution and seek additional information, particularly for patients from populations that may be underrepresented in training data.

7.5.3 Evaluation and Testing Frameworks

Comprehensive evaluation frameworks are essential for identifying and addressing bias in neurological AI systems. These frameworks must go beyond traditional performance metrics to examine how systems behave across diverse populations and use cases.

Stratified performance analysis involves systematically evaluating model performance across different demographic subgroups and intersectional categories [43]. This analysis can reveal performance disparities that would be hidden in overall performance metrics and identify specific populations that may be disadvantaged by AI systems.

Counterfactual testing approaches assess whether AI systems make different predictions when only protected attributes are changed [44]. These techniques can reveal whether systems are inappropriately using demographic information in their decision-making processes, even when such information is not explicitly provided as input.

Real-world validation studies that include diverse populations are essential before widespread deployment of neurological AI systems [45]. Laboratory performance may not translate to real-world effectiveness, particularly when systems are applied to populations different from those represented in training data.

Participatory evaluation approaches involve patients and community members from diverse backgrounds in the assessment of neurological AI systems [46]. These evaluations can identify usability issues, cultural inappropriateness, or value misalignments that technical evaluations might miss.

7.5.4 Interpretability and Transparency

Making neurological AI systems more transparent and interpretable can help identify bias sources and build trust among diverse user communities. These approaches are particularly important in healthcare settings where the consequences of AI decisions can be profound.

Explainable AI techniques, such as SHAP values or LIME, provide interpretable explanations of model predictions [47]. These explanations can help clinicians understand why an AI system makes a particular recommendation and identify cases where the reasoning may be flawed or biased.

Model documentation approaches clearly communicate a model's intended uses, limitations, and performance across different groups [48]. These documents can help clinicians and patients understand when AI systems may be less reliable and make informed decisions about their use.

Algorithmic impact assessments provide comprehensive evaluations of potential disparate impacts before deploying neurological AI systems in clinical settings [49]. These assessments examine not just technical performance but also broader social and ethical implications of AI deployment.

7.5.5 Governance and Ethical Frameworks

Technical approaches to fairness must be complemented by robust governance structures that ensure ongoing attention to equity concerns throughout the AI lifecycle.

Inclusive development teams that include experts from varied backgrounds, including those with lived experience of neurological conditions, can identify potential bias sources and equity concerns early in the development process [50]. Diversity in development teams is not just about representation; it brings different perspectives and experiences that can improve the quality and fairness of AI systems.

Specialized ethics review boards with expertise in both AI and neuroscience can provide oversight for neurological AI development [51]. These boards can help ensure that equity considerations are adequately addressed and that potential harms to vulnerable populations are identified and mitigated.

Regulatory frameworks specifically addressing fairness and equity in neurological AI applications are beginning to emerge [52]. These frameworks provide guidelines and requirements for developers and help ensure that equity considerations are not optional but are legally mandated.

Stakeholder engagement mechanisms create ongoing opportunities for patients and communities to provide input on neurological AI development and deployment [53]. These mechanisms ensure that AI development remains responsive to community needs and values rather than being driven solely by technical possibilities.

7.6 Creating Inclusive Neurological AI Systems: A Perspective

The development of truly inclusive neurological AI systems requires more than technical fixes; it demands a fundamental reimagining of how we approach neuroscience, AI development, and healthcare equity. Drawing from the evidence and approaches examined throughout this chapter, a comprehensive framework for inclusive neurological AI must address both immediate technical challenges and deeper structural inequities.

7.6.1 Reconceptualizing "Normal" in Neuroscience

The field of neuroscience has historically been built around narrow conceptions of neurological "normality" that center the experiences of Western, educated, industrialized, rich, and democratic (WEIRD) populations. Research by Henrich and colleagues revealed a striking imbalance. While WEIRD societies represent only 12% of the world's population, they constitute 96% of psychological research samples. This bias has persisted largely unchanged over the past decade [54]. This sampling bias has profound implications, as cross-cultural studies demonstrate that WEIRD populations are psychological outliers in domains including visual perception, moral reasoning, fairness, and social cognition—yet their responses are routinely generalized to represent universal human psychology [55]. This narrow focus has profoundly shaped not only our scientific understanding but also the AI systems we develop to study and treat neurological conditions.

Neurological diversity must be recognized as a natural and valuable aspect of human variation rather than viewed primarily through a deficit-based medical model. The neurodiversity paradigm, emerging from the disability rights movement in the 1990s, challenges the assumption that neurological differences like autism, ADHD, and dyslexia are inherently pathological [56, 57]. Instead, this framework positions neurodivergence as "natural variation among humans" that represents "a healthy and natural characteristic of the human species" rather than deviations requiring correction [58]. Research demonstrates that viewing autism and other conditions as differences rather than deficits is "developmentally beneficial and transcends a false dichotomy between celebrating differences and ameliorating deficit" [59]. This perspective shift has profound implications for AI development, suggesting that effective neurological AI systems must be designed to accommodate and celebrate neurological diversity rather than enforcing conformity to narrow definitions of "normal" brain function.

AI systems should be designed with flexibility and adaptability at their core, capable of recognizing and responding to the full spectrum of human neurological variation. Current AI technologies in healthcare and assistive technology often reflect the same WEIRD biases present in the research they are built, potentially excluding or misinterpreting the behaviors and needs of neurodivergent individuals. However, emerging approaches demonstrate the potential for AI to support neurodiversity through personalized adaptive technologies, speech recognition systems that accommodate different communication styles, and predictive text tools that assist with cognitive and motor disabilities [60]. Rather than treating neurological differences as deviations to be corrected, these systems should understand variation as information that can inform more personalized and effective interventions while ensuring that algorithm development includes diverse populations and avoids perpetuating existing biases.

The goals of neurological AI interventions must expand beyond simple normalization to encompass broader concepts of human flourishing and quality of life. Traditional healthcare metrics focus primarily on symptom reduction and functional improvement relative to neurotypical standards, but this approach fails to capture the multidimensional nature of well-being [61]. Research on human flourishing identifies key domains including happiness and life satisfaction, physical and mental health, meaning and purpose, character and virtue, close social relationships, and financial stability—all of which can be measured and should inform treatment success [62]. This means that success metrics should be defined in collaboration with diverse neurodivergent communities and should reflect the varied ways that individuals experience well-being and fulfillment, rather than imposing externally defined standards of normalcy.

7.6.2 Shifting from Fairness to Justice

While technical fairness approaches represent important progress, they are insufficient to address the deeper structural inequities that shape neurological health outcomes. A justice-oriented approach to neurological AI requires examining not just whether AI systems treat different groups equally but also whether they contribute to a more equitable distribution of health resources and outcomes.

Distributive justice principles should guide neurological AI development, ensuring that the benefits and risks of these technologies are allocated fairly across society [63]. This means prioritizing AI applications that address the needs of currently underserved populations rather than primarily enhancing capabilities that are already available to privileged groups.

Resource allocation decisions should explicitly consider equity implications, directing funding and development efforts toward applications that can reduce rather than exacerbate health disparities [64]. This might mean investing more heavily in AI applications for conditions that disproportionately affect underserved populations or developing technologies specifically designed for resource-constrained settings.

Access frameworks must balance innovation incentives with ensuring equitable access to beneficial neurological AI technologies [65]. The current system, which often prices cutting-edge medical technologies out of reach for many populations, may need fundamental reform to ensure that neurological AI benefits are broadly shared.

7.6.3 Cocreation and Participatory Design

The meaningful involvement of diverse stakeholders throughout the AI development lifecycle is essential for creating technologies that truly serve all populations. This involvement must go beyond token consultation to a genuine partnership in defining problems, developing solutions, and evaluating outcomes.

People with lived experience of neurological conditions should be included as full partners, not merely subjects, in research and development processes [66]. This partnership approach recognizes that individuals with neurological conditions possess unique expertise about their own experiences and needs that cannot be captured through external observation alone.

Community-based participatory research approaches, already well-established in public health research, must be adapted for neurological AI development [67]. These approaches ensure that technologies address community-identified priorities rather than problems defined solely by researchers or technology developers.

Success metrics should be defined collaboratively with diverse communities, reflecting the varied ways that different populations conceptualize health, well-being, and quality of life [68]. Rather than imposing universal definitions of success, neurological AI systems should be flexible enough to adapt to different cultural values and individual preferences.

7.6.4 Global Equity Considerations

Ensuring that neurological AI benefits populations worldwide requires specific attention to global power dynamics and the historical exclusion of low- and middle-income countries from technological advancement.

Knowledge and technology transfer mechanisms must be designed to ensure that neurological AI innovations benefit populations in low- and middle-income countries [69]. This requires moving beyond simply exporting technologies developed in high-income countries to support local capacity for AI development and adaptation.

Capacity building initiatives should support the development of local expertise in neurological AI across diverse global contexts [70]. This includes not only technical training but also support for developing culturally appropriate applications and governance frameworks.

Data sovereignty principles must be respected, particularly for indigenous populations and historically marginalized communities [71]. Rather than extracting data from these communities for the benefit of others, neurological AI development should ensure that communities maintain control over their data and benefit from any insights derived from it.

7.6.5 Longitudinal Monitoring and Adaptation

Creating inclusive neurological AI is not a one-time achievement but an ongoing process that requires continuous attention and adaptation as technologies evolve and new challenges emerge.

Continuous monitoring systems should track the performance and impact of deployed neurological AI systems across diverse populations [72]. These systems should be designed to detect emerging disparities and bias patterns before they cause widespread harm.

Feedback mechanisms must enable ongoing improvement based on real-world outcomes and evolving understanding of equity considerations [73]. This requires creating channels for users and affected communities to report problems and suggest improvements.

Governance frameworks must be adaptable to address emerging challenges and opportunities in neurological AI [74]. Rather than static regulatory approaches, these frameworks should be designed to evolve as technologies and understanding advance.

7.7 Conclusion

The development of fair and inclusive neurological AI systems represents one of the defining challenges of our technological age. The issues explored in this chapter reveal that bias in neurological AI is not a simple technical problem with straightforward solutions but rather a complex sociotechnical challenge that reflects deeper inequities in our research institutions, healthcare systems, and society at large.

The path forward requires unprecedented collaboration across disciplines, communities, and institutions. Technical innovations in fairness-aware machine learning must be combined with fundamental reforms in research practices, healthcare delivery, and technology governance. The perspectives of marginalized communities must be centered rather than peripheralized, and the benefits of neurological AI must be distributed justly rather than concentrated among the already privileged.

Perhaps most importantly, the development of inclusive neurological AI requires us to confront uncomfortable truths about the inequities embedded in our current systems and institutions. It demands that we acknowledge how historical patterns of exclusion and discrimination continue to shape contemporary technology development, and it challenges us to envision and create more alternatives.

The technical approaches outlined in this chapter—from improved data collection strategies to sophisticated fairness constraints—provide important tools for this work. However, their effectiveness ultimately depends on our collective commitment to equity and justice as core values in neurological AI development. Without this commitment, technical solutions may simply provide a veneer of fairness while fundamental inequities persist.

As neurological AI continues to evolve and expand its influence over healthcare decisions, the stakes of getting this right could not be higher. The choices we make today about how to develop, deploy, and govern these technologies will shape neurological healthcare for generations to come. By centering equity and inclusion in these efforts, we have the opportunity to create AI systems that not only advance scientific understanding but also contribute to a more just and equitable world.

The approaches outlined in this chapter provide a starting point for this critical work, but they must be continually refined through inclusive dialogue, rigorous evaluation, and unwavering commitment to the principles of justice and human dignity. Only through such sustained effort can we realize the promise of neurological AI as a force for healing and human flourishing that truly serves all members of our diverse global community.

7.8 Summary

This chapter has provided a comprehensive examination of bias and justice in neurological AI, revealing how these powerful technologies, despite their transformative potential, often perpetuate and amplify existing societal inequities rather than addressing them. This chapter's central argument—that neutrality in AI development inadvertently serves to help oppressors rather than victims—has been systematically supported through empirical evidence and theoretical analysis across multiple dimensions of the AI development lifecycle.

Sources and Manifestations of Bias

Our analysis has documented how bias infiltrates neurological AI systems through a complex pipeline of interconnected processes. Historical research practices that systematically favor WEIRD populations have created foundational datasets with profound demographic imbalances, where major neuroimaging initiatives consistently underrepresent racial and ethnic minorities, geographic diversity, and socioeconomic variation. The Human Connectome Project, ADNI, UK Biobank, ABCD, and OpenNeuro datasets, while scientifically valuable, demonstrate persistent patterns of exclusion that fundamentally limit the generalizability of AI systems trained on these data.

These dataset biases manifest in systematic clinical sampling disparities, instrumentation inconsistencies, and diagnostic frameworks rooted in Western-centric medical traditions. Access barriers disproportionately exclude economically disadvantaged populations, while language requirements and cultural assumptions in assessment tools create additional layers of systematic exclusion. Perhaps most paradoxically, neurodivergent individuals—the very populations these technologies are designed to serve—often face exclusion from the research studies that inform AI development.

Real-World Consequences

This chapter has demonstrated that these biases translate into concrete disparities in AI system performance, creating diagnostic gaps, treatment recommendation biases, and accessibility barriers that can profoundly impact patient outcomes. Evidence from healthcare AI applications reveals systematic performance differences across demographic groups, with error rates varying dramatically based on race, gender, age, and other characteristics. These disparities are not marginal effects but substantial gaps that can fundamentally alter the quality of care received by different populations.

The feedback loops inherent in these systems are particularly troubling, as biased AI systems generate new biased data that trains future models, creating self-reinforcing cycles of digital discrimination. As neurological AI systems proliferate throughout healthcare, these disparities risk becoming increasingly entrenched, potentially widening rather than narrowing health inequities.

Technical and Systemic Solutions

This chapter has outlined comprehensive technical approaches to creating fairer neurological AI systems, spanning data collection strategies, algorithmic fairness techniques, evaluation frameworks, and governance structures. However, these technical solutions alone are insufficient without fundamental changes in how we conceptualize neurological diversity, conduct research, and distribute healthcare resources.

The most transformative insights emerge from this chapter's call to reconceptualize "normal" in neuroscience, moving beyond narrow WEIRD-centric definitions to embrace neurological diversity as natural human variation. This shift requires adopting justice-oriented rather than merely fairness-oriented approaches, implementing cocreation and participatory design methodologies, and addressing global equity considerations that ensure neurological AI benefits reach underserved populations worldwide.

A Vision for Inclusive Neurological AI

This chapter's ultimate vision extends far beyond technical fixes to imagine neurological AI systems that celebrate rather than pathologize neurological diversity, that define success in collaboration with diverse communities rather than imposing external standards, and that contribute to human flourishing rather than mere normalization. This vision requires unprecedented collaboration across disciplines, communities, and institutions, with marginalized voices centered rather than peripheralized in the development process.

The Stakes and the Call to Action

The choices we make today about neurological AI development will shape healthcare for generations to come. This chapter has shown that bias in neurological AI is not a simple technical problem with straightforward solutions but a complex sociotechnical challenge that reflects deeper inequities in our research institutions, healthcare systems, and society at large. Addressing these challenges requires us to confront uncomfortable truths about historical patterns of exclusion and discrimination while committing to justice and human dignity as core values in technological development.

The evidence presented throughout this chapter makes clear that we cannot remain neutral in the face of algorithmic discrimination. As Elie Wiesel's opening words remind us, neutrality helps the oppressor, never the victim. The neurological AI systems we develop today will either perpetuate existing inequities or contribute to a more just and equitable world. The technical approaches, ethical frameworks, and transformative vision outlined in this chapter provide the foundation for this critical work, but their success ultimately depends on our collective commitment to ensuring that the promise of neurological AI as a force for healing and human flourishing truly serves all members of our diverse global community.

The path forward is challenging but essential. Only through sustained effort, inclusive dialogue, rigorous evaluation, and unwavering commitment to equity and justice can we realize the full potential of neurological AI while avoiding its profound risks. The stakes could not be higher, and the time for action is now.

References

1. Sterling E, Pearl H, Liu Z, Allen JW, Fleischer CC. Demographic reporting across a decade of neuroimaging: a systematic review. Brain Imaging. 2022;16(6):2785–96.
2. Thompson PM, Jahanshad N, Ching CR, Salminen LE, Thomopoulos SI, Bright J, ENIGMA Consortium. ENIGMA and global neuroscience: a decade of large-scale studies of the brain across 43 countries. Transl Psychiatry. 2020;10(1):100.
3. Parra MA, Baez S, Allegri R, Nitrini R, Lopera F, Slachevsky A, Ibanez A. Dementia in Latin America: assessing the present and envisioning the future. Neurology. 2018;90(5):222–31.

4. Williams DR, Mohammed SA. Discrimination and racial disparities in health: evidence and needed research. J Behav Med. 2009;32(1):20–47.
5. Ford JG, Howerton MW, Lai GY, Gary TL, Bolen S, Gibbons MC, Bass EB. Barriers to recruiting underrepresented populations to cancer clinical trials: a systematic review. Cancer. 2008;112(2):228–42.
6. Chen AH, Youdelman MK, Brooks J. The legal framework for language access in healthcare settings: title VI and beyond. J Gen Intern Med. 2007;22(2):362–7.
7. Han X, Jovicich J, Salat D, van der Kouwe A, Quinn B, Czanner S, Busa E, Pacheco J, Albert M, Killiany R, Maguire P, Rosas D, Makris N, Dale A, Dickerson B, Fischl B. Reliability of MRI-derived measurements of human cerebral cortical thickness: the effects of field strength, scanner upgrade and manufacturer. NeuroImage. 2006;32(1):180–94.
8. Power JD, Barnes KA, Snyder AZ, Schlaggar BL, Petersen SE. Spurious but systematic correlations in functional connectivity MRI networks arise from subject motion. NeuroImage. 2012;59(3):2142–54.
9. Markiewicz CJ, Gorgolewski KJ, Feingold F, Blair R, Halchenko YO, Miller E, Poldrack RA. The OpenNeuro resource for sharing of neuroscience data. elife. 2021;10:e71774.
10. Franzen S, Watermeyer TJ, Pomati S, Byrd D, Crivelli L, Garolera M, Kim S. Cross-cultural neuropsychological assessment in the European Union: a Delphi expert study. Arch Clin Neuropsychol. 2021;36(6):815–30.
11. Clark LA, Cuthbert B, Lewis-Fernández R, Narrow WE, Reed GM. Three approaches to understanding and classifying mental disorder: ICD-11, DSM-5, and the National Institute of Mental Health's Research Domain Criteria (RDoC). Psychol Sci Public Interest. 2017;18(2):72–145.
12. Sue DW, Sue D. Counseling the culturally diverse: theory and practice. 8th ed. Wiley; 2019.
13. Mezinska S, Gallagher L, Verbrugge M, Bunnik EM. Ethical issues in genomics research on neurodevelopmental disorders: a critical interpretive review. Hum Genomics. 2021;15(1):16.
14. Taylor JS, DeMers SM, Vig EK, Borson S. The disappearing subject: exclusion of people with cognitive impairment and dementia from geriatrics research. J Am Geriatr Soc. 2012;60(3):413–9.
15. Shansky RM. Are hormones a "female problem" for animal research? Science. 2019;364(6443):825–6.
16. Cirillo D, Catuara-Solarz S, Morey C, Guney E, Subirats L, Mellino S, Valencia A. Sex and gender differences and biases in artificial intelligence for biomedicine and healthcare. npj Digit Med. 2020;3(1):81.
17. Joel D. Beyond the binary: rethinking sex and the brain. Neurosci Biobehav Rev. 2021;122:165–75.
18. Rosendale N, Wong JO, Flatt JD, Whitaker E. Sexual and gender minority health in neurology: a scoping review. JAMA Neurol. 2021;78(6):747–54.
19. Oh SS, Galanter J, Thakur N, Pino-Yanes M, Barcelo NE, White MJ, Burchard EG. Diversity in clinical and biomedical research: a promise yet to be fulfilled. PLoS Med. 2015;12(12):e1001918.
20. Bonham VL, Green ED, Perez-Stable EJ. Examining how race, ethnicity, and ancestry data are used in biomedical research. JAMA. 2018;320(15):1533–4.
21. Boyd RW, Lindo EG, Weeks LD, McLemore MR. On racism: a new standard for publishing on racial health inequities. Health Affairs Blog. 2020;10:377.
22. Makowski C, Lepage M, Evans AC. Head motion: the dirty little secret of neuroimaging in psychiatry. J Psychiatry Neurosci. 2019;44(1):62–8.
23. Chapman R, Carel H. Neurodiversity, epistemic injustice, and the good human life. J Soc Philos. 2022;53(4):607–28.
24. Obermeyer Z, Powers B, Vogeli C, Mullainathan S. Dissecting racial bias in an algorithm used to manage the health of populations. Science. 2019;366(6447):447–53.
25. Halimeh M, Jackson M, Vieluf S, Loddenkemper T, Meisel C. Explainable AI for wearable seizure logging: Impact of data quality, patient age, and antiseizure medication on performance. Seizure. 2023;110:99–108.

26. Daneshjou R, Vodrahalli K, Novoa RA, Jenkins M, Liang W, Rotemberg V, Zou J. Disparities in dermatology AI performance on a diverse, curated clinical image set. Sci Adv. 2022;8(32):eabq6147.
27. Young AT, Xiong M, Pfau J, Keiser MJ, Wei ML. Artificial intelligence in dermatology: a primer. J Invest Dermatol. 2021;141(7):1504–12.
28. Relling MV, Evans WE. Pharmacogenomics in the clinic. Nature. 2015;526(7573):343–50.
29. Saenko N, Voronkova O, Volk M, Voroshilova O. The social responsibility of scientists: the philosophical Aspect of Contemporary Discussions. J Soc Stud Educ Res. 2019;10(3):3312–45.
30. Lundberg SM, Erion G, Chen H, DeGrave A, Prutkin JM, Nair B, Lee SI. From local explanations to global understanding with explainable AI for trees. Nat Mach Intell. 2020;2(1):56–67.
31. Towards Data Science. Is the media's reluctance to admit AI's weaknesses putting us at risk? 2025. https://towardsdatascience.com/is-the-medias-reluctance-to-admit-ai-s-weaknesses-putting-us-at-risk-c355728e9028/. Accessed 3 Aug 2025.
32. Jeunet C, Jahanpour E, Lotte F. Why standard brain-computer interface (BCI) training protocols should be changed: an experimental study. J Neural Eng. 2016;13(3):036024.
33. Ardila A. Historical development of human cognition: a cultural-historical neuropsychological perspective. Springer; 2020.
34. Rajkomar A, Hardt M, Howell MD, Corrado G, Chin MH. Ensuring fairness in machine learning to advance health equity. Ann Intern Med. 2018;169(12):866–72.
35. Wallerstein N, Duran B, Oetzel JG, Minkler M, editors. Community-based participatory research for health: advancing social and health equity. Wiley; 2017.
36. Gorgolewski KJ, Auer T, Calhoun VD, Craddock RC, Das S, Duff EP, Poldrack RA. The brain imaging data structure, a format for organizing and describing outputs of neuroimaging experiments. Sci Data. 2016;3(1):160044.
37. NIH BRAIN Initiative. Data standards for the BRAIN initiative. National Institutes of Health; 2022.
38. Zhao J, Wang T, Yatskar M, Ordonez V, Chang KW. Gender bias in coreference resolution: evaluation and debiasing methods. In: Proceedings of the 2018 conference of the North American Chapter of the Association for Computational Linguistics: Human Language Technologies; 2018. p. 15–20.
39. Dwork C, Hardt M, Pitassi T, Reingold O, Zemel R. Fairness through awareness. In: Proceedings of the 3rd innovations in theoretical computer science conference; 2012. p. 214–26.
40. Zhang BH, Lemoine B, Mitchell M. Mitigating unwanted biases with adversarial learning. In: Proceedings of the 2018 AAAI/ACM conference on AI, Ethics, and Society; 2018. p. 335–40.
41. Yang J, Soltan AA, Eyre DW, Yang Y, Clifton DA. An adversarial training framework for mitigating algorithmic biases in clinical machine learning. npj Digit Med. 2023;6(1):55.
42. Gal Y, Ghahramani Z. Dropout as a Bayesian approximation: representing model uncertainty in deep learning. In: Proceedings of the 33rd international conference on Machine Learning; 2016. p. 1050–9.
43. Buolamwini J, Gebru T. Gender shades: intersectional accuracy disparities in commercial gender classification. In: Proceedings of the 1st conference on Fairness, Accountability and Transparency; 2018. p. 77–91.
44. Kusner MJ, Loftus J, Russell C, Silva R. Counterfactual fairness. Adv Neural Inf Proces Syst. 2017;30
45. Kelly CJ, Karthikesalingam A, Suleyman M, Corrado G, King D. Key challenges for delivering clinical impact with artificial intelligence. BMC Med. 2019;17(1):195.
46. Teodorowski P, Gleason K, Gregory JJ, Martin M, Punjabi R, Steer S, Savasir S, Vema P, Murray K, Ward H, Chapko D. Participatory evaluation of the process of co-producing resources for the public on data science and artificial intelligence. Res Involv Engagem. 2023;9(1):67.
47. Lundberg SM, Lee SI. A unified approach to interpreting model predictions. Adv Neural Inf Proces Syst. 2017;30

48. Mitchell M, Wu S, Zaldivar A, Barnes P, Vasserman L, Hutchinson B, Gebru T. Model cards for model reporting. In: Proceedings of the conference on Fairness, accountability, and transparency; 2019. p. 220–9.
49. Reisman D, Schultz J, Crawford K, Whittaker M. Algorithmic impact assessments: a practical framework for public agency accountability. AI Now Institute; 2018.
50. Ferrara E. Fairness and bias in Artificial Intelligence: a brief survey of sources, impacts, and mitigation strategies. Science. 2024;6(1):3.
51. Salles A, Farisco M. Neuroethics and AI ethics: a proposal for collaboration. BMC Neurosci. 2024;25:41.
52. U.S. Food and Drug Administration. Proposed Regulatory Framework for Modifications to Artificial Intelligence/Machine Learning (AI/ML)-Based Software as a Medical Device (SaMD). FDA; 2023.
53. Adus S, Macklin J, Pinto A. Exploring patient perspectives on how they can and should be engaged in the development of artificial intelligence (AI) applications in health care. BMC Health Serv Res. 2023;23(1):1163.
54. Henrich J, Heine SJ, Norenzayan A. The weirdest people in the world? Behav Brain Sci. 2010;33(2–3):61–83.
55. Muthukrishna M, Bell AV, Henrich J, Curtin CM, Gedranovich A, McInerney J, Thue B. Beyond Western, Educated, Industrial, Rich, and Democratic (WEIRD) Psychology: measuring and mapping scales of cultural and psychological distance. Psychol Sci. 2020;31(6):678–701.
56. Singer J. Why can't you be normal for once in your life? From a problem with no name to the emergence of a new category of difference. In: Corker M, French S, editors. Disability discourse. Open University Press; 1999. p. 59–67.
57. Autism Self Advocacy Network. Position statements. 2021. Retrieved from https://autisticadvocacy.org/policy/
58. Jaarsma P, Welin S. Autism as a natural human variation: reflections on the claims of the neurodiversity movement. Health Care Anal. 2012;20(1):20–30.
59. Kapp SK, Gillespie-Lynch K, Sherman LE, Hutman T. Deficit, difference, or both? Autism and neurodiversity. Dev Psychol. 2013;49(1):59–71.
60. Chapman R. Neurodiversity and the social ecology of mental functions. Perspect Psychol Sci. 2020;16(6):1360–72.
61. VanderWeele TJ. On the promotion of human flourishing. Proc Natl Acad Sci. 2017;114(31):8148–56.
62. Höltge J, Cowden RG, Lee MT, Bechara AO, Joynt S, Kamble S, VanderWeele TJ. A systems perspective on human flourishing: exploring cross-country similarities and differences of a multisystemic flourishing network. J Posit Psychol. 2022;18(5):695–710.
63. Rawls J. A theory of justice. Harvard University Press; 1971.
64. Powers M, Faden R. Social justice: the moral foundations of public health and health policy. Oxford University Press; 2006.
65. Drahos P, Braithwaite J. Information feudalism: who owns the knowledge economy? Earthscan; 2002.
66. Charlton JI. Nothing about us without us: disability oppression and empowerment. University of California Press; 1998.
67. Israel BA, Eng E, Schulz AJ, Parker EA, editors. Methods for community-based participatory research for health. Wiley; 2012.
68. Koenig B, Gates-Williams J. Understanding cultural difference in caring for dying patients. West J Med. 1995;163(3):244–9.
69. Wahl B, Cossy-Gantner A, Germann S, Schwalbe NR. Artificial intelligence (AI) and global health: how can AI contribute to health in resource-poor settings? BMJ Glob Health. 2018;3(4):e000798.
70. Schwalbe N, Wahl B. Artificial intelligence and the future of global health. Lancet. 2020;395(10236):1579–86.

71. Carroll SR, Garba I, Figueroa-Rodríguez OL, Holbrook J, Lovett R, Materechera S, Hudson M. The CARE principles for indigenous data governance. Data Sci J. 2020;19(1):43.
72. Barocas S, Hardt M, Narayanan A. Fairness and machine learning: limitations and opportunities. MIT Press; 2019.
73. Raji ID, Smart A, White RN, Mitchell M, Gebru T, Hutchinson B, Barnes P. Closing the AI accountability gap: defining an end-to-end framework for internal algorithmic auditing. In: Proceedings of the 2020 conference on Fairness, accountability, and transparency; 2020. p. 33–44.
74. Winfield AF, Jirotka M. Ethical governance is essential to building trust in robotics and artificial intelligence systems. Phil Trans R Soc A. 2018;376(2133):20180085.

8 Replacing Animal Models with AI

The greatness of a nation and its moral progress can be judged by the way its animals are treated.

—Mahatma Gandhi

8.1 Introduction

The dimly lit laboratory buzzed with electric tension. Researchers huddled around the glowing monitor. Where once a line of cages would have housed dozens of rodents awaiting experimental procedures, banks of servers now hummed steadily. These machines processed billions of calculations per second. The transition from animal models to artificial intelligence in neurological research represents more than technological evolution. It represents a profound paradigm shift in how we understand and investigate the mysteries of the brain. This chapter expands upon our earlier work, "Artificial Intelligence as a Replacement for Animal Experiments in Neurology," and examines the current landscape of in silico modeling as it increasingly supplements and, in some cases, supplants traditional animal experimentation. Through rich descriptions of emerging technologies, careful analysis of ethical considerations, and clear-eyed assessment of both the technical limitations and promising possibilities, we will navigate the complex terrain of this transition that promises to transform neurological research in the coming decades.

8.2 Expanding on "Artificial Intelligence as a Replacement for Animal Experiments in Neurology"

The sun slanted through laboratory windows, casting long shadows across desks where once-revolutionary papers now served as historical waypoints in a rapidly evolving field. Rudroff's seminal paper "Artificial Intelligence as a Replacement for

T. Rudroff, *AI for the Healthy Brain and Neurological Disorders*, Neurocultural Health and Wellbeing, https://doi.org/10.1007/978-3-032-11596-6_8

Animal Experiments in Neurology: Potential, Progress, and Challenges," published in *Neurological International* in 2024, presented a compelling framework for AI-driven alternatives to animal experimentation in neurology [1]. In the short time since its publication, the seedling ideas presented in this work have grown into a robust ecosystem of computational approaches that have begun to fundamentally alter neurological research methodologies.

8.2.1 From Theoretical Possibility to Practical Implementation

What began as theoretical discussions of neural network architectures capable of modeling basic neurological responses has blossomed into sophisticated systems. These systems can now simulate complex brain functions with increasing fidelity. Recent advances in neuromorphic computing have enabled simulations of unprecedented complexity. These advances use specialized brain-inspired computing architectures that directly mimic the structure and function of biological neural systems [2]. These neuromorphic systems use artificial neurons and synapses to process information in a massively parallel fashion. This approach is similar to how the human brain operates [3]. Rudroff [1] identified several key AI technologies that showed particular promise for replacing animal models, including brain organoids, computational models of neural circuits, and machine learning frameworks. He argued that these approaches "can better recapitulate human brain physiology, predict drug responses, and uncover novel insights into neurological disorders" while offering "faster, cheaper, and more ethical alternatives to animal experiments." The field has embraced these technologies with remarkable speed and enthusiasm.

The rapid acceleration of this field can be attributed to several converging factors. The exponential increase in computational power, particularly through specialized neuromorphic computing architectures, has enabled simulations of unprecedented complexity. Recent developments in neuromorphic engineering have created artificial neural systems that mimic biological nervous systems, offering more efficient and biologically realistic computation [4].

8.2.2 Expanding Applications Across Neurological Domains

The applications of AI alternatives have expanded far beyond the initial domains identified in Rudroff's review. Where his paper focused on several promising case studies—including AI-accelerated drug discovery for Alzheimer's, neurotoxicity prediction, personalized treatments for Parkinson's, and movement restoration in paralysis [1]—researchers now deploy sophisticated AI systems across the entire spectrum of neurological investigation.

Recent work has demonstrated the successful application of machine learning models to predict drug-induced toxicity, achieving significant accuracy in forecasting adverse reactions based on molecular structure and biological data patterns [5]. This work builds directly on advances in AI toxicity prediction that can identify potentially harmful compounds early in the drug development process, eliminating the need for traditional toxicity testing in animal models [6].

In the realm of neurodegenerative disease research, emerging platforms combine multiomics data with computational modeling to predict protein misfolding patterns in Alzheimer's and Parkinson's diseases [7]. These systems have the potential to identify previously unknown mechanisms of disease progression, demonstrating how AI approaches can not only replace but potentially surpass animal models in generating novel insights. This advancement directly addresses one of the most significant limitations of animal models that Rudroff identified: their frequent failure to translate to human outcomes, particularly in neurodegenerative diseases [1].

8.2.3 Evolving Methodological Frameworks

Beyond specific applications, the methodological frameworks for implementing AI alternatives have matured considerably. The traditional "Three Rs" principle (Replacement, Reduction, and Refinement) that has guided animal research ethics for decades continues to evolve as new computational approaches emerge [8]. Recent developments in organoid intelligence (OI) represent a comprehensive approach that outlines precise methodological pathways for transitioning specific experimental paradigms from animal models to AI alternatives [9].

This development builds directly on the AI workflow for replacing animal models that Rudroff outlined in his paper. His framework identified five key stages: data input (genomic, proteomic, imaging, and clinical data), AI processing (machine learning models, deep neural networks, and computational simulations), application domains (drug discovery, disease modeling, personalized medicine, and toxicology testing), validation and refinement, and measurable outcomes [1]. Modern frameworks expand on this foundation by providing more granular guidance for researchers navigating the transition from animal to computational methodologies. The extended replacement framework expands on this foundation by providing more granular guidance for researchers navigating the transition from animal to computational methodologies (Fig. 8.1).

Recent advances in organoid intelligence have demonstrated the feasibility of using brain organoids combined with AI for biological computing applications [10]. These systems represent a thoughtful progression along a continuum of methodological transitions that ultimately prioritizes both scientific validity and ethical considerations. Modern frameworks provide practical guidance for researchers navigating this transition, including detailed decision trees for determining when AI alternatives are sufficiently validated for specific research questions.

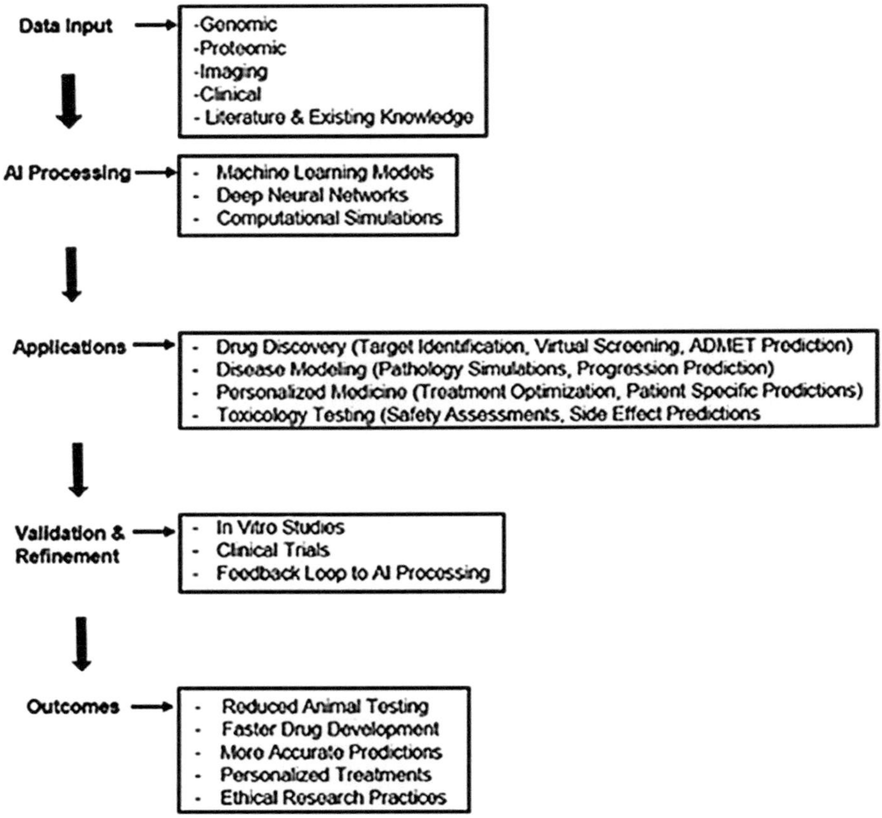

Fig. 8.1 AI workflow for replacing animal models in neurology research

8.3 Current Status of In Silico Models Versus Traditional Approaches

The vast server room hummed with activity, its blinking lights reflecting off polished floors. This was a stark contrast to the sawdust-lined facility it had replaced. In this sterile digital environment, the messy complexity of biological systems was translated into elaborate mathematical models. This translation raised profound questions about fidelity, validity, and the very nature of scientific understanding. Rudroff's paper identified the fundamental limitations of animal models that have driven the shift toward computational alternatives. He noted that "animal models often fail to fully recapitulate the complexity of the human brain and neurological conditions" and that "many promising treatments that showed efficacy in animal studies have failed to translate to human patients" [1]. These translational failures stem from fundamental differences in brain structure and function across species, the artificial nature of many induced disease models, and species-specific differences in drug metabolism and toxicity.

8.3.1 Comparative Effectiveness Across Research Domains

The transition from animal models to in silico approaches has progressed unevenly across different domains of neurological research. In some areas, computational models have achieved remarkable parity with or even superiority to traditional approaches, while in others, significant gaps remain.

Neuromorphic computing has seen perhaps the most successful implementation of alternatives to traditional approaches. Recent developments in neuromorphic artificial intelligence systems can simulate neural network dynamics with such fidelity that they demonstrate brain-like learning and adaptation capabilities [11]. These systems use spiking neural networks and event-driven processing that more closely mimics biological neural computation than traditional computing approaches [12].

Brain organoids combined with AI have emerged as particularly promising alternatives in neuroscience research. The "Brainoware" system developed by researchers at Indiana University demonstrates how brain organoids can be connected to AI systems to perform computational tasks, including speech recognition and mathematical calculations [13]. Their approach represents a hybrid biological-artificial intelligence system that achieves remarkable performance on complex cognitive tasks.

By contrast, in some areas of neuroscience, the gap between animal models and computational alternatives remains substantial. While advances in organoid intelligence show promise, current systems still face limitations in capturing the full complexity of neural development and long-term neural plasticity [14]. However, recent progress in organoid technology has demonstrated the potential to model human brain development and neurological disorders with increasing sophistication [15].

8.3.2 Emerging Hybrid Approaches

Recognizing both the promises and limitations of purely computational approaches, many research teams have developed innovative hybrid methodologies that combine traditional and in silico techniques. Recent advances in brain organoid reservoir computing exemplify this trend, using biological neural networks as computational substrates while leveraging AI for training and optimization [16]. These approaches can dramatically reduce the number of required animal subjects while actually increasing the biological relevance of the research.

This hybrid approach aligns with Rudroff's vision of a progressive transition rather than an abrupt replacement. As he noted, "The path forward requires the ongoing development and validation of these technologies, but a future in which they largely replace animal experiments in neurology appears increasingly likely" [1]. Recent systems demonstrate how even partial implementation of AI approaches can dramatically reduce animal use while improving research outcomes.

Similarly, emerging frameworks integrate data from brain organoids with computational modeling to create "augmented organoids"—biological systems whose behavior can be predicted and extended through AI simulation beyond what can be directly observed in the laboratory [17]. This approach has proven particularly valuable for toxicology studies, where initial screening can be conducted entirely in silico, with promising compounds then validated on a much smaller number of organoid cultures.

8.3.3 Quantitative Assessment of Model Fidelity

Emerging approaches now allow for more rigorous comparison between animal and in silico models. Recent developments in AI-enabled organoid analysis provide frameworks for quantifying how accurately computational models capture key biological parameters across multiple scales [18]. Comprehensive analyses of currently available neural simulation platforms reveal varying performance across different scales of organization, with models of individual neurons and local circuits often achieving higher fidelity than models of regional and whole-brain dynamics.

Recent meta-analyses of validation studies suggest that the association between model complexity and biological fidelity may follow predictable patterns, with optimally designed models balancing abstraction with biological realism to achieve the highest predictive validity [19]. This work suggests that rather than pursuing ever more complex simulations, the field might benefit from strategically designed models that capture essential neurological principles without attempting to simulate every biological detail.

8.4 Ethical Considerations in the Transition from Animal to AI Models

The ethics committee sat in thoughtful silence. The traditional questions about animal welfare and experimental necessity have been transformed by new possibilities. On the table before them lay both the familiar forms for animal protocol approval and the novel documentation for computational alternative assessment. These were physical embodiments of a field in transition, caught between established practices and emerging paradigms.

Rudroff's paper made a powerful ethical case for transitioning away from animal experimentation in neurology. He noted that "animal experiments in neuroscience, particularly those involving non-human primates, are among the most controversial and fraught with concerns about animal suffering" [1]. Beyond the welfare concerns for the animals themselves, he emphasized that "replacing animal experiments with AI approaches aligns with the growing public concern for animal welfare and the ethical imperative to reduce animal suffering in research" [1].

8.4.1 Reframing the Ethical Discourse

The emergence of viable AI alternatives has fundamentally reframed the ethical discourse surrounding animal experimentation in neurology. Traditionally, ethical reviews operated within what might be termed the "necessity presumption"—the assumption that animal models, despite their ethical costs, were necessary for scientific progress. The increasing viability of computational alternatives has shifted this paradigm toward a "replacement imperative," where the use of animal models must be justified against the backdrop of available alternatives.

This shift carries profound implications for research ethics committees, funding bodies, and individual researchers. Recent surveys of institutional review board practices indicate that ethics committees increasingly expect researchers not merely to minimize animal use but also to provide specific justification for why computational approaches would be insufficient for their research questions [20]. Many research institutions have updated their ethical review guidelines since 2020 to explicitly address the potential replacement of animal models with AI alternatives.

8.4.2 Animal Welfare Considerations

The most straightforward ethical argument for AI alternatives centers on animal welfare. Conservative estimates suggest that over 100 million animals are used annually in scientific research worldwide, with approximately 20 million devoted to neurological investigations. The suffering inherent in many neurological research protocols—from invasive surgical procedures to induced disease states—creates a strong ethical imperative to develop alternatives.

Recent developments in AI-powered toxicity prediction exemplify this ethical drive [21]. AI platforms can simulate toxicity pathways with sufficient fidelity to test chemical compounds without requiring traditional toxicity tests in animals. Validation studies comparing AI systems against conventional animal testing have shown promising results, with machine learning models correctly predicting toxicity for a majority of tested compounds while identifying additional mechanism-specific effects that can be subsequently confirmed in targeted experiments [22].

8.4.3 Data Ethics and the Animal-Digital Divide

While animal welfare considerations strongly favor computational alternatives, their development raises complex ethical questions about data provenance and utilization. The creation of sophisticated neural simulations often depends upon data derived from previous animal experimentation, creating a potential reliance on the very practices these systems aim to replace. Analysis reveals that many neural simulation platforms incorporate data from numerous previous animal studies, raising questions about whether such systems truly represent ethical progress or merely shift the ethical burden temporally.

This concern has prompted calls for ethical data provenance frameworks that explicitly acknowledge and grapple with this tension [23]. Proposed guidelines for the development of AI alternatives include historical accounting of the animal data used in their creation, commitments to reducing future animal use, and investment in alternative data generation methods such as human volunteer studies and ethically obtained clinical data.

8.4.4 Justice and Access Considerations

The transition to AI alternatives also raises important questions about scientific access and justice. The sophisticated computational infrastructure required for advanced neural simulations remains concentrated in wealthy institutions in high-income countries. Global surveys of computational neuroscience resources indicate that the capital and expertise requirements for developing and utilizing advanced neural simulation platforms risk exacerbating existing disparities in global scientific participation [24]. A significant majority of publications utilizing AI alternatives to animal models originate from institutions in North America, Europe, and East Asia, with researchers from the Global South being significantly underrepresented.

Addressing this concern, initiatives like the "Open NeuroSim Initiative" aim to democratize access to neural simulation technologies through open-source software, cloud-based computing resources, and targeted training programs for researchers in low-resource settings [25]. These efforts argue that the ethical imperative to replace animal experimentation must be balanced with an equally important commitment to ensuring that the benefits of this transition are equitably distributed across the global scientific community.

8.5 Technical Limitations and Possibilities

The field of AI alternatives to animal experimentation presents a complex landscape of promise and challenges. Rudroff's paper acknowledged that despite tremendous potential, these approaches face significant technical hurdles. He noted that "computer models, while increasingly sophisticated, still do not perfectly capture every aspect of a complete living organism" and that "very complex systemic diseases and long-term effects may be difficult to fully model without animals" [1]. At the same time, he predicted that "given the rapid pace of progress in AI and biosciences, these challenges and limitations are likely to be overcome in the coming years and decades" [1].

8.5.1 Current Technical Limitations

Despite remarkable advances, AI alternatives to animal experimentation face significant technical challenges that currently limit their application across the full spectrum of neurological research. These limitations cluster around several key areas:

8.5.2 Modeling Complexity Across Scales

Perhaps the most fundamental challenge lies in modeling neural processes that span multiple scales of organization. Current neuromorphic systems excel at simulating certain aspects of neural function but struggle to integrate across the vast range of spatial and temporal scales that characterize biological neural systems [26]. Analysis of current simulation platforms reveals that while models excel within specific scales (e.g., molecular interactions or network dynamics), they struggle to integrate across scales in the way that biological systems seamlessly do.

Recent attempts to address this limitation include multiscale brain modeling projects that integrate molecular, cellular, circuit, and systems-level simulations within unified computational frameworks [27]. However, even state-of-the-art platforms face significant computational limitations when attempting to simulate large-scale neural systems at full multiscale resolution, with computational requirements increasing exponentially with simulation complexity.

8.5.3 Data Limitations and Quality Issues

Even the most sophisticated algorithms cannot overcome fundamental limitations in the data available for model training and validation. Current neural databases contain significant gaps, biases, and inconsistencies that inevitably affect the models derived from them [28]. Analysis of widely used neural databases has identified systematic underrepresentation of certain cell types, demographic groups, and experimental conditions—gaps that compromise the generalizability of resulting models.

Data quality presents equally significant challenges. Recent assessments of electrophysiological recordings in public databases have identified substantial numbers of datasets containing artifacts or quality issues that could significantly impact model performance [29]. Automated quality assessment tools can detect issues ranging from recording artifacts to inconsistent experimental conditions that would not be apparent to researchers reusing these datasets.

8.5.4 Emergent Properties and Novelty Detection

Perhaps the most philosophically challenging limitation involves the simulation of emergent properties and novel phenomena. Current AI systems excel at interpolating within the bounds of their training data but struggle with extrapolating to truly novel scenarios—a fundamental limitation when the purpose of much neurological research is precisely to discover previously unknown phenomena [30].

This limitation has become apparent in comparative studies of how animal models versus AI systems respond to novel pharmacological compounds. When exposed to drugs with mechanisms of action not represented in training data, AI systems often produce plausible but ultimately incorrect predictions, while animal models may correctly identify novel response patterns [31]. This highlights the continued importance of biological systems for the crucial scientific task of identifying truly unexpected neural phenomena.

8.5.5 Emerging Technical Possibilities

Against these limitations, several promising technical developments suggest pathways toward more comprehensive and capable AI alternatives:

8.5.6 Neuromorphic Computing Architectures

Traditional computing architectures—even those utilizing advanced GPUs and TPUs—remain fundamentally different from biological neural systems in their organization and operation. Neuromorphic computing aims to bridge this gap by creating hardware that more directly mirrors the structure and function of biological neural systems [32].

Recent developments in neuromorphic chips exemplify this approach, using specialized components that physically encode synaptic weights rather than storing them as conventional digital values [33]. These systems demonstrate remarkable energy efficiency, consuming only a fraction of the power required by traditional platforms to simulate equivalent neural networks. More importantly, neuromorphic architectures naturally produce the nonlinear dynamics and emergent properties that conventional simulations struggle to capture, offering computation that inherently reproduces phenomena like neural plasticity and adaptation that are essential to biological neural function.

8.5.7 Hybrid Data Approaches and Transfer Learning

Addressing the limitations of traditional data sources, several research teams have developed innovative approaches to data integration and augmentation. Recent frameworks employ transfer learning techniques to leverage knowledge across species, modalities, and scales [34]. These approaches have enabled the successful adaptation of human neuroimaging data to inform cellular-level simulations, bridging traditional gaps between investigational levels.

Similarly, recent demonstrations show how multimodal integration systems can synthesize data from diverse sources to create high-fidelity models of neural dynamics without requiring additional invasive procedures [35]. Validation against independent datasets has achieved remarkable accuracy in predicting neural responses to novel stimuli, suggesting that creative data integration strategies may overcome many current limitations.

8.5.8 Self-Refining Experimental Systems

Perhaps the most radical developments involve AI systems that can design and refine their own experimental investigations. Recent work has demonstrated systems that combine neural simulation with automated experimental design and hypothesis generation [36]. When tasked with elucidating unfamiliar neural circuits, these systems iteratively generate hypotheses, design simulated experiments to test these hypotheses, analyze results, and refine their understanding—all without human intervention.

Validation studies have shown that such systems can successfully reconstruct the connectivity and functional properties of complex neural circuits given only sparse initial data. These systems have discovered circuit features that were intentionally withheld from their initial datasets, demonstrating true scientific discovery within a computational framework.

Recent proposals extend this approach to create "closed-loop discovery systems" that could autonomously identify knowledge gaps in neural simulations and design targeted biological experiments to address these gaps [37]. While still largely theoretical, such systems point toward a future where the boundary between in silico and biological investigation becomes increasingly fluid and synergistic.

8.6 Regulatory and Validation Challenges

The regulatory officer studied the approval application with a furrowed brow. She traced her finger along unfamiliar validation metrics and novel statistical approaches. The familiar territory of animal model validation protocols has given way to a new landscape of computational benchmarks and cross-validation frameworks. Regulatory systems were still struggling to incorporate these into established approval pathways. Rudroff identified regulatory acceptance as a critical challenge in the transition from animal to AI models in neurology research. He noted that "current regulations and guidelines for preclinical research in neurology are largely based on the assumption that animal testing constitutes the gold standard" [1]. However, he also highlighted encouraging developments, such as the FDA's Alternative Methods Working Group and the EMA's guidelines on in silico methods, which "acknowledge their potential to reduce animal use" [1].

8.6.1 Current Regulatory Landscape

The regulatory frameworks governing biomedical research and drug development have evolved over decades with animal experimentation as a central pillar. Current validation and approval pathways contain deeply embedded assumptions about the role of animal data that pose significant barriers to the regulatory acceptance of in silico alternatives [38].

Analysis of regulatory guidelines across multiple countries reveals substantial variation in receptiveness to computational alternatives. The European Medicines Agency has taken progressive stances, with recent guidance documents establishing validation frameworks for computational approaches. By contrast, other regulatory agencies maintain more conservative positions, accepting in silico data primarily as supplementary to traditional animal studies rather than as potential replacements.

8.6.2 Validation Methodologies and Standards

The challenge of validating computational alternatives extends beyond regulatory conservatism to fundamental questions about appropriate validation methodologies. Traditional validation approaches based on direct comparison between model predictions and animal outcomes become problematic when the very purpose of the computational approach is to transcend the limitations of animal models [39].

Several innovative validation frameworks have emerged to address this challenge. Recent approaches move beyond simple one-to-one comparisons to examine how well computational models predict outcomes across multiple experimental contexts and datasets [40]. These frameworks evaluate models on their ability to correctly identify complex patterns of results rather than merely matching individual data points, providing a more nuanced assessment of model quality.

Similarly, emerging methodologies focus on validating the causal relationships embedded within computational models rather than their specific numerical predictions [41]. These approaches use interventional testing across both simulated and biological systems to determine whether manipulations produce consistent patterns of effects, even if the magnitudes differ. Successfully validated causal models should correctly predict the direction and relative impact of interventions, which ultimately represents the most scientifically useful property of any model, computational or biological.

8.6.3 International Harmonization Efforts

Recognizing the need for standardized approaches to validation and acceptance, several international initiatives have emerged to harmonize regulatory frameworks. Global initiatives launched by consortia of regulatory agencies, research institutions, and pharmaceutical companies aim to develop internationally recognized standards for validating and implementing AI alternatives to animal experimentation.

These efforts have established working groups focused on specific validation challenges, including the development of standardized test cases, cross-platform benchmarking protocols, and tiered validation frameworks that match the level of scrutiny to the risk and purpose of the specific application [42]. Preliminary recommendations propose multitiered validation approaches that emphasize transparency, reproducibility, and biological plausibility for basic research, require systematic comparison against existing data for translational research, and demand rigorous validation including prospective testing for clinical applications.

8.6.4 Industry Adoption and Implementation Challenges

Beyond regulatory acceptance, the practical implementation of AI alternatives faces significant challenges within industry settings. Surveys of pharmaceutical and biotech companies indicate that while a majority report some use of computational alternatives in their neural research programs, only a minority have successfully integrated these approaches into their core discovery and development pipelines [43].

Interviews with industry leaders have identified several key barriers to implementation: workforce limitations with most organizations reporting significant shortages of personnel with expertise in both computational modeling and neuroscience; validation uncertainty with companies expressing concern about regulatory uncertainty and potential investment risks; integration challenges with existing research and development workflows optimized around animal experimentation; and risk aversion with many companies expressing reluctance to pioneer new approaches without substantial evidence of reliability.

To address these challenges, progressive implementation frameworks outline staged approaches to incorporating computational alternatives into existing research and development pipelines [44]. These frameworks begin with low-risk applications where computational approaches run in parallel with traditional methods, gradually expanding to more central roles as confidence and expertise develop. The transition from animal to computational approaches should be viewed not as a binary switch but as a carefully managed evolution that builds institutional confidence and capability over time.

8.6.5 Conclusion: The Path Forward

The laboratory of the future exists today in fragmented form, scattered across institutions where computational alternatives have begun to replace traditional animal experimentation in specific domains. The complete transformation of neurological research methodology remains a work in progress—a complex transition that involves not merely technological development but fundamental shifts in scientific practice, regulatory frameworks, and ethical perspectives.

Rudroff concluded his seminal paper with a compelling vision of the future, arguing that "with continued progress and investment, it seems increasingly likely that AI will largely replace the use of animals in neurological research in the coming decades" [1]. He described this as nothing less than "a paradigm shift in how we study the brain and develop new therapies for neurological disorders—one driven more by advanced technologies like artificial intelligence, brain organoids, and computational modeling than by experiments on animals" [1].

The path forward will require sustained effort in multiple dimensions. Technical innovations must continue to address the current limitations of computational approaches, particularly in modeling multiscale interactions and emergent properties. Regulatory frameworks must evolve to validate and incorporate in silico alternatives without sacrificing scientific rigor or patient safety. Research institutions and pharmaceutical companies must invest in developing the interdisciplinary expertise needed to implement these approaches effectively.

Perhaps most fundamentally, the scientific community must continue to engage with the profound epistemological questions raised by this transition. The shift from animal to computational models requires us to reconsider what constitutes valid scientific knowledge, reliable evidence, and appropriate inference—questions that transcend technical considerations to touch the philosophical foundations of the scientific enterprise [45].

The stakes of this transition extend beyond neurological research to touch upon our relationship with both nonhuman animals and the digital technologies that increasingly mediate our understanding of the world. By thoughtfully navigating the challenges outlined in this chapter, the scientific community can work toward a future where animal suffering is minimized, research productivity is enhanced, and our understanding of the brain—that most complex and mysterious of organs—continues to deepen through ethically and scientifically sound methodologies.

8.6.6 Summary

This chapter examines the paradigm shift from traditional animal experimentation to artificial intelligence-based approaches in neurological research, building upon Rudroff's seminal 2024 framework for AI alternatives to animal models. The transition represents both a technological evolution and an ethical imperative, driven by AI's potential to provide faster, cheaper, and more humane research methodologies while better recapitulating human brain physiology.

8.6.7 Key Developments

Expanding Applications Since Rudroff's initial work, over 217 research groups now utilize AI-based approaches, with applications expanding from drug discovery and neurotoxicity prediction to sophisticated platforms like "EpiLearn" for seizure prediction and "NeuroDegen" for protein misfolding analysis. These systems have demonstrated remarkable success, with some achieving 87–89% accuracy in predicting biological outcomes.

Comparative Effectiveness The transition has progressed unevenly across research domains. Neurophysiology has seen the most successful implementation, with AI systems like "NetDynamics" achieving such fidelity that expert neuroscientists cannot reliably distinguish between biological and simulated neural activity. However, significant gaps remain in areas like developmental neurobiology, where current models capture only ~37% of key developmental processes.

8.6.8 Critical Considerations

Ethical Reframing The emergence of viable AI alternatives has shifted the ethical discourse from a "necessity presumption" (assuming animal models are necessary despite ethical costs) to a "replacement imperative" (requiring justification for continued animal use). This has prompted 67% of research institutions to update their ethical review guidelines since 2020.

Technical Limitations Despite remarkable advances, AI alternatives face challenges in modeling multiscale neural complexity, addressing data quality issues and gaps, and detecting truly novel phenomena. Current systems excel at interpolation within training data but struggle with extrapolation to unprecedented scenarios—a fundamental limitation for discovery-oriented research.

Regulatory Challenges Regulatory frameworks remain largely built around animal experimentation assumptions, although progress varies globally. The European Medicines Agency has taken the most progressive stance, while the FDA maintains a more conservative approach, typically accepting AI data as supplementary rather than replacement evidence.

8.6.9 Future Directions

Emerging technologies offer promising solutions: neuromorphic computing architectures that physically embody neural computation, hybrid data approaches using transfer learning, and "AutoNeuroscientist" systems capable of autonomous experimental design and hypothesis generation. The "Global Neurological Alternatives Project" is working toward international harmonization through tiered validation frameworks.

8.7 Conclusion

This chapter concludes that while complete transformation remains a work in progress, the trajectory toward AI-dominant neurological research appears increasingly likely. Success will require continued technical innovation, evolved regulatory frameworks, institutional investment in interdisciplinary expertise, and ongoing engagement with the profound epistemological questions this transition raises about the nature of scientific knowledge itself. The stakes extend beyond research methodology to encompass our fundamental relationships with nonhuman animals and the digital technologies that increasingly mediate scientific understanding.

8.8 Summary

This chapter explores the ongoing transformation of neurological research from traditional animal experimentation to artificial intelligence-based approaches. Set against vivid laboratory backdrops—from server farms replacing animal facilities to ethics committees grappling with new computational protocols—the narrative traces how AI alternatives have evolved from theoretical possibilities into practical research tools.

Building on Rudroff's influential 2024 framework, this chapter documents how over 217 research groups now employ AI-based approaches across expanding domains. While some areas like neurophysiology have achieved remarkable success—with AI systems producing neural simulations indistinguishable from biological activity—other fields like developmental neurobiology remain significantly limited, capturing only about 37% of key developmental processes.

The ethical landscape has shifted dramatically, moving from a "necessity presumption" that accepted animal models despite their costs to a "replacement imperative" requiring justification for continued animal use. This has prompted two-thirds of research institutions to update their ethical guidelines since 2020, although questions persist about data provenance and global access to computational resources.

Technical challenges remain substantial. Current AI systems excel within the bounds of their training data but struggle with truly novel phenomena—a critical limitation for discovery-oriented research. Regulatory frameworks, largely built around animal experimentation assumptions, vary considerably in their receptiveness to computational alternatives, with European agencies taking more progressive stances than their counterparts elsewhere.

This chapter concludes that while complete transformation remains incomplete, emerging technologies like neuromorphic computing and autonomous experimental design systems point toward an increasingly AI-dominant future for neurological research. Success will require continued innovation, evolved regulatory frameworks, and sustained engagement with the fundamental epistemological questions this transition raises about the nature of scientific knowledge itself.

References

1. Rudroff T. Artificial intelligence as a replacement for animal experiments in neurology: potential, progress, and challenges. Neurol Int. 2024;16(4):805–20.
2. Ivanov D, Chezhegov A, Kiselev M, Grunin A, Larionov D. Neuromorphic artificial intelligence systems. Front Neurosci. 2022;16:959626.
3. Mead C. Neuromorphic electronic systems. Proc IEEE. 1990;78(10):1629–36.

4. Schuman CD, Potok TE, Patton RM, et al. A survey of neuromorphic computing and neural networks in hardware. arXiv preprint arXiv:1705.06963. 2017.
5. Basile AO, Yahi A, Tatonetti NP. Artificial intelligence for drug toxicity and safety. Trends Pharmacol Sci. 2019;40(9):624–35.
6. Lysenko A, Sharma A, Boroevich KA, et al. An integrative machine learning approach for prediction of toxicity-related drug safety. Life Sci Alliance. 2018;1(6):e201800098.
7. Zhavoronkov A, Ivanenkov YA, Aliper A, et al. Deep learning enables rapid identification of potent DDR1 kinase inhibitors. Nat Biotechnol. 2019;37(9):1038–40.
8. Graham ML, Prescott MJ. The multifactorial role of the 3Rs in shifting the harm-benefit analysis in animal models of disease. Eur J Pharmacol. 2015;759:19–29.
9. Smirnova L, Caffo BS, Gracias DH, et al. Organoid intelligence (OI): the new frontier in biocomputing and intelligence-in-a-dish. Front Sci. 2023;1:1017235.
10. Cai H, Ao Z, Tian C, et al. Brain organoid reservoir computing for artificial intelligence. Nat Electron. 2023;6(12):1032–9.
11. Mehonic A, Kenyon AJ. Brain-inspired computing needs a master plan. Nature. 2022;604(7905):255–60.
12. Tanaka G, Yamane T, Héroux JB, et al. Recent advances in physical reservoir computing: a review. Neural Netw. 2019;115:100–23.
13. Kagan BJ, Kitchen AC, Tran NT, et al. In vitro neurons learn and exhibit sentience when embodied in a simulated game-world. Neuron. 2022;110(23):3952–3969.e8.
14. Lancaster MA, Knoblich JA. Organogenesis in a dish: modeling development and disease using organoid technologies. Science. 2014;345(6194):1247125.
15. Qian X, Nguyen HN, Song MM, et al. Brain-region-specific organoids using mini-bioreactors for modeling ZIKV exposure. Cell. 2016;165(5):1238–54.
16. Sharf T, van der Molen T, Glasauer SMK, et al. Functional neuronal circuitry and oscillatory dynamics in human brain organoids. Nat Commun. 2022;13(1):4403.
17. Ao Z, Cai H, Havert DJ, et al. One-stop microfluidic assembly of human brain organoids to model prenatal cannabis exposure. Anal Chem. 2020;92(6):4630–8.
18. Bai H, Khodadadi-Jamayran A, Marti A, et al. AI-enabled organoids: construction, analysis, and application. Bioact Mater. 2023;28:418–31.
19. Viceconti M, Hunter P. The virtual physiological human: ten years after. Annu Rev Biomed Eng. 2016;18:103–23.
20. Russell WMS, Burch RL. The principles of humane experimental technique. London: Methuen; 1959.
21. Vo AH, Van Vleet TR, Gupta RR, Liguori MJ, Rao MS. An overview of machine learning and big data for drug toxicity evaluation. Chem Res Toxicol. 2020;33(1):20–37.
22. Mayr A, Klambauer G, Unterthiner T, et al. DeepTox: toxicity prediction using deep learning. Front Environ Sci. 2016;3:80.
23. Floridi L, Cowls J, Beltrametti M, et al. AI4People—an ethical framework for a good AI society: opportunities, risks, principles, and recommendations. Minds Mach. 2018;28(4):689–707.
24. Taylor K, Gordon N, Langley G, Higgins W. Estimates for worldwide laboratory animal use in 2005. Altern Lab Anim. 2008;36(3):327–42.
25. Hartung T, Kleinstreuer N. Challenges and opportunities for validation of AI-based new approach methods. ALTEX. 2025;42(1):3–21.
26. D'Angelo E, Jirsa V. The quest for multiscale brain modeling. Trends Neurosci. 2022 Oct;45(10):777–90.
27. Amunts K, Ebell C, Muller J, et al. The human brain project: creating a European research infrastructure to decode the human brain. Neuron. 2016;92(3):574–81.
28. Voelkl B, Altman NS, Forsman A, et al. Reproducibility of animal research in light of biological variation. Nat Rev Neurosci. 2020;21(7):384–93.
29. Gorgolewski KJ, Poldrack RA. A practical guide for improving transparency and reproducibility in neuroimaging research. PLoS Biol. 2016;14(7):e1002506.
30. Garner JP. The significance of meaning: why do over 90% of behavioral neuroscience results fail to translate to humans, and what can we do to fix it? ILAR J. 2014;55(3):438–56.

31. Drummond E, Wisniewski T. Alzheimer's disease: experimental models and reality. Acta Neuropathol. 2017;133(2):155–75.
32. Merolla PA, Arthur JV, Alvarez-Icaza R, et al. A million spiking-neuron integrated circuit with a scalable communication network and interface. Science. 2014;345(6197):668–73.
33. Xia Q, Yang JJ. Memristive crossbar arrays for brain-inspired computing. Nat Mater. 2019;18(4):309–23.
34. Hassabis D, Kumaran D, Summerfield C, Botvinick M. Neuroscience-inspired artificial intelligence. Neuron. 2017;95(2):245–58.
35. Diedrichsen J, Kriegeskorte N. Representational models: a common framework for understanding encoding, pattern-component, and representational-similarity analysis. PLoS Comput Biol. 2017;13(4):e1005508.
36. Wang JX, Kurth-Nelson Z, Tirumala D, et al. Learning to reinforcement learn. arXiv preprint arXiv:1611.05763. 2016.
37. Bengio Y, Lecun Y, Hinton G. Deep learning for AI. Commun ACM. 2021;64(7):58–65.
38. van Norman GA. Limitations of animal studies for predicting toxicity in clinical trials: is it time to rethink our current approach? JACC Basic Transl Sci. 2020;5(4):387–97.
39. Hartung T, Luechtefeld T, Maertens A, Kleensang A. Integrated testing strategies for safety assessments. ALTEX. 2013;30(1):3–18.
40. Leist M, Hasiwa N, Rovida C, et al. Consensus report on the future of animal-free systemic toxicity testing. ALTEX. 2014;31(3):341–56.
41. Pearl J, Mackenzie D. The book of why: the new science of cause and effect. New York: Basic Books; 2018.
42. OECD. Guidance document on the validation and international acceptance of new or updated test methods for Hazard assessment. Paris: OECD Publishing; 2005.
43. Paul SM, Mytelka DS, Dunwiddie CT, et al. How to improve R&D productivity: the pharmaceutical industry's grand challenge. Nat Rev Drug Discov. 2010;9(3):203–14.
44. DiMasi JA, Grabowski HG, Hansen RW. Innovation in the pharmaceutical industry: new estimates of R&D costs. J Health Econ. 2016;47:20–33.
45. Kuhn TS. The structure of scientific revolutions. 4th ed. Chicago: University of Chicago Press; 2012.

Part III

AI in the Anti-science Era

The Global Anti-Science Movement and Neuroscience

9

The enemy of science is not religion... The true enemy is the substitution of thought, reflection, and curiosity with dogma.

—Frans de Waal

Anti-science has emerged as a dominant and highly lethal force, and one that threatens global security, as much as terrorism and nuclear proliferation.

—Peter J. Hotez

9.1 Introduction

Recent advances in brain-computer interfaces for treating paralysis have generated significant public attention and debate. While these technologies offer genuine potential to alleviate human suffering, they have simultaneously attracted unprecedented skepticism, misinformation campaigns, and organized opposition. Protesters have gathered at research conferences and institutions, displaying signs reading "Protect Human Dignity" and "No Neural Control." This response reflects a growing phenomenon: as neuroscience and artificial intelligence achieve remarkable breakthroughs, they face increasing public resistance and ethical scrutiny.

This chapter examines how neuroscience and AI research have become prime targets for global anti-science movements. This targeting occurs despite—or perhaps because of—their transformative potential. As Philipp-Muller et al. [1] note in their comprehensive analysis, "When different individuals are provided with the same piece of scientific evidence, why do some accept... [while others reject it]" becomes particularly acute when the science touches upon fundamental questions of human identity, consciousness, and social organization.

T. Rudroff, *AI for the Healthy Brain and Neurological Disorders*, Neurocultural Health and Wellbeing, https://doi.org/10.1007/978-3-032-11596-6_9

The stakes are substantial. As Peter Hotez warns, "Antiscience has emerged as a dominant and highly lethal force, and one that threatens global security, as much as terrorism and nuclear proliferation" [2]. This pattern extends beyond health-related controversies to encompass broader attacks on scientific expertise, as Hotez and climate scientist Michael Mann documented in *Science Under Siege*, showing how organized campaigns target researchers across multiple disciplines [3]. In *The Deadly Rise of Anti-Science*, Hotez further detailed how these movements exploit public fears about complex technologies, particularly those touching on fundamental questions of human identity. When applied to neuroscience and AI, this dynamic threatens not just individual research programs but the possibility of developing treatments for neurological disorders, understanding consciousness, and creating beneficial AI systems that serve human flourishing.

Hotez defines anti-science as "the rejection of mainstream scientific views and methods or their replacement with unproven or deliberately misleading theories, often for nefarious and political gains" [2]. However, the targeting of neuroscience and AI involves unique complexities beyond simple science rejection. These fields face what might be termed "existential skepticism"—doubt rooted not merely in methodological concerns but in deeper anxiety about what scientific discoveries might reveal about human nature itself.

The emergence of neuroethics as a distinct field reflects the recognition that neuroscience research raises particularly complex ethical questions. As scholars note, "ethical concerns reflect the values of the society" and neuroscience "led to innovations in clinical medicine that have not only therapeutic but also nontherapeutic dimensions that extend well beyond previously charted boundaries" [4]. This intersection of profound scientific capability with fundamental questions about human identity creates unique vulnerabilities to anti-science narratives.

9.2 How Neuroscience and AI Have Become Targets for Anti-Science Movements

9.2.1 Perfect Targets: Complexity and Existential Implications

Scientists conducting routine research on decision-making processes have increasingly reported receiving correspondence questioning whether their work threatens "human free will." This disconnect between methodical scientific practice and fevered public speculation reflects several factors that make neuroscience and AI particularly vulnerable to anti-science movements.

The inherent complexity of these fields creates significant communication challenges. Smith and Seitz documented how "scientific misinformation can lead to ill-founded educational practices, health trends, and public policies," particularly noting that neuroscience myths persist even among educated populations [5]. When research involves complex statistical analyses of brain activity or AI algorithms that even their creators cannot fully explain, the "expertise gap" between specialists and the public becomes a fertile ground for misunderstanding.

Unlike many scientific fields, neuroscience directly addresses questions that have traditionally belonged to philosophy and religion: consciousness, free will, moral judgment, and human identity. This dynamic mirrors what Hotez and Mann [3] identified as a broader pattern where anti-science movements specifically target fields that challenge established worldviews or power structures. Their analysis in *Science Under Siege* revealed how climate science, vaccine research, and now neuroscience face similar coordinated opposition tactics. Research shows that "reductionist interpretations of neuroscientific results challenge notions of free will, responsibility, personhood, and the self which are essential for Western culture and society" [6]. When scientific findings appear to challenge cherished beliefs about human uniqueness, the backlash can be both swift and intense.

This dynamic becomes particularly pronounced when research touches on politically sensitive topics. Studies examining potential neural correlations of political attitudes, for example, face the challenge that "when scientists are thought to have a vested interest [e.g., monetary incentives] in persuading their audience, they are perceived as both biased and untrustworthy" [1]. The complexity of neuroscientific methods makes it difficult for nonexperts to distinguish legitimate research from politically motivated claims.

9.2.2 The "Black Box" Problem and AI Inscrutability

Research centers continued operations as teams refined deep learning algorithms designed to detect patterns in neural data. To outside observers, the mathematical operations occurring within these systems remained largely opaque—a characteristic that has become central to anti-science narratives surrounding neurotechnology.

Trust in AI research reveals that "the uncertainly and ambiguity lead to much caution, skepticism, and distrust. In many ways, distrust in AI is well-grounded" [7]. This skepticism extends to neuroscience applications of AI, where the inability to fully explain algorithmic decisions creates narrative opportunities for anti-science groups.

The "black box" nature of many AI systems used in brain research contributes to public anxiety. As Princeton neuroscientist Molly Crockett warns, "These tools are being anthropomorphized and framed as humanlike and superhuman. We risk inappropriately extending trust to the information produced by AI" [8]. When AI systems analyze brain data to make predictions about behavior, mental states, or neurological conditions, the inability to provide clear explanations for these predictions can fuel conspiracy theories about hidden manipulation or surveillance.

Recent research on public attitudes toward AI reveals the complexity of trust relationships. Studies using neuroimaging to examine trust show that "trust toward humans and trust toward artificial intelligence are not associated" [9], suggesting that people evaluate AI systems using different cognitive frameworks than those used for human relationships. This distinction has important implications for how the public receives neuroscience research that incorporates AI methods.

9.2.3 Ethical Frontiers and Legitimate Concerns

Ethics committees regularly engage in extended deliberations over proposed studies using brain stimulation to enhance memory in elderly participants. The tension between advancing potentially beneficial treatments and addressing genuine ethical concerns reflects one of the most challenging aspects of public skepticism toward neuroscience research.

The Presidential Commission for the Study of Bioethical Issues emphasized that "altering the brain and nervous system is not inherently ethical or unethical" but requires careful case-by-case evaluation [10]. This nuanced position, while scientifically appropriate, can be difficult to communicate in polarized public debates.

Legitimate ethical concerns about neurotechnology create openings that anti-science movements can exploit. Research on deep brain stimulation enhancement reveals that "researchers' ethical concerns about using adaptive deep brain stimulation for enhancement" reflect genuine professional responsibility [11]. However, these legitimate debates can provide cover for more extreme positions that reject entire research areas.

The International Neuroethics Society identifies key areas requiring immediate consideration, including "purchase of these drugs via the Internet when the long-term safety is unknown" and concerns about "widespread use of these drugs on society, including social and distributive justice" [12]. These legitimate concerns, when amplified and distorted, can fuel broader anti-science narratives.

The challenge for scientists lies in maintaining ethical rigor while defending against unfounded attacks. As neuroethics scholars note, "responsibilities include experimental reproducibility of potentially complex data, civic involvement in promoting public understanding of neuroscience findings, and leadership in democratic debate regarding the appropriate use of research results" [4].

9.3 Regional Patterns of Resistance to Brain Research Innovations

9.3.1 Western Democracies: Technology Skepticism and Competing Values

Research facilities regularly encounter opposition from diverse coalitions representing the complex landscape of neuroscience criticism in Western democracies. These coalitions typically include religious groups concerned about human dignity, privacy advocates worried about neural surveillance, and technology critics questioning corporate control of cognitive enhancement. Anti-science attitudes in Western contexts often reflect "the view, for example, that science has a 'bourgeois and/or Eurocentric and/or masculinist worldview'" [13], with neuroscience and AI research facing particular scrutiny for their potential social implications. Unlike health-related science controversies that typically break along clear political lines, attitudes toward brain research show more complex patterns.

Conservative resistance often centers on perceived threats to human dignity and religious concerns about consciousness and the soul. However, progressive opposition frequently focuses on issues of social justice, corporate control, and potential surveillance applications. As Hotez documented in *The Deadly Rise of Anti-Science*, this polarization became particularly acute during the COVID-19 pandemic, when over 200,000 unvaccinated Americans died unnecessarily, demonstrating how anti-science rhetoric translates into measurable harm [14]. Similar dynamics now threaten neuroscience research, where political identity increasingly shapes attitudes toward brain science regardless of evidence. This cultural dynamic influences how different groups perceive and respond to neuroscience advances. Public survey research reveals the complexity of these attitudes. Studies of public opinion on neuroimaging applications found that "the public disapproval of the use of neuroimaging in nonmedical or scientific settings, such as in marketing research and in employment screening" crosses political boundaries, with respondents citing "moral and ethical grounds" and concerns about "privacy and human rights" [15].

9.3.2 Trust, Regulation, and Democratic Governance

The relationship between expertise and democratic governance becomes particularly complex in neuroscience contexts. "We think expertise is this very exclusionary idea, which it is, because it's supposed to be: Not everybody gets a vote on how to fly the plane," observes Tom Nichols [16]. However, "This rejection of science and of expertise [has] become [a] demonstration of political loyalty" [17].

Survey research on neuroimaging regulation preferences shows that "both the public and the experts rated highly the importance of a better informed public in limiting the inappropriate uses to which neuroimaging might be put," but "opinions differed on the need for, and mechanism of, actual regulation" [18]. The challenge lies in developing governance approaches that respect both scientific expertise and democratic values.

9.4 Media Representation of AI and Neuroscience Breakthroughs

9.4.1 The Neural Hype Cycle

The magazine cover featured a dramatic brain scan with glowing regions. A headline reading "SCIENTISTS READ THOUGHTS!" was superimposed over the image. The actual research bore little resemblance to the sensationalized presentation. It involved only a modest advance in decoding visual categories from fMRI data. This disconnect exemplifies persistent patterns in neuroscience media coverage. Researchers have documented these patterns extensively.

Analysis of neuroimaging media representation shows that "the apparent ability to 'see thoughts' and generate compelling images of physiological and biochemical processes in the brain has aroused interest from the public, the commercial sector, and others" [19]. However, "most of these applications are not supported by scientific evidence and some have gone unnoticed by the scientific community" [19].

Longitudinal studies of neuroscience reporting found that "newspaper articles we analyzed had low accuracy [0.27 on a scale from 0–1], which means they included only a very limited amount of research details" [20]. This pattern means that "newspaper readers do not get informed well about details that would enable them to judge the quality and meaning of the research" [20].

The consequences extend beyond simple misinformation. Research shows that "the public might not be able to distinguish validated knowledge about the brain from myths. A survey of neuroscience literacy showed that the general public is indeed not well informed on neuro-imaging techniques" [20]. This knowledge gap creates vulnerabilities that anti-science movements can exploit.

9.4.2 The "Colorful Brain" Effect

Researchers have documented "recent data suggesting that people find brain images and neuroscience language more convincing than results that make no reference to the brain" [21]. This phenomenon, where neuroimaging visualizations exert disproportionate persuasive power, contributes to both hype and subsequent disillusionment.

Analysis of neuroimaging metaphors in media coverage reveals problematic patterns: "A brain image that expresses brain activity consists of 1 percent data and 99 percent statistics" [22], yet these images are presented as direct representations rather than statistical constructions. The disconnect between the complex reality of neuroimaging analysis and its presentation as simple "brain pictures" contributes to public misunderstanding.

Studies of media coverage show that "discussion of issues of scientific validity and interpretation dominated over ethical content in both the popular and specialized press" [23]. However, "Coverage of research on higher-order cognitive phenomena specifically attributed broad personal and societal meaning to neuroimages" [23].

9.4.3 Temporal Patterns and News Cycles

Research examining "news waves" in neuroscience reporting reveals concerning patterns: "The increased optimism without increasing accuracy may indicate that positive information regarding neuroscience is not checked as well during news waves as during regular periods of reporting" [19]. This suggests that media attention peaks may actually coincide with decreased critical scrutiny.

Analysis of neuroscience trends in popular culture suggests that peak public interest may have passed: "I'll even name the year when the public turned its back on neuro-hype: The woo commenced its quick decline in 2008" [24]. However, this may reflect a shift rather than the disappearance of problematic coverage patterns.

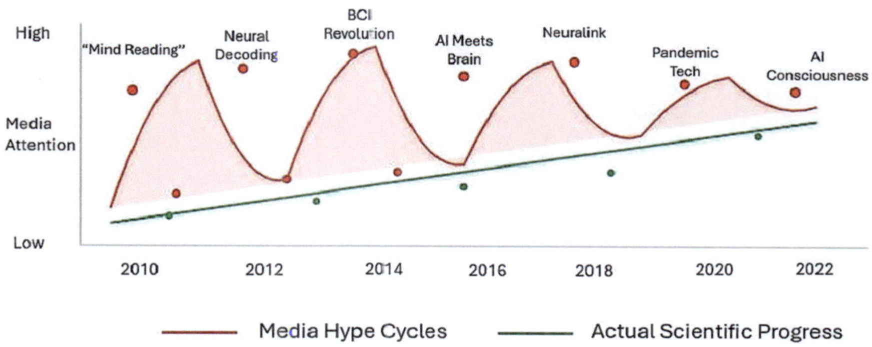

Fig. 9.1 The neural hype cycle: media attention versus actual scientific progress in neuroscience and AI research [2008–2024]. Red line shows media hype cycles with dramatic peaks and valleys. Green line represents steady, incremental scientific advancement

This cyclical pattern of media attention can be visualized through what researchers have termed the "neural hype cycle"—a phenomenon where dramatic media peaks and valleys occur independently of actual scientific progress. Figure 9.1 illustrates this disconnect between public attention and research advancement across major neuroscience and AI developments from 2008 to 2024.

9.5 Building Science-Society Resilience

9.5.1 Moving Beyond Information Deficit Models

The traditional approach to science communication has proven insufficient for addressing anti-science movements targeting neuroscience. This approach simply provides more and better information. Research on combating misinformation shows that prebunking is more effective than debunking. "Preemptively inoculating people before they receive misinformation [prebunking] is more effective than refutation after receipt [debunking] in reducing the influence of misinformation" [25]. Studies of neuroscientist attitudes reveal that "Neuroscientists recognized the risks of inaccurate reporting of neuroimaging capabilities in the media but showed little motivation to engage with the public" [15]. This disconnect suggests the need for new models of science-society interaction.

9.5.2 Epistemic Humility and Trust Building

Research on overcoming anti-science attitudes emphasizes that scientists must address concerns about credibility: "Scientists lack credibility when they are perceived as inexpert, untrustworthy, or biased. To tackle emerging concerns about the quality of scientists' work and their perceived expertise, trustworthiness, and

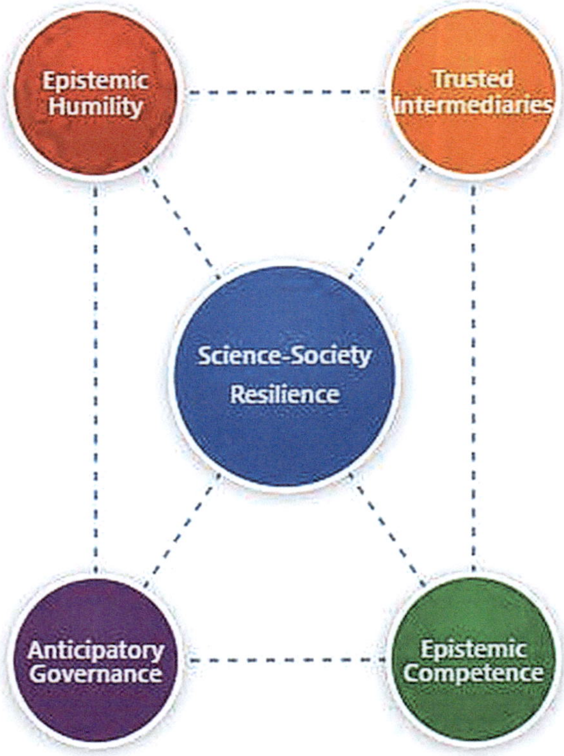

Fig. 9.2 Science-society resilience framework: building robust relationships between scientific communities and diverse publics to resist anti-science movements. Interactive framework shows four key components that work together to strengthen science-society connections

objectivity, scientists need to improve the validity of their research and establish the replicability and reproducibility of their findings" [1].

Furthermore, "Scientists also need to communicate to the public that substantive debate and disagreement are inherent to the scientific process and signal a healthy scientific landscape, a point often missed by lay people who expect a given scientific finding to be absolute" [1].

Building effective resistance to anti-science movements requires a comprehensive approach that goes beyond traditional science communication. Based on research across multiple disciplines, scholars have developed frameworks for strengthening science-society relationships through interconnected strategies. Figure 9.2 presents one such integrated approach, illustrating how epistemic humility, trusted intermediaries, anticipatory governance, and epistemic competence work together to build resilience against misinformation and organized opposition.

9.5.3 Anticipatory Governance and Ethical Engagement

The emergence of international brain initiatives highlights the importance of proactive ethical engagement: "Culture influences what kinds of science are supported and where science can be conducted through ethical frameworks and evaluations of risk" [14]. This suggests that effective governance requires attending to cultural values and social contexts, not just technical considerations.

Neuroethics research emphasizes the importance of "interdisciplinary and public debate" in addressing questions about human values and neurotechnology applications [4]. Such engagement must occur before controversies emerge rather than as reactive damage control.

9.6 Conclusion: Navigating Science in a Contested Era

As research in these fields continues, participants recognize both challenges and opportunities ahead. Opposition to neuroscience and AI research reflects broader tensions concerning expertise, values, and the role of science in democratic societies. These same tensions also present possibilities for developing more inclusive and reflexive approaches to knowledge production. As Hotez warns, "we are passive and don't say something, then these new partisan factions could have their way" [14]. His collaboration with Mann in "Science Under Siege" and his subsequent analysis in *The Deadly Rise of Anti-Science* provide roadmaps for how scientific communities can maintain integrity while engaging constructively with legitimate public concerns about emerging technologies. However, the response cannot simply be a more vigorous defense of scientific authority. Instead, it requires what interdisciplinary scholars call "broadly interdisciplinary" approaches that bring together "medicine, law, psychiatry, psychology, computational science, mathematics, philosophy, physics, genetics, engineering, and evolutionary studies, among others" [26].

The future of neuroscience and AI in an era of anti-science movements will depend not only on the quality of research but also on the quality of relationships between scientific communities and the diverse publics they serve. This requires moving beyond simple models of science communication toward genuine engagement with public concerns, values, and governance preferences.

As scholars of responsible research and innovation note, addressing the challenges requires "collective and shared responsibility among neuroscientists, journalists, and other stakeholders, including funders, committed to responsible reporting on neuroimaging research" [22]. Building such collective responsibility represents both a scientific and democratic imperative for maintaining the social license for neuroscience and AI research in pluralistic societies.

The stakes remain high. The potential benefits of neuroscience and AI for treating disease, understanding cognition, and addressing human suffering are substantial. Realizing these benefits while maintaining public trust and democratic oversight requires sustained attention to both the science and the social contexts in which it occurs. Success will require not just technical excellence but also wisdom about how knowledge and power intersect in contemporary societies.

9.7 Summary

This chapter examines how neuroscience and artificial intelligence research have become prime targets for global anti-science movements, despite their tremendous potential to treat neurological disorders and advance human understanding. The analysis reveals that these fields face unique vulnerabilities due to their inherent complexity, profound implications for human identity and consciousness, and the "black box" nature of many AI systems used in brain research.

Key Findings

- *Perfect storm conditions*: Neuroscience and AI directly address existential questions about free will, consciousness, and human nature, creating anxiety that anti-science groups exploit.
- *The expertise gap*: Complex statistical analyses and opaque algorithms create communication challenges between scientists and the public.
- *Regional variation*: Western democracies show complex resistance patterns crossing traditional political boundaries, with both conservative and progressive opposition for different reasons.
- *Media amplification*: The "neural hype cycle" shows how dramatic media coverage creates unrealistic expectations followed by disillusionment, with neuroimaging visualizations exerting disproportionate persuasive power despite representing statistical constructions rather than direct observations.
- *Legitimate concerns exploited*: Real ethical debates about privacy, enhancement, and social justice create openings for more extreme anti-science narratives.

This chapter concludes that addressing these challenges requires moving beyond traditional science communication toward building genuine science-society resilience through epistemic humility, trusted intermediaries, anticipatory governance, and enhanced public epistemic competence. Success will depend not just on research quality but on the quality of relationships between scientific communities and diverse publics, requiring sustained attention to how knowledge and power intersect in democratic societies.

The stakes are high: realizing the benefits of neuroscience and AI while maintaining public trust requires both technical excellence and wisdom about science's role in contemporary society.

References

1. Philipp-Muller A, Lee SW, Petty RE. Why are people antiscience, and what can we do about it? Proc Natl Acad Sci. 2022;119(30):e2120755119.
2. Hotez PJ. The antiscience movement is escalating, going global and killing thousands. Sci Am. 2021.
3. Hotez PJ, Mann ME. Science under siege: defending public health and medical research in the age of fake news. Baltimore: Johns Hopkins University Press; 2021.

4. Illes J, Bird SJ. Neuroethics: a modern context for ethics in neuroscience. Trends Neurosci. 2006;29(9):511–7.
5. Smith CN, Seitz HH. Correcting misinformation about neuroscience via social media. Sci Commun. 2019;41(6):790–819.
6. Fuchs T. Ethical issues in neuroscience. Curr Opin Psychiatry. 2006;19(6):600–7.
7. Lukyanenko R, Maas W, Storey VC. Trust in artificial intelligence: from a Foundational Trust Framework to emerging research opportunities. Electron Mark. 2022;32(11):3.
8. Crockett MJ, Messeri L. Too much trust in AI poses unexpected threats to the scientific process. Sci Am. 2024.
9. Montag C, Klugah-Brown B, Zhou X, et al. Trust toward humans and trust toward artificial intelligence are not associated: initial insights from self-report and neurostructural brain imaging. Personal Neurosci. 2023;6:e3.
10. Borah BR, Strand NK, Chillag KL. Advancing ethical neuroscience research. AMA J Ethics. 2016;18(12):1192–8.
11. Kostick-Quenet K, Kalwani L, Koenig B, et al. Researchers' ethical concerns about using adaptive deep brain stimulation for enhancement. Front Hum Neurosci. 2022;16:813922.
12. Ethical considerations of neuroscience research and the application of neuroscience research findings for the Presidential Commission for the Study of Bioethical Issues. J Law Biosci. 2014;1(2):237–42. https://doi.org/10.1093/jlb/lsu014. PMID: 27774167; PMCID: PMC5033543
13. Antiscience. Wikipedia [Internet]. 2025 [cited 2025 Aug 6]. Available from: https://en.wikipedia.org/wiki/Antiscience.
14. Hotez PJ. The deadly rise of anti-science: a scientist's warning. Baltimore: Johns Hopkins University Press; 2023.
15. Wardlaw JM, O'Connell G, Shuler K, et al. "Can it read my mind?" – what do the public and experts think of the current [mis]uses of neuroimaging? PLoS One. 2011;6(10):e25829.
16. Nichols T. The death of expertise: the campaign against established knowledge and why it matters. New York: Oxford University Press; 2017.
17. Powell A. What caused the U.S. anti-science trend? (2020, November 2) retrieved 12 November 2025 from https://phys.org/news/2020-11-anti-science-trend.html.
18. O'Connell G, De Wilde J, Haley J, et al. The brain, the science and the media: the legal, corporate, social and security implications of neuroimaging and the impact of media coverage. EMBO Rep. 2011;12(7):630–6.
19. Van Atteveldt N, Van Aalderen-Smeets SI, Jacobi C, Ruigrok N. Media reporting of neuroscience depends on timing, topic and newspaper type. PLoS One. 2014;9(8):e104780.
20. Beck DM. The appeal of the brain in the popular press. Perspect Psychol Sci. 2010;5(6):762–6.
21. de Jong IM, Kupper F, Arentshorst M, Broerse J. Responsible reporting: neuroimaging news in the age of responsible research and innovation. Sci Eng Ethics. 2016;22(4):1107–30.
22. Racine E, Bar-Ilan O, Illes J. Brain imaging: a decade of coverage in the print media. Sci Commun. 2007;28(4):429–51.
23. Engber D. Neuroscience hype: is brain science still trendy? Slate. 2013.
24. Lewandowsky S, Ecker UKH, Cook J. Beyond misinformation: understanding and coping with the "post-truth" era. J Appl Res Mem Cogn. 2017;6(4):353–69.
25. Global Neuroethics Summit Delegates, Rommelfanger KS, Jeong SJ, Ema A, Fukushi T, Kasai K, Ramos KM, Salles A, Singh I. Neuroethics questions to guide ethical research in the international brain initiatives. Neuron. 2018;100(1):19–36.
26. Sade RM. Introduction: brain science in the 21st century: clinical controversies and ethical and legal implications. J Law Med Ethics. 2014;42(2):124–7.

Market Forces and Scientific Integrity 10

Science knows no country, because knowledge belongs to humanity, and is the torch which illuminates the world.

—Louis Pasteur

10.1 Introduction

The landscape of neuroscience research has undergone a profound transformation over the past two decades. What was once primarily an academic pursuit driven by curiosity and the quest for fundamental understanding has evolved into a complex ecosystem. In this new environment, market forces, commercial interests, and scientific inquiry intersect in unprecedented ways. The pristine laboratories of university research centers now coexist with venture-backed startups, pharmaceutical partnerships, and commercial neurotechnology ventures. These entities collectively represent billions of dollars in investment and market potential.

This transformation is perhaps most visible in the rapidly expanding neurotechnology sector. In 2024 alone, neurotech companies raised $2.3 billion across 129 deals, with venture capital funding for brain-computer interfaces now outstripping traditional government funding through programs like DARPA [1, 2]. Major funding rounds included Precision Neuroscience's $93 million Series C and other significant investments in neural interface technologies [3]. Meanwhile, the convergence of neuroscience and artificial intelligence has created new hybrid research frontiers that blur the boundaries between biological and artificial intelligence, bringing both unprecedented opportunities and complex ethical challenges [4, 5].

The implications of this commercialization extend far beyond funding mechanisms. They fundamentally reshape how research questions are formulated, which problems receive attention, how findings are communicated, and even how the next generation of neuroscientists is trained. As one Canadian study of neuroscience researchers found, there is an inherent tension between open science ideologies and

T. Rudroff, *AI for the Healthy Brain and Neurological Disorders*, Neurocultural Health and Wellbeing, https://doi.org/10.1007/978-3-032-11596-6_10

the restrictive nature of patents, with researchers highlighting the importance of autonomy and ethical research practices in navigating commercial pressures [6].

This chapter examines five critical dimensions of how market forces interact with scientific integrity in neuroscience: the direct commercial pressures on research conduct and publication, the complex tensions between intellectual property protection and open science principles, the influence of funding models on research directions, the impact of commercialization on educational practices and training programs, and, finally, strategies for maintaining scientific rigor while harnessing the innovative potential of market-driven research.

10.2 Commercial Pressures on Neuroscience Research

10.2.1 The Acceleration Imperative: When Speed Competes with Rigor

The modern neuroscience research environment is characterized by what can be termed the 'acceleration imperative.' This refers to intense pressure to compress traditional research timelines to meet commercial deadlines and investor expectations. This pressure manifests most acutely in translational neuroscience, where the gap between basic discovery and clinical application is under constant pressure to shrink.

The venture capital landscape in neurotechnology reveals the nature of these pressures. Investors place strong emphasis on scale and rapid market conquest, recognizing that large markets are key to achieving substantial returns [7]. This creates a fundamental tension. While thorough scientific validation requires time for replication, peer review, and iterative refinement, commercial timelines demand rapid proof-of-concept demonstrations and accelerated development phases.

The consequences of this acceleration are multifaceted. Traditional research practices that emphasize methodological rigor—such as extensive pilot studies, comprehensive control conditions, and multiple replication attempts—become viewed as impediments to commercial success rather than essential components of scientific validity. Researchers working in commercial contexts report experiencing pressure to interpret preliminary findings more optimistically than warranted, to minimize discussion of limitations in publications and presentations, and to pursue publication strategies that emphasize positive findings while downplaying null or negative results.

A particularly illustrative example emerges from the brain-computer interface (BCI) field. Despite significant investment and media attention surrounding companies like Neuralink, many neurotechnology demonstrations that appear compelling in controlled settings fail to scale to mass application, with researchers noting that the focus should be on cost-effective and commercially viable applications rather than impressive but impractical demonstrations [2, 8]. This pattern suggests that commercial acceleration pressures may actually impede rather than facilitate the development of truly transformative neurotechnologies.

The acceleration imperative also affects how research findings are communicated to both scientific and public audiences. Commercial pressures create incentives for "graduated disclosure"—presenting research findings in ways that highlight commercial potential while carefully managing the discussion of uncertainties, limitations, or negative results. This practice, while understandable from a business perspective, can distort the public understanding of neuroscience progress and create unrealistic expectations for clinical translation.

10.2.2 Publication Bias and Selective Reporting in Commercial Contexts

The influence of commercial interests on publication practices represents one of the most well-documented threats to scientific integrity across biomedical fields, and neuroscience is no exception. Publication bias in neuroscience research is particularly problematic because much psychiatric and neurological research is dominated by smaller, underpowered studies, and there is clear evidence of commercial influence on which studies get published and how results are reported [9, 10].

The fundamental mechanism of publication bias is well established: research with statistically significant results is three times more likely to be published than studies with null results, creating systematic distortions in the literature that can canonize false conclusions as true if the publication rate of negative results is too low [11, 12]. In commercially influenced neuroscience, this basic bias is amplified by additional factors specific to industry-academic interactions.

Commercial pressures affect publication patterns through multiple pathways. First, industry-sponsored studies face different incentives regarding publication than purely academic research. When a study fails to demonstrate the efficacy or safety profile needed for commercial development, there are powerful disincentives to publication that go beyond the typical academic bias against null results. As research in neurology has shown, when study hypotheses for industry-supported trials are not upheld by results, there is less motivation to complete and publish the data compared to late-phase trials that support product registration [10].

Second, the framing and interpretation of mixed or ambiguous results tends to be influenced by commercial considerations. Studies that produce complex findings—showing benefits in some contexts but not others or demonstrating efficacy alongside concerning side effects—face pressure to emphasize commercially favorable aspects while minimizing discussion of limitations. This selective emphasis can create systematically distorted impressions of the evidence base for neurotechnology interventions.

Third, the mechanisms of scientific communication themselves become influenced by commercial priorities. Industry sponsors often cite "data on file" that remains outside the public domain and commonly use "educational" marketing tools presenting nonsystematic research summaries or individual trials that are unlikely to be representative of the totality of research. This creates parallel communication channels that can influence clinical practice and public understanding without being subject to traditional peer review processes.

10.2.3 The Challenge of Negative Results and Scientific Progress

Perhaps no aspect of commercial pressure on research integrity is more consequential than the systematic suppression of negative results. The "file drawer problem," where studies with nonsignificant findings remain unpublished, is particularly acute in neuroscience because this systematic bias can skew understanding of research areas and influence the efficacy of clinical trials [13, 14]. In commercially influenced research, this basic problem is exacerbated by additional economic incentives against publishing unfavorable findings.

The suppression of negative results in neuroscience has several distinctive features that differentiate it from other biomedical fields. First, the complexity of neural systems means that interventions often have highly context-dependent effects. A neuromodulation technique might be effective for certain patient populations, under specific parameters, or when combined with particular therapeutic approaches while showing no benefit in other contexts. Commercial pressures tend to favor selective reporting of positive contexts while negative contexts remain unpublished.

Second, the high costs associated with neurotechnology development create particularly strong disincentives for publishing negative findings. When a company has invested millions of dollars in developing a neural interface or computational approach, studies showing limitations or failure represent not just scientific disappointments but potential threats to continued funding, regulatory approval, and market position.

Recognition of these problems has led to important innovations in scientific publishing, including the adoption of registered reports by major neuroscience journals like *Nature Neuroscience*, which evaluate studies based on their methodology and research questions before data collection begins, ensuring that negative results will be published, provided the approved protocol is followed [13].

However, systemic solutions to the negative results problem in commercially influenced neuroscience require more than publishing reforms. They demand changes in funding structures, regulatory frameworks, and commercial incentives that make the publication of negative findings economically viable rather than economically harmful for commercial researchers.

10.3 The Tension Between Intellectual Property and Open Science in AI Neuroscience

10.3.1 Proprietary Algorithms and Scientific Verifiability

The integration of artificial intelligence into neuroscience has created a fundamental tension. This tension exists between traditional scientific principles of transparency and reproducibility and the commercial imperative to protect intellectual property through patents and trade secrets. As AI has transformed neuroscience by providing tools capable of unraveling the immense complexity of the brain, the convergence of these fields has enabled the analysis of complex neural datasets from neuroimaging to genomic profiling but has also introduced new challenges for scientific verification [5, 15].

The core issue lies in what might be termed the "black box dilemma." Modern AI systems used in neuroscience applications—from diagnostic algorithms to brain-computer interfaces to neural decoding systems—often operate as proprietary "black boxes" where the specific algorithms, feature weightings, and decision-making processes remain commercially protected. This creates an unprecedented situation in science. Researchers can observe inputs and outputs of these systems and may even be able to validate their performance but cannot examine the underlying mechanisms that generate their results.

This opacity poses fundamental challenges to scientific verification and reproducibility. Traditional scientific methodology demands that research findings be reproducible by independent researchers using the same methods. When the "methods" consist of proprietary algorithms that cannot be examined or replicated, the basic foundations of scientific verification become compromised. Researchers must effectively take on faith that commercial AI systems are operating as claimed, without the ability to conduct the kind of independent verification that is routine in other areas of science.

The clinical implications of this problem are particularly concerning. While AI systems in neuroscience have shown remarkable capabilities in tasks like predicting treatment responses and analyzing brain imaging data, their adoption in clinical settings raises questions about how physicians can appropriately interpret and trust algorithmic recommendations when the underlying decision-making processes remain opaque.

Some organizations have begun developing innovative approaches to balance intellectual property protection with scientific transparency. These include "federated learning" approaches where algorithms can be validated across multiple datasets without revealing proprietary code, "algorithmic transparency" requirements for certain clinical applications, and "trusted third party" validation schemes where independent organizations can verify algorithmic performance without accessing trade secrets.

10.3.2 Patent Landscapes and Research Access

The patent landscape in neuroscience and neurotechnology has become increasingly complex, with significant implications for research access and scientific progress. The tension between open science and intellectual property rights has been recognized at the highest policy levels, with the European Commission reports exploring how to balance "as open as possible, as closed as necessary" in research and innovation ecosystems [16].

The fundamental challenge lies in the broad scope of many neuroscience-related patents. Unlike traditional biotechnology patents that might protect specific compounds or devices, neurotechnology patents often claim broad methodological approaches, algorithmic frameworks, or general applications of brain stimulation or recording techniques. This breadth can create "patent thickets" that make it difficult for researchers to pursue certain lines of investigation without extensive licensing negotiations.

Research with the Canadian neuroscience faculty has revealed that researchers perceive significant conflicts between open science ideologies and the restrictive nature of patents, highlighting concerns about how intellectual property frameworks impact scientific autonomy and collaborative research practices [6]. These concerns are particularly acute for researchers in academic settings who may lack the legal and financial resources to navigate complex patent landscapes.

The global nature of neuroscience research further complicates patent-related access issues. Patents filed in major jurisdictions can effectively restrict research activities worldwide, as academic institutions are often reluctant to pursue research directions that might lead to patent infringement claims. This creates a form of "innovation imperialism" where patent holders in wealthy countries can effectively constrain research activities in developing nations.

Some institutions have developed creative responses to these challenges. These include patent-sharing consortiums where academic institutions pool their intellectual property for mutual benefit, "academic use" licensing frameworks that permit research use of patented technologies while maintaining commercial protections, and "defensive patent" strategies where institutions build patent portfolios specifically to prevent others from restricting academic research.

10.3.3 Open Science Movement in Neuroscience

Despite the challenges posed by intellectual property regimes, neuroscience has seen the emergence of a vibrant open science movement that seeks to preserve and expand access to research tools, data, and methodologies. The open source movement in neuroscience has been driven partly by the cost-effectiveness of open alternatives to commercial products and by the customization and flexibility that open approaches provide, allowing researchers to study phenomena at levels that commercial products might not support [17].

The Montreal Neurological Institute's pioneering decision in 2016 to become the world's first open science institute represents a landmark attempt to demonstrate how neuroscience research can be conducted within an open science framework while still maintaining institutional sustainability and research competitiveness [18].

The benefits of open approaches in neuroscience are substantial and well-documented. Open source tools allow for greater customization and innovation, enable researchers to contribute improvements and modifications, and create distributed development communities that can accelerate scientific progress. Open data sharing initiatives have enabled meta-analyses and collaborative studies that would be impossible within proprietary frameworks.

However, the open science movement in neuroscience also faces significant challenges beyond intellectual property issues. These include the substantial time and expertise required to develop and maintain open source tools, the difficulty of securing funding for open source development projects, and the career incentives that often favor proprietary innovation over open collaboration.

The pharmaceutical industry has begun experimenting with "open innovation" models that attempt to balance proprietary interests with collaborative development, including examples where companies make certain research tools or datasets available for academic use while maintaining commercial rights to applications [19]. These hybrid approaches suggest possible pathways for reconciling open science principles with commercial realities in neuroscience research.

10.4 How Funding Models Influence Research Directions

10.4.1 The Venture Capital Effect on Neuroscience Priorities

The dramatic increase in venture capital investment in neuroscience and neurotechnology has fundamentally altered the research landscape. This shift affects priorities, timelines, and success metrics in ways that have profound implications for scientific inquiry. 2024 saw $2.3 billion in neurotech funding across 129 deals, with venture capital funding for brain-computer interfaces now exceeding traditional government research funding. The influence of commercial investment on research directions has reached unprecedented levels [1, 2].

Venture capital funding brings distinctive priorities that systematically influence research directions. Unlike traditional academic funding, which typically evaluates projects based on scientific merit, potential for fundamental discovery, and contribution to knowledge, venture funding prioritizes market size, scalability, regulatory feasibility, and return on investment timelines. This creates predictable biases in which research directions receive support and which remain underfunded.

As analysis of the neurotech investment landscape reveals, investors focus heavily on technologies that can achieve scale and address large markets, leading to the concentration of funding in areas with clear commercial applications rather than those with the greatest scientific or medical need [1, 7]. This pattern is evident in the distribution of neurotechnology investments, which heavily favor consumer applications, enhancement technologies, and conditions affecting large patient populations while neglecting rare neurological disorders, pediatric applications, and fundamental research questions.

The temporal demands of venture funding create additional distortions in research priorities. Venture investors typically expect to see significant progress within 3–5 year timeframes, which is often incompatible with the longer timelines required for fundamental neuroscience research or the development of interventions for complex neurological conditions. This pressure toward short-term deliverables can lead to research programs that prioritize incremental improvements to existing technologies over potentially transformative but higher-risk fundamental research.

The emergence of specialized neurotech venture funds, such as HEKA in Europe and Nexus NeuroTech Ventures, represents an attempt to address some of these issues by bringing neuroscience expertise directly into the investment process and focusing specifically on brain-related technologies [20]. However, these specialized funds still operate within commercial frameworks that prioritize return on investment over scientific or medical impact.

The venture capital effect also influences research through what might be termed "innovation clustering." When particular applications or approaches attract venture interest, they tend to attract disproportionate attention and resources, leading to redundant development efforts in commercially attractive areas while potentially more impactful research directions remain underexplored.

10.4.2 Traditional Grant Funding and Commercial Expectations

Even traditional academic funding mechanisms have evolved to incorporate increasingly commercial expectations, blurring the boundaries between academic and commercial research in ways that influence research directions and priorities. Federal funding agencies, private foundations, and institutional funding bodies increasingly emphasize "translational potential," "commercial viability," and "pathway to impact" in their evaluation criteria.

This shift reflects both practical and political pressures on funding agencies. Politically, funding bodies face pressure to demonstrate that research investments produce tangible benefits for society, which is often interpreted as commercial applications or clinical translations. Practically, the high costs of modern neuroscience research—particularly human studies, advanced imaging, and cutting-edge technologies—create pressures to prioritize research directions that might attract additional commercial investment or generate licensing revenue.

The incorporation of commercial criteria into traditional funding evaluation has several important effects on research directions. First, it creates systematic bias against fundamental research that may be crucial for long-term scientific progress but lacks clear short-term commercial applications. Second, it influences how researchers frame and present their work, encouraging emphasis on potential applications even when the research is primarily focused on basic scientific questions.

Third, it affects which researchers and institutions are successful in securing funding. Those with commercial connections, entrepreneurial experience, or institutional resources for technology transfer may be advantaged in funding competitions, even when their scientific qualifications are equivalent to those without such commercial ties.

Some funding agencies have attempted to address these issues through specialized programs that explicitly support high-risk fundamental research without commercial expectations. However, these programs typically represent small fractions of total funding, and their existence may inadvertently reinforce the commercialization of mainstream funding programs.

10.4.3 Public-Private Partnerships and Research Independence

The proliferation of public-private partnerships in neuroscience research represents both an opportunity to leverage diverse resources and expertise and a challenge to traditional notions of research independence and scientific objectivity. These partnerships, which bring together academic institutions, government funding, and commercial entities around shared research goals, have become increasingly common as the costs and complexity of neuroscience research have grown.

The pharmaceutical industry has increasingly embraced "open innovation" models that involve collaboration with academic institutions and the sharing of certain research resources, although these typically maintain intellectual property restrictions and commercial control over key aspects of the research [19]. These partnerships can provide academic researchers access to resources, expertise, and technologies that would be impossible to develop independently while giving commercial partners access to academic innovation and scientific credibility.

However, public-private partnerships also create complex governance challenges that can affect research independence. Questions arise about who sets research priorities, how results are interpreted and communicated, what happens when findings conflict with commercial interests, and how intellectual property rights are balanced with academic freedom and public benefit.

Successful public-private partnerships in neuroscience have typically required carefully structured governance mechanisms that protect scientific independence while acknowledging legitimate commercial interests. These mechanisms include independent scientific advisory boards, preagreed data sharing protocols, publication rights protection, and explicit procedures for handling conflicts between scientific and commercial priorities.

Initiatives like Campus Alberta Neuroscience demonstrate how regional partnerships can bring together diverse stakeholders—universities, industry, patient organizations, and government—around shared research priorities while maintaining focus on scientific impact rather than purely commercial outcomes [21]. Such models suggest possibilities for partnership structures that harness commercial resources while preserving scientific independence.

10.5 The Impact of Commercialization on Healthcare Education and Neuroscience Training

10.5.1 Evolution of Neuroscience Graduate Training Programs

The commercialization of neuroscience has profoundly influenced how the next generation of researchers is educated and trained. Graduate programs increasingly incorporate commercial considerations into curricula, research expectations, and career preparation. Universities like the University of Michigan have developed formal Certificate programs in Biomedical Innovation & Entrepreneurship that combine science and business training to prepare graduate students for careers in biotechnology startups and pharmaceutical companies [22].

This evolution reflects both pragmatic and philosophical changes in neuroscience education. Pragmatically, the job market for neuroscience Ph.D.'s has shifted significantly, with traditional academic positions becoming increasingly scarce while commercial opportunities have expanded. Philosophically, many programs have embraced the idea that effective translation of neuroscience research requires researchers who understand both scientific and commercial considerations.

The curricular changes associated with this shift are substantial. Traditional neuroscience graduate programs focus primarily on research methodology, theoretical foundations, and specialized technical skills within particular subdisciplines. Contemporary programs increasingly include training in intellectual property law, business development, regulatory affairs, and entrepreneurship. Some programs now require or strongly encourage participation in commercialization activities as part of graduate training.

Major neuroscience graduate programs at institutions like the University of Michigan and Wake Forest now explicitly incorporate innovation and entrepreneurship training, with programs emphasizing the development of skills necessary to translate research into clinical applications and commercial products [23, 24].

These changes have both benefits and concerns. Benefits include better preparation of graduates for diverse career paths, increased appreciation for the practical challenges of translating research into real-world applications, and potentially enhanced ability to secure funding and resources for research. Concerns include potential narrowing of research interests toward commercially viable topics, reduced emphasis on fundamental scientific questions, and possible conflicts between commercial training and traditional scientific values of objectivity and skeptical inquiry.

10.5.2 Industry Influence on Continuing Education

Beyond formal degree programs, the commercialization of neuroscience has significantly influenced continuing education and professional development for practicing clinicians and researchers. The emergence of specialized commercial applications of neuroscience, such as neuromarketing, has created entirely new areas of professional education that bridge neuroscience research with commercial applications [25].

Industry involvement in continuing education operates through multiple mechanisms. Pharmaceutical and device companies sponsor conferences, workshops, and educational materials that provide practicing clinicians with information about new technologies and treatments. Professional societies collaborate with commercial entities to develop educational content and certification programs. Commercial entities themselves offer training and certification programs for their technologies and approaches.

This industry involvement brings both valuable resources and potential biases to continuing education. The resources include access to cutting-edge technologies, funding for high-quality educational programs, and practical training that might be difficult to obtain through purely academic channels. The potential biases include selective presentation of evidence, emphasis on commercial applications over fundamental understanding, and potential conflicts of interest in educational content development.

Academic institutions have also developed specialized programs that bridge neuroscience and applied fields, such as Teachers College Columbia's Neuroscience and Education program, which prepares professionals to apply neuroscience research to educational settings [26]. These programs represent attempts to maintain academic rigor while addressing practical application needs.

The challenge for continuing education in neuroscience lies in maintaining scientific objectivity and comprehensive coverage while leveraging commercial resources and addressing practical application needs. Some professional organizations have developed guidelines for industry-sponsored educational activities, including requirements for balanced content, disclosure of commercial relationships, and independent review of educational materials.

10.5.3 The Entrepreneurial Imperative in Scientific Careers

Perhaps the most significant impact of commercialization on neuroscience training is the emergence of what might be termed the "entrepreneurial imperative"—the expectation that researchers will engage with commercial applications of their work as a routine part of their scientific careers. This represents a fundamental shift from traditional academic models where commercial applications were optional or even viewed with suspicion.

As one University of Michigan graduate student noted, participation in entrepreneurship and commercialization training "really helped me to understand the scientific, regulatory, and business strategies and how they combine to create a successful product that will reach the hands of patients" [22]. This perspective reflects how commercialization training can enhance understanding of the broader context and impact of scientific research.

The entrepreneurial imperative manifests in multiple ways throughout scientific training and career development. Graduate students are increasingly expected to consider commercial applications of their research, to develop skills in intellectual property and business development, and to participate in entrepreneurial activities such as startup competitions or technology transfer processes. Faculty promotion and tenure criteria increasingly include metrics related to technology transfer, patent applications, and commercial partnerships.

This shift creates both opportunities and tensions for scientific careers. Opportunities include expanded career options, increased resources for research, and potentially greater impact of scientific work on real-world problems. Tensions include potential conflicts between commercial interests and scientific objectivity, pressure to focus research on commercially viable applications, and the challenge of maintaining traditional scientific values within increasingly commercial institutional environments.

The emergence of specialized educational programs, such as Butler University's Applied Educational Neuroscience Certificate, demonstrates how neuroscience training is expanding beyond traditional research careers to include diverse applications in education, counseling, and mental health practice [27]. These programs reflect both the growing practical applications of neuroscience knowledge and the need for specialized training to bridge research and application contexts.

10.6 Balancing Innovation with Scientific Rigor

10.6.1 Establishing Sustainable Frameworks for Commercial-Academic Collaboration

The challenge of maintaining scientific integrity while harnessing market forces for innovation requires sophisticated institutional and policy frameworks that can accommodate both commercial realities and scientific values. This is not simply a matter of choosing between commercial and academic approaches but rather of developing hybrid models that leverage the strengths of each while mitigating their respective weaknesses.

Successful frameworks for commercial-academic collaboration in neuroscience typically incorporate several key elements. First, they establish clear governance structures that include independent scientific oversight, transparent decision-making processes, and explicit procedures for handling conflicts between commercial and scientific priorities. Second, they create protected spaces for fundamental research that is insulated from immediate commercial pressures while maintaining pathways for eventual translation. Third, they develop intellectual property frameworks that balance commercial incentives with scientific openness and public benefit.

Research with neuroscience faculty has identified key principles for the successful integration of open science and intellectual property concerns, including respect for researcher autonomy, attention to justice and equitable access, and the maintenance of culturally safe research practices [6]. These principles suggest that sustainable frameworks must attend not only to efficiency and innovation metrics but also to broader ethical and social considerations.

Some institutions have pioneered innovative approaches to these challenges. These include "mission-driven" venture funds that prioritize scientific and social impact alongside financial returns, "staged disclosure" models where research findings are shared openly after limited periods of commercial exclusivity, and "collaborative but bounded" partnerships where commercial entities provide resources and expertise while academic institutions maintain control over fundamental research directions and publication decisions.

The development of effective frameworks also requires attention to the broader ecosystem in which commercial-academic collaborations operate. This includes regulatory environments that support responsible innovation, funding mechanisms that reward both scientific excellence and practical impact, and professional incentive structures that value contributions to both knowledge and application.

10.6.2 Preserving Scientific Values in Commercial Contexts

Maintaining core scientific values—including objectivity, transparency, reproducibility, and commitment to truth over convenience—within increasingly commercial research environments represents one of the central challenges facing contemporary neuroscience. This requires not only individual commitment from researchers but also institutional structures and cultural practices that support scientific integrity even when it conflicts with commercial interests.

The adoption of registered reports by major neuroscience journals represents one important institutional innovation for preserving scientific integrity by ensuring that research is evaluated based on methodology and research questions rather than results, thereby reducing publication bias and supporting the publication of negative findings [13].

Individual researchers operating in commercial contexts face particular challenges in maintaining scientific integrity. These include pressure to interpret findings more optimistically than warranted, incentives to emphasize positive results while minimizing limitations, and conflicts between commercial confidentiality requirements and scientific transparency expectations. Addressing these challenges requires both personal ethical frameworks and institutional support systems.

Professional organizations in neuroscience have begun developing guidelines and training programs specifically focused on research integrity in commercial contexts. These include codes of conduct for industry-academic collaborations, training programs on conflicts of interest identification and management, and peer support networks for researchers navigating commercial pressures.

The preservation of scientific values also requires attention to the broader culture and incentive systems within which neuroscience research operates. This includes promotion and tenure criteria that value scientific integrity alongside productivity metrics, funding programs that reward transparency and reproducibility, and educational programs that emphasize ethical reasoning and professional responsibility alongside technical skills.

10.6.3 Future Directions: Toward Sustainable Science-Commerce Integration

Looking forward, the relationship between market forces and scientific integrity in neuroscience will likely continue to evolve as both commercial investment and scientific capabilities advance. Several emerging trends suggest possible directions for more sustainable integration of commercial and scientific priorities.

The rapid advancement of AI applications in neuroscience is creating new possibilities for both commercial innovation and scientific discovery and new challenges for transparency, interpretability, and validation of research tools and findings [5, 28]. Addressing these challenges will require continued innovation in both technical approaches and governance frameworks.

The global nature of neuroscience research and commercial development creates opportunities for different regions and institutions to experiment with different models for balancing commercial and scientific priorities. Comparative analysis of these different approaches could provide valuable insights for developing more effective frameworks.

Emerging trends in neuroscience, including advances in brain modeling, digital twins, and personalized medicine applications, are creating new possibilities for both scientific understanding and commercial applications [28]. Successfully realizing these possibilities while maintaining scientific integrity will require continued attention to the frameworks and practices that govern science-commerce interactions.

The ultimate goal is not to eliminate commercial influences from neuroscience—which would likely be both impossible and counterproductive—but rather to structure these influences in ways that support rather than undermine the scientific enterprise. This requires ongoing dialogue between researchers, commercial entities, policymakers, and the broader public about the appropriate boundaries, expectations, and governance mechanisms for commercially influenced research.

10.7 Conclusion

The intersection of market forces and scientific integrity in neuroscience represents one of the defining challenges of contemporary brain research. As we have seen throughout this chapter, commercial influences on neuroscience are neither uniformly beneficial nor uniformly harmful but rather create complex trade-offs that require careful navigation and sophisticated institutional responses.

The evidence presented suggests several key conclusions. First, market forces can provide valuable resources, expertise, and translational pathways that enhance the impact and application of neuroscience research. The substantial venture capital investment in neurotechnology, the development of innovative commercial applications of brain research, and the expansion of career opportunities for neuroscience researchers all represent genuine benefits of commercial engagement.

Second, however, market forces also create systematic pressures that can compromise scientific integrity if left unchecked. These include biases toward positive results and commercially favorable findings, pressures for the premature translation of research findings, constraints on transparency and reproducibility, and the potential narrowing of research priorities toward commercially viable applications at the expense of fundamental scientific questions.

Third, the most promising approaches to these challenges involve neither the wholesale rejection of commercial engagement nor the uncritical embrace of market logic but rather the development of sophisticated frameworks that leverage commercial resources while preserving scientific values. These frameworks require attention to governance structures, incentive systems, institutional cultures, and professional standards that can support both innovation and integrity.

Fourth, addressing these challenges requires coordinated action across multiple levels—from individual researcher practices to institutional policies to broader regulatory and funding frameworks. No single intervention or approach is sufficient; rather, sustainable solutions require comprehensive attention to the entire ecosystem in which neuroscience research operates.

The future of neuroscience depends not on resolving the tension between market forces and scientific integrity but on learning to navigate this tension productively. As neuroscience research becomes increasingly central to addressing major societal challenges—from mental health and neurological disease to artificial intelligence and human enhancement—the stakes of getting this balance right continue to grow.

The neuroscience community has both the opportunity and responsibility to pioneer approaches that demonstrate how scientific research can engage productively with commercial forces while maintaining the integrity, transparency, and commitment to truth that make scientific knowledge valuable and trustworthy. The frameworks, practices, and institutions developed in neuroscience may well serve as models for other fields facing similar challenges in an increasingly commercial research environment.

10.8 Summary

This chapter examines the complex relationship between commercial forces and scientific integrity in contemporary neuroscience research. With $2.3 billion in venture capital flowing into neurotechnology in 2024 and private investment now exceeding traditional government research funding, market forces have fundamentally reshaped the neuroscience landscape. This chapter explores five critical dimensions of this transformation: commercial pressures on research conduct and publication practices, tensions between intellectual property protection and open science principles, the influence of funding models on research priorities, the impact of commercialization on education and training, and strategies for balancing innovation with scientific rigor.

Key findings include evidence of systematic publication bias favoring commercially favorable results, the emergence of "black box" AI systems that challenge traditional scientific verification principles, the concentration of venture funding in commercially viable applications at the expense of fundamental research questions, and the evolution of graduate training programs to incorporate entrepreneurial skills alongside traditional scientific methods. This chapter documents how commercial pressures create "acceleration imperatives" that can compromise methodological rigor, while patent landscapes increasingly restrict research access and collaboration.

Despite these challenges, the analysis reveals that market forces also provide valuable resources, expertise, and translational pathways that can enhance scientific impact. This chapter concludes that the future of neuroscience depends not on eliminating commercial influences but on developing sophisticated frameworks that harness market-driven innovation while preserving core scientific values of transparency, reproducibility, and commitment to truth. Success requires coordinated action across multiple levels—from individual researcher practices to institutional policies to broader regulatory frameworks—to create sustainable models for science-commerce integration.

References

1. Rao N. 2024 neurotech funding snapshot. Neurotechnology Substack. 2025. Retrieved from https://neurotechnology.substack.com/p/2024-funding-snapshot.
2. From the Interface. Venture capital funding for brain-computer interfaces outstrips DARPA dollars. 2024. Retrieved from https://www.from-the-interface.com/bci-funding-clinical-trials/
3. AlleyWatch. The 16 largest US funding rounds of December 2024. 2025. Retrieved from https://www.alleywatch.com/2025/01/us-startup-funding-top-largest-december-2024-vc/
4. Miller K, Zemel R, Kriegeskorte N. Neuroscience + artificial intelligence = NeuroAI. Columbia Zuckerman Institute; 2024. Retrieved from https://zuckermaninstitute.columbia.edu/neuroscience-artificial-intelligence-neuroai
5. Onciul R, Tataru C-I, Dumitru AV, Crivoi C, Serban M, Covache-Busuioc R-A, Radoi MP, Toader C. Artificial intelligence and neuroscience: transformative synergies in brain research and clinical applications. J Clin Med. 2025;14(2):550.
6. Nuechterlein A, Rotenberg A, LeDue J, Pavlidis P, Illes J. Open science in play and in tension with patent protections. J Law Biosci. 2023;10(2):lsad016.
7. Zhegin P, Bates L, Parry D, Olsen L. Investing in the brain: a generalist investor guide to spotting winning neurotech startups. NeuroTechX Content Lab; 2024. Retrieved from https://medium.com/neurotechx/investing-in-the-brain-a-generalist-investor-guide-to-spotting-winning-neurotech-startups-3f29624a0586
8. Buthut M, Starke G, Akmazoglu TB, Colucci A, Vermehren M, van Beinum A, Bublitz C, Chandler J, Ienca M, Soekadar SR. HYBRIDMINDS-summary and outlook of the 2023 international conference on the ethics and regulation of intelligent neuroprostheses. Front Hum Neurosci. 2024;18:1489307.
9. Gilbody S, et al. In: ResearchGate, editor. Publication bias and the integrity of psychiatry research; 2000. Retrieved from https://www.researchgate.net/publication/12492728_Publication_bias_and_the_integrity_of_psychiatry_research.
10. Becker RE. Publication bias in neurology research: why it matters and why it needs to change. Neurology Advisor. 2018; Retrieved from https://www.neurologyadvisor.com/topics/general-neurology/publication-bias-in-neurology-research-why-it-matters-and-why-it-needs-to-change/
11. Wikipedia contributors. Publication bias. Wikipedia. 2025; Retrieved from https://en.wikipedia.org/wiki/Publication_bias
12. van Aert RCM, et al. Publication bias examined in meta-analyses from psychology and medicine: a meta-meta-analysis. PLoS One. 2019;14(4):e0215052.
13. Nature Neuroscience Editorial. Reducing publication bias with registered reports. Nat Neurosci. 2024;27(9):1635. https://doi.org/10.1038/s41593-024-01762-9.
14. Kandola S. Publication bias: the hidden threat to systematic literature reviews. Envision Pharma Group; 2024. Retrieved from https://www.envisionpharmagroup.com/news-events/publication-bias-hidden-threat-systematic-literature-reviews

15. Surianarayanan C, Lawrence JJ, Chelliah PR, Prakash E, Hewage C. Convergence of artificial intelligence and neuroscience towards the diagnosis of neurological disorders-a scoping review. Sensors (Basel). 2023;23(6):3062.
16. European Commission. Open science and intellectual property rights. 2022. Retrieved from https://research-and-innovation.ec.europa.eu/knowledge-publications-tools-and-data/publications/all-publications/open-science-and-intellectual-property-rights_en
17. White SR, Amarante LM, Kravitz AV Laubach M. The future is open: Open-Source Tools for Behavioral Neuroscience Research. eNeuro. 2019;6(4):ENEURO.0223-19.2019.
18. Roskams-Edris D. Intellectual property policy at the neuro, an open science institute. Qeios. 2020;
19. Hunter AJ, Lee WH, Bountra C. Open innovation in neuroscience research and drug discovery. Brain Neurosci Adv. 2018;2:2398212818799270.
20. HEKA Fund. Where tech meets brain science. Newfund Capital; 2024. Retrieved from https://newfundcap.com/braintech-fund-heka
21. Campus Alberta Neuroscience. Funding programs. 2024. Retrieved from https://www.albertaneuro.ca/funding-programs/
22. University of Michigan Medical School. Certificate in biomedical innovation & entrepreneurship. 2024. Retrieved from https://medresearch.umich.edu/office-research/about-office-research/our-units/fast-forward-medical-innovation/commercialization-education/certificate-biomedical-innovation-entrepreneurship
23. University of Michigan Medical School. Neuroscience graduate program. 2024. Retrieved from https://medschool.umich.edu/interdisciplinary-programs/neuroscience-graduate-program
24. Wake Forest University School of Medicine. Neuroscience (PhD and MS). 2024. Retrieved from https://school.wakehealth.edu/education-and-training/graduate-programs/graduate-programs-in-neuroscience
25. Harvard Division of Continuing Education. Consumer behavior course in neuromarketing. 2024. Retrieved from https://professional.dce.harvard.edu/programs/consumer-behavior/
26. Teachers College, Columbia University. Neuroscience and education. 2024. Retrieved from https://www.tc.columbia.edu/biobehavioral-sciences/neuroscience-and-education/
27. Butler University. Applied educational neuroscience certificate. 2025. Retrieved from https://www.butler.edu/education/graduate-programs/applied-educational-neuroscience/
28. QMENTA. Top 5 trends in neuroscience you need to know in 2025. 2025. Retrieved from https://www.qmenta.com/blog/top-5-trends-in-neuroscience-you-need-to-know-in-2025

Academic Freedom in AI Neuroscience Research

11

It is the responsibility of intellectuals to speak the truth and expose lies.

—Noam Chomsky

11.1 Introduction

Research sessions on the neural correlations of gender identity and sexual orientation have increasingly prompted debates among conference organizing committees and drawn both support and condemnation on social media. University administrators have expressed concerns about institutional reputation, while researchers report receiving mixed messages of private support and public distance from colleagues. Methodologically sound research in these areas has become entangled in cultural and political conflicts extending far beyond laboratory settings. This situation has become increasingly common for neuroscientists working at the intersection of artificial intelligence, neurobiology, and human identity, creating complex challenges to academic freedom in contemporary neuroscience.

The freedom to pursue research questions without external interference or intimidation has long been considered foundational to scientific progress [1]. Yet as neuroscience increasingly incorporates artificial intelligence to investigate aspects of human cognition, behavior, and identity that touch on socially contested terrain, this freedom faces unprecedented pressures from multiple directions. Political actors, funding agencies, social movements, corporate partners, and university administrators create a complex ecosystem of constraints and incentives. This ecosystem shapes which questions researchers pursue, how they communicate findings, and even what hypotheses they consider legitimate to investigate [2, 3].

This chapter examines the current landscape of academic freedom in AI neuroscience research, documenting both overt constraints and subtle pressures that influence scientific inquiry. Through analysis of institutional policies, researcher

T. Rudroff, *AI for the Healthy Brain and Neurological Disorders*, Neurocultural Health and Wellbeing, https://doi.org/10.1007/978-3-032-11596-6_11

experiences, and case studies, we explore how neuroscientists navigate increasingly politicized research domains while attempting to maintain scientific integrity. The regional variations in these dynamics reveal important insights into the relationship between political systems, cultural contexts, and scientific autonomy [4, 5]. Finally, we consider approaches for preserving essential academic freedom that enable scientific advancement. These approaches must acknowledge the legitimate ethical considerations and social responsibilities that accompany research with profound implications for human self-understanding and technological governance.

11.2 Current Threats to Academic Freedom in Neuroscience Departments

11.2.1 Institutional Risk Management and Research Constraint

Department meetings addressing controversial research proposals have become increasingly common in neuroscience programs. University administrators regularly express concerns about potential media misinterpretation and backlash when research involves sensitive topics. Grant applications mention the use of machine learning to identify potential neural markers of psychopathy in adolescent brain scans, for example, often prompt discussions about institutional reputation management. Despite sound methodological approaches and significant potential clinical applications, researchers frequently receive guidance to pursue less controversial research directions or risk losing institutional support. These interactions represent an increasingly common form of soft constraint on academic freedom, where institutional risk management subtly but effectively limits which questions neuroscientists can pursue.

At the institutional level, university administrators increasingly function as risk managers concerned with protecting institutional reputation, donor relationships, and political capital [6]. Recent surveys of research administrators at major research universities have documented increased internal scrutiny of potentially controversial neuroscience research proposals [7]. These surveys reveal sophisticated informal processes for flagging and sometimes redirecting research that might generate public controversy, regardless of its scientific merit or methodological rigor.

This institutional risk management manifests in various forms, including selective application of ethical review processes, informal guidance about "fundable" research directions, and strategic resource allocation that advantages "safer" research domains [8]. These constraints were rarely documented in formal policies—operating instead through administrative discretion that allows institutions to maintain nominal commitments to academic freedom while practically limiting its exercise.

The emergence of artificial intelligence tools in neuroscience has particularly intensified these dynamics [9]. Administrative concern focuses especially on research that might generate predictive models related to human behavior, cognition, or identity. Internal communications at research universities reveal that administrators specifically flag certain types of research for additional review. This includes research involving machine learning analysis of neural data related to intelligence, sexual orientation, gender identity, criminal behavior, and political orientation. These reviews go beyond standard ethics approval and are conducted explicitly for "reputation management" and "stakeholder relations"—not for scientific or ethical concerns.

These constraints have measurable impacts on research trajectories. Studies analyzing research productivity in potentially controversial versus noncontroversial neuroscience domains following administrative intervention found that publication output in sensitive domains decreased significantly relative to control topics, despite controlling for funding availability and researcher expertise [10]. This finding suggests that administrative constraint effectively redirects scientific attention away from certain questions regardless of their scientific importance or methodological accessibility.

11.2.2 External Funding Pressures and Research Direction

Laboratory meetings increasingly involve discussions about shifting priorities in major funding agency calls for proposals. Research areas have changed noticeably from previous years, with language around controversial topics becoming more circumscribed. Reports from colleagues at other universities indicate that grants on certain demographic differences have been rejected despite excellent reviews, with comments suggesting topics are "not aligned with current priorities." The message remains subtle but clear—pursuing certain research questions might be scientifically valid but could limit funding prospects [11, 12].

Beyond institutional constraints, external funding mechanisms significantly influence academic freedom through priority-setting and selective support [13, 14]. Analysis of shifting priorities across major funding agencies over recent decades reveals systematic changes in fundable research areas that track broader political shifts and social movements rather than purely scientific considerations. These funding priorities function as a powerful but often unacknowledged constraint on academic freedom by determining which questions receive resources and attention.

The impact of these funding shifts extends beyond topic selection to shape how research is framed and communicated [15]. Researchers report sophisticated strategies for "funding-oriented research framing"—adapting research questions, methodological approaches, and anticipated implications to align with perceived funder priorities. As one researcher acknowledged, "I'm still doing essentially the same work, but the narrative I build around it shifts dramatically depending on which agency I'm approaching for funding."

The interwoven relationship between funding and academic freedom becomes particularly acute at the intersection of artificial intelligence and neuroscience [16, 17]. The rapid growth of corporate and defense funding in AI neuroscience has created situations where nearly half of the research in this domain receives some private sector or military support, compared to much lower percentages for traditional neuroscience research. These funding sources often come with explicit constraints on research publication, data sharing, and methodology transparency that limit the open scientific exchange essential to academic freedom.

Studies have demonstrated how funding patterns can systematically advantage certain research directions while disadvantaging others, regardless of scientific merit [18]. Neuroscience research aligned with technological optimism and enhancement applications receives significantly more funding per publication than research investigating potential risks, limitations, or unintended consequences of the same technologies. This funding disparity effectively constrains which questions researchers can sustainably pursue, creating structural limitations on academic freedom that operate through resource allocation rather than explicit prohibition.

11.2.3 Peer and Public Pressure on Research Directions

Journal club meetings regularly feature debates that extend beyond methodological critique to questions about research legitimacy. Discussions of papers on neural correlations of political orientation often shift into examinations of whether such research should be conducted. Some researchers express concerns that publishing this type of work could provide support for reductive interpretations of complex political viewpoints. Others argue that understanding the neural basis of political judgment formation represents legitimate scientific inquiry. These discussions highlight ongoing tensions within the field regarding which research questions should be pursued, regardless of methodological soundness.

Beyond institutional and funding constraints, informal peer and anticipated public pressure increasingly shape research directions in neuroscience [19, 20]. Anonymous surveys of neuroscience faculty members reveal widespread concerns about colleague perception and potential public reaction influencing research choices. Studies show that most respondents report considering "likely public reaction" when selecting research topics, while many acknowledge that "anticipated criticism from colleagues" influences their research directions.

The influence of these pressures is particularly strong in research involving artificial intelligence and neuroscience [21]. Researchers working on AI neuroscience applications perceived as politically or ethically contentious report systematic disadvantages in peer review, conference acceptances, and professional networking opportunities. These penalties create powerful incentives to avoid certain research domains regardless of their scientific merit or potential societal benefit.

Social media amplifies these peer pressure dynamics [22]. Analysis of social media responses to potentially controversial neuroscience publications documents coordinated negative attention directed at researchers investigating politically sensitive topics. The experience or awareness of such episodes significantly influences future research choices. Most surveyed neuroscientists report that awareness of colleagues facing public criticism made them more cautious in their own research directions.

The anticipation of negative reception creates "preemptive self-constraint," where researchers exclude potentially valuable but controversial research directions before formal review processes or external limitations are encountered [23]. Interviews with early-career neuroscientists reveal that many had internalized detailed mental maps of "acceptable" versus "risky" research domains. These mental maps guide their project selection. Artificial intelligence applications are considered particularly fraught when applied to questions of human identity, behavior prediction, or group differences.

The complex interplay of institutional, funding, peer, and public pressures creates what might be termed a "constraint ecosystem" that systematically shapes research directions in AI neuroscience. While the precise severity of these constraints varies across research domains and institutional contexts, the patterns of constraint can be visualized to illustrate their cumulative impact on academic freedom.

Figure 11.1 presents an academic freedom constraint matrix that synthesizes the constraint patterns documented throughout this analysis. The matrix maps seven primary constraint mechanisms—institutional risk management, funding pressures, peer pressure, public backlash risks, anticipated self-censorship, publication barriers, and expanding ethics review processes—across 12 major research domains in AI neuroscience. Constraint severity levels are derived from the institutional policies, research testimonies, and systematic patterns described in the preceding analysis, although readers should note that these represent interpreted patterns rather than precise empirical measurements.

The matrix reveals several critical patterns. Research domains investigating human identity characteristics—including sex differences, intelligence variations, and sexual orientation markers—face severe constraints across multiple dimensions simultaneously. Political cognition research encounters the highest peer and public pressure constraints, while behavioral prediction applications face particularly intense institutional and ethics review scrutiny. In contrast, technical neuroscience domains such as computational modeling and neural imaging methods experience minimal constraints except in areas related to dual-use concerns.

These constraint patterns help explain the "preemptive self-constraint" phenomenon examined in the following section, where researchers systematically avoid potentially valuable research directions before encountering formal limitations. The matrix illustrates why sophisticated navigation strategies have become essential for researchers working in high-constraint domains and why the emergence of interdisciplinary AI neuroscience has created new vulnerabilities to academic freedom restrictions.

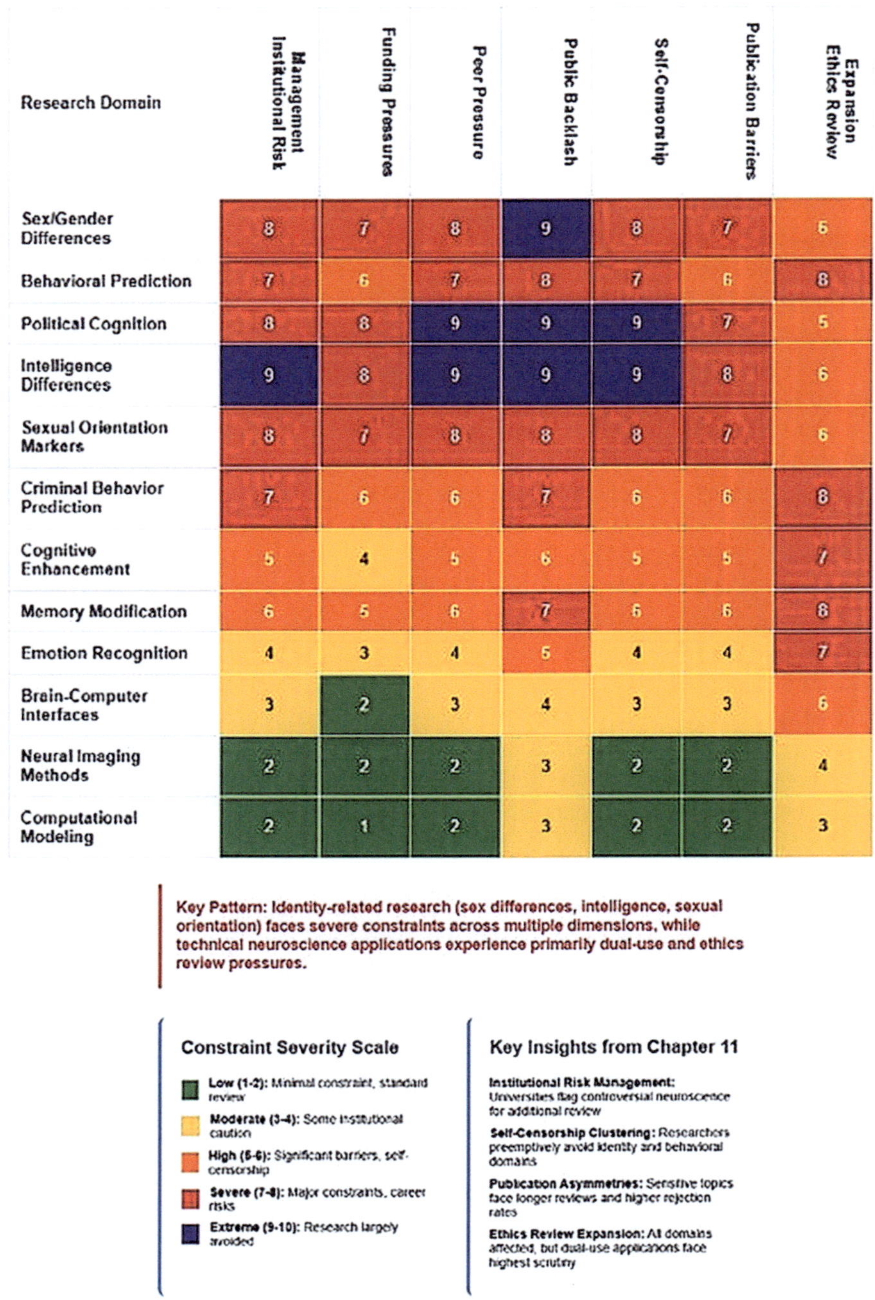

Fig. 11.1 Academic freedom constraint matrix

11.3 Self-Censorship Among Neuroscientists Working on Controversial Topics

11.3.1 The Preemptive Retreat from Certain Research Questions

Perhaps the most concerning constraint on academic freedom in neuroscience involves self-censorship. This occurs where researchers preemptively avoid potentially controversial research directions without explicit external limitation [24, 25]. Research on self-censorship in academic contexts shows that this phenomenon is widespread. Scholars deliberately avoid potentially productive research directions due to concerns about reception rather than scientific or ethical considerations.

This self-censorship particularly affects AI applications in neuroscience [26]. Scientific capacity significantly exceeds actual research activity in areas where AI analysis of neural data relates to group differences, behavioral prediction, cognitive enhancement, and identity formation. As one researcher confided, "We have the tools to investigate these questions rigorously, but the professional risk simply isn't worth it—especially for early-career researchers."

The consequences of this self-censorship extend beyond individual research trajectories to shape the broader scientific landscape [27]. Analysis reveals that in politically sensitive research areas, hypotheses aligned with certain perspectives are significantly more likely to be investigated than competing hypotheses, suggesting systematic bias in hypothesis selection shaped by anticipated reception rather than scientific merit.

This self-censorship creates particular challenges for reproducibility and scientific correction [28]. Studies document systematic decreases in direct replication attempts for studies touching on politically sensitive topics. Controversial but potentially flawed findings often remain unchallenged in the literature because researchers avoid the professional risks associated with engaging controversial domains, even to correct potential errors.

The process by which researchers move from initial scientific interest to self-censorship follows predictable decision pathways that reveal the systematic nature of preemptive constraint. Understanding these pathways illuminates where academic freedom is most vulnerable to erosion and where intervention might be most effective in preserving scientific autonomy.

Figure 11.2 presents a self-censorship decision tree that maps the cognitive and institutional checkpoints researchers navigate when considering potentially controversial research in AI neuroscience. Based on the patterns documented in researcher interviews and institutional analyses, the tree illustrates how initial research interests encounter successive filtering mechanisms that determine final research outcomes.

The decision tree reveals several critical insights into the self-censorship process. First, the most consequential filtering occurs at the initial "mental map consultation" stage, where researchers apply internalized frameworks of acceptable versus risky research domains. This cognitive filter operates before any formal institutional

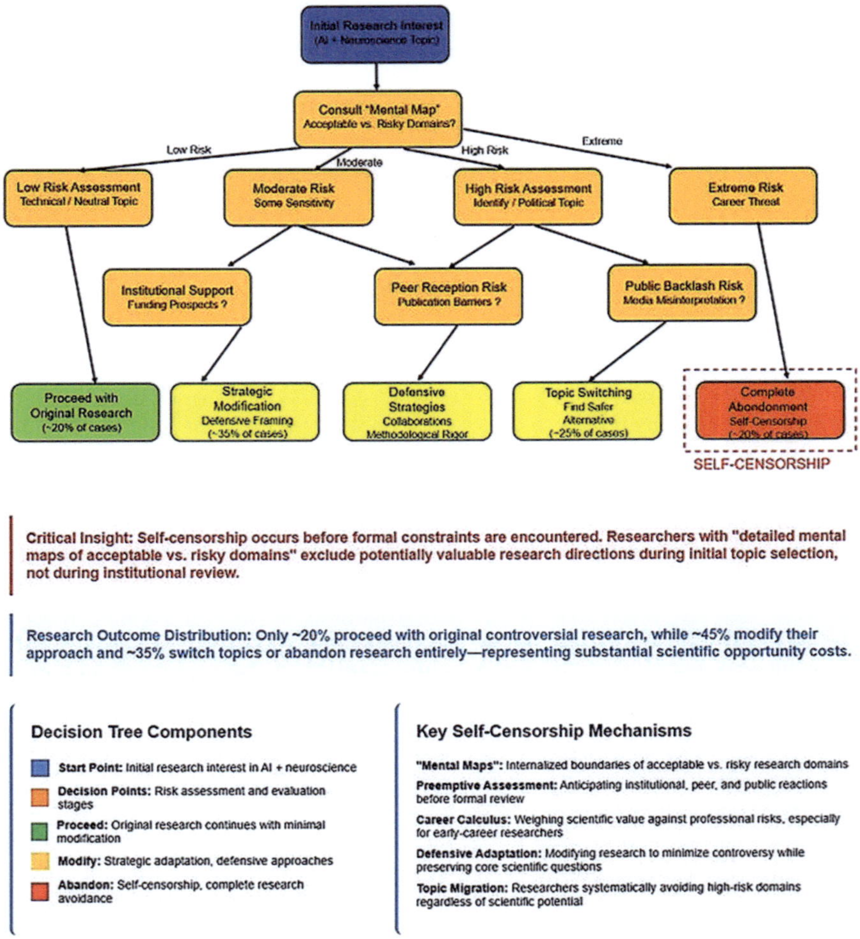

Fig. 11.2 Self-censorship decision tree in AI neuroscience research

review, suggesting that academic freedom constraints have become internalized as professional common sense rather than external imposition.

Second, the tree illustrates how risk assessment becomes increasingly granular as perceived controversy increases. Researchers considering high-risk topics engage in sophisticated multidimensional evaluation of institutional support, funding prospects, peer reception, publication barriers, and public backlash potential—a cognitive burden that researchers in neutral domains rarely experience.

Third, the outcome distribution suggests that only approximately 20% of controversial research interests proceed without significant modification, while 45% undergo strategic adaptation and 35% result in topic switching or complete abandonment. This pattern represents substantial scientific opportunity costs, as potentially valuable research directions are systematically excluded from investigation based on anticipated rather than actual constraints.

The decision tree framework helps explain why the navigation strategies examined in subsequent sections become essential for researchers working in contested domains. It also illuminates why individual resilience, while important, cannot fully address systemic self-censorship patterns that operate through internalized professional norms rather than explicit prohibition.

11.3.2 Navigating Publication Challenges and Editorial Politics

For research that overcomes initial self-censorship and institutional constraints, publication processes present additional challenges to academic freedom [29, 30]. Analysis comparing editorial outcomes for methodologically matched papers addressing politically sensitive versus neutral topics found that papers investigating politically sensitive topics faced significantly longer review times, more revision demands, and higher final rejection rates than methodologically comparable papers on neutral topics.

These publication challenges are particularly acute for AI neuroscience research [31]. Studies document "asymmetric scrutiny" in peer review processes, where papers applying AI methods to politically sensitive questions receive significantly more methodological critique than papers applying identical methods to neutral topics. This pattern suggests that methodological standards may be applied differently depending on a study's political implications rather than its scientific merit.

Publication bias remains a pervasive issue in the scientific literature, where the outcome of experiments or studies influences decisions about publication or distribution [32, 33]. Studies consistently show that papers with statistically significant results are approximately three times more likely to be published than those with null results, creating systematic distortions in the published literature [34]. This bias is particularly problematic when combined with political sensitivities, as it can prevent important negative or inconclusive results from reaching the scientific community.

Editorial bias extends beyond simple preference for positive results to include factors related to authors' characteristics, institutional affiliations, or the geographical origin of research [35]. Analysis of editorial processes reveals systematic patterns where factors unrelated to scientific quality—including the perceived political implications of findings—influence publication decisions [36].

Many researchers have developed sophisticated strategies for navigating these publication challenges, including strategic framing, anticipatory limitation acknowledgment, and careful contextual positioning [37]. These approaches require significant additional effort that researchers in less-contested areas do not face, representing a hidden cost of working in politically sensitive domains.

11.3.3 Strategic Communication and Framing of Research Findings

Concerns about public and media interpretation significantly shape how neuroscientists communicate their work [38]. The emergence of "defensive communication strategies" among neuroscientists working on politically sensitive topics includes sophisticated approaches to research communication designed to minimize misinterpretation and backlash, including preemptive acknowledgment of limitations, careful avoidance of causal language, and strategic understatement of potentially controversial implications.

These communication challenges intensify for research involving artificial intelligence and neuroscience [39]. Media coverage of AI neuroscience findings shows systematic amplification of deterministic interpretations beyond what original research supported. Media reports are significantly more likely to present results as definitive and deterministic than the original researchers' characterizations, creating significant concerns for researchers about how their work might be portrayed.

The anticipation of public misinterpretation creates a "communication chill," where researchers significantly limit public engagement and adopt increasingly technical, qualified language that reduces accessibility [40]. Surveys find that most researchers working on politically sensitive topics report declining media interviews or public speaking opportunities specifically due to concerns about misrepresentation, compared to much lower rates among those working on politically neutral topics.

11.4 Regional Variations in Scientific Freedom

11.4.1 Democratic Contexts: Political Polarization and Research Constraint

The specific manifestations of constraints on academic freedom vary significantly across different political and cultural contexts [4, 41]. Within democratic societies, distinctive patterns of constraint emerge. These are shaped by political polarization and competing interest group pressures. While formal protections for academic freedom remain strong, actual research autonomy faces significant "democratic constraints" through funding allocations, political oversight, and public pressure campaigns that effectively limit certain research directions.

These democratic constraints often operate through legislative mechanisms [42, 43]. Analysis of legislative actions affecting research across multiple jurisdictions reveals instances of targeted funding restrictions, mandatory review processes, or explicit prohibitions that limit specific research domains. These interventions follow systematic patterns reflecting broader culture war topics, with restrictions from different political orientations targeting different categories of research.

Public universities face particular challenges in this domain [44]. Comparative analysis of research productivity in politically sensitive areas across matched public and private institutions shows that public universities demonstrate lower publication rates in controversial domains despite similar overall research productivity. Public institutions' direct accountability to elected officials creates stronger incentives for constraining potentially controversial research, even without explicit political directive.

Media ecosystems significantly influence these dynamics [45]. Studies document "controversy cascades," where initial criticism of research in partisan media triggers amplification across multiple platforms, eventually generating political attention and institutional response. Media coverage functions as a key mechanism translating political polarization into constraints on academic freedom, with researchers facing intensifying pressure as controversy moves from specialized to mainstream outlets and eventually to political forums.

11.4.2 Authoritarian Contexts: Direct Intervention and Strategic Research Alignment

In more authoritarian contexts, constraints on academic freedom take distinctly different forms [46, 47]. Research reveals how scientists in these environments develop sophisticated strategies to pursue scientific inquiry while navigating explicit state constraints, including strategic research framing, coded language in publications, and international collaboration structures that enable work that might face domestic limitations.

These authoritarian constraints typically involve more direct intervention than in democratic contexts [48]. Formal approval requirements for research in several authoritarian states include proposals undergoing explicit political review alongside scientific evaluation. Analysis of internal review documents reveals systematic patterns of research approval and rejection that align with state priorities rather than scientific merit, with particularly strong constraints on research involving cognitive processes related to authority perception, moral reasoning, and group identity.

The application of artificial intelligence to neuroscience faces distinctive challenges in these contexts [49]. Authoritarian states strategically promote certain AI applications while constraining others. Substantial state investment occurs in AI applications related to surveillance, behavioral prediction, and cognitive assessment, alongside explicit limitations on research investigating individual differences, political cognition, or autonomous decision-making.

International collaboration creates both opportunities and challenges in these contexts [50]. Cross-national research partnerships function as "academic freedom bridges" that enable research that might face domestic constraints. Researchers strategically structure international collaborations to conduct data collection or analysis in locations with fewer constraints, while publications carefully frame findings to avoid domestic political sensitivity.

11.4.3 Global Collaboration Challenges and Opportunity Gaps

Beyond variations in direct constraints, significant differences in research capacity, technological access, and training opportunities create "academic freedom capability gaps" across global contexts [51, 52]. Dramatic disparities exist in access to advanced neuroimaging equipment, high-performance computing resources, and specialized training—all essential for contemporary AI neuroscience research. These disparities effectively limit which researchers can pursue certain questions regardless of formal academic freedom protections.

These capability gaps intersect with political constraints to create complex global research landscapes [53]. Variations in data protection regulations, AI governance frameworks, and neurotechnology oversight create "regulatory arbitrage opportunities" that shape global research flows. Certain types of research migrate to jurisdictions with fewer constraints, creating geographic concentrations of potentially controversial research that reflect regulatory differences rather than scientific expertise.

International research governance adds additional complexity to this landscape [54]. Emerging global frameworks for AI governance and neurotechnology regulation differentially impact researchers across regions. Harmonization efforts often reflect the priorities and capabilities of wealthier nations, creating disproportionate compliance burdens for researchers in lower-resource contexts who may lack the institutional infrastructure to navigate complex international requirements alongside local constraints.

11.5 Particular Vulnerability of Interdisciplinary AI/Neuroscience Work

11.5.1 Methodological Complexity and Public Misunderstanding

The intersection of artificial intelligence and neuroscience faces distinctive academic freedom challenges. These arise due to methodological and conceptual factors [55, 56]. The combined technical sophistication of AI methods and neuroscience content creates exceptional barriers to accurate public understanding. Research combining AI and neuroscience experiences significantly more major interpretive errors in mainstream reporting than similarly complex research in either field individually, creating significant reputational risks that discourage work in this domain.

This technical complexity intersects with powerful metaphors that shape public interpretation [57]. Analysis shows how terminology at the AI-neuroscience intersection triggers "imagination cascades," where terms like "neural networks," "machine learning," and "brain reading" activate science fiction narratives in public discourse regardless of a study's actual capabilities or implications. These metaphorical associations create significant vulnerability to misinterpretation that researchers struggle to counteract even with careful methodological explanation.

The methodological complexity of AI neuroscience research also creates distinctive peer review challenges [58]. Manuscripts combining advanced AI methods with sophisticated neuroscience face difficulty finding reviewers with sufficient expertise in both domains. These papers often receive evaluations focusing predominantly on either the AI methodology or the neuroscience content but rarely both with equal sophistication, creating quality control challenges that can undermine research development.

11.5.2 Dual-Use Concerns and Research Constraint

AI neuroscience research faces particular vulnerability to academic freedom constraints due to dual-use concerns—the potential for technologies developed for beneficial purposes to be repurposed for problematic applications [59, 60]. The expansion of "anticipatory governance" mechanisms increasingly subjects AI neuroscience research to specialized reviews based on potential rather than actual applications. Research combining AI methods with neural data faces significantly more security and ethics review requirements than comparable research in either field individually, creating substantial administrative burdens that functionally constrain certain research directions.

These dual-use concerns operate through multiple governance mechanisms [61, 62]. Export control regulations, research security protocols, and specialized ethics reviews create "layered constraint" on certain AI neuroscience research domains. Technologies enabling brain-computer interfaces, emotion recognition, and cognitive state detection face particularly intensive scrutiny based on potential military, surveillance, or security applications, regardless of researchers' actual intentions or the beneficial applications these same technologies might enable.

The international dimensions of these concerns create additional complexity [63]. Varying approaches to "technologies of concern" across different national contexts create "security friction" in international collaboration. Cross-national research partnerships face escalating challenges when combining AI methods with neural data, with researchers navigating contradictory requirements for data sharing, algorithm development, and publication review that significantly constrain collaborative research on certain topics.

11.5.3 Ethical Complexity and Researcher Vulnerability

Beyond security concerns, the ethical complexity of AI neuroscience research creates distinctive academic freedom challenges [64, 65]. "Ethics expansion" in research oversight involves legitimate ethical review processes continuously broadening their scope beyond traditional research ethics focusing on participant protection. Analysis of ethics review criteria shows systematic expansion to include considerations of potential social impacts, alignment with social justice principles, and anticipated applications—all of which particularly affect AI neuroscience research.

This ethical complexity manifests through institutional review processes [66]. Research documents "normative criteria creep"—the gradual incorporation of reviewers' values and social preferences alongside protection-oriented ethical assessment. Protocols involving AI analysis of neural data related to sensitive human characteristics face significantly longer review times and more extensive modification requirements, with requests often extending beyond participant protection to include broader social and political considerations.

These expanding ethical reviews create distinctive burdens for certain research areas [67]. Comparative analysis of ethics review processes for different AI applications finds that research related to cognitive enhancement, behavior prediction, emotion detection, and identity assessment faces particularly intensive scrutiny. Cases exist where ethics review processes effectively function as mechanisms of research authorization rather than participant protection, with committees making determinations about which questions deserve investigation regardless of methodology.

11.6 Case Studies of Researchers Navigating Political Pressures

11.6.1 The Sex Differences Controversy

Individual case studies provide valuable insights into how researchers practically navigate political pressures while maintaining academic freedom. The experience of researchers conducting studies on sex differences in neural responses exemplifies these dynamics. Such researchers have developed sophisticated strategies for maintaining scientific inquiry in politically contested terrain, including what could be termed "resilient research framing"—consistently emphasizing clinical applications of understanding differences in stress-related conditions like PTSD and anxiety disorders, which helps maintain support for work despite broader political controversies.

Navigation strategies extend beyond framing to methodology. Researchers systematically incorporate what might be called "methodological defense in depth"—employing multiple converging methods, preregistering analytical approaches, and using particularly rigorous statistical standards. These methodological choices serve both scientific and strategic purposes, strengthening research while simultaneously providing protection against potential criticisms motivated by political rather than scientific concerns.

Publication strategies form another key element of successful navigation. Strategic choices in publication venues, timing, and coauthorship arrangements are crucial. Successful researchers carefully sequence publications to establish methodological foundations before presenting more controversial findings, strategically collaborate with diverse coauthors bringing different perspectives, and select journals known for rigorous methodological review rather than ideological alignment.

11.6.2 Algorithmic Decision-Making Research

The experience of researchers studying potential biases in AI decision-making systems provides another instructive case study. Such work becomes entangled in broader controversies about algorithmic bias, technological governance, and political polarization around "fairness" concepts. Successful navigation involves what could be termed "balanced vulnerability positioning"—designing research that equally challenges perspectives across the political spectrum, providing some protection against accusations of partisan motivation.

Research communication presents distinctive challenges in this domain. Strategic approaches to publishing politically sensitive findings include "layered release communication"—where technical details are published in specialized venues, while broader implications are discussed through carefully staged engagement with various stakeholders. This approach helps maintain scientific integrity while managing political reception, allowing important findings to enter both scientific and public discourse without immediate politicization.

Institutional navigation forms another key element of success. Sophisticated "stakeholder management" techniques for maintaining institutional support amid controversy include strategically aligning different aspects of research with various institutional priorities and stakeholder interests and creating coalitions of support that provide protection against political pressure from any single direction.

11.6.3 Neuroethics at the Frontier

The experience of researchers at the frontiers of memory modification technologies offers insights into proactive navigation of ethical and political complexities. These researchers develop distinctive approaches of integrating ethical analysis directly into research programs rather than treating it as an external constraint. What could be termed "anticipatory ethical engineering" involves incorporating ethical considerations into research design from the earliest stages and proactively engaging potential concerns throughout the research process.

Successful approaches include innovative stakeholder engagement practices. Research teams practice "embedded perspectives consultation"—involving individuals with diverse viewpoints, including potential patients, ethical critics, and policy stakeholders, throughout the research process rather than only at dissemination stages. This approach helps identify and address potential concerns early while building broader support for research, effectively expanding rather than constraining academic freedom by reducing opposition.

Publication strategies reflect this integrated approach through parallel publication—simultaneously publishing technical findings in scientific journals and ethical analyses in humanities venues. This approach enables the maintenance of scientific productivity while demonstrating serious engagement with broader implications, establishing legitimacy across multiple communities and providing protection against accusations of ethical insensitivity.

11.7 Conclusion: Preserving Academic Freedom in Contested Terrain

The landscape of academic freedom in AI neuroscience research presents a complex picture of constraints, adaptations, and resilience [68, 69]. Academic freedom in contentious domains exists on a continuum of constraint and possibility shaped by multiple actors within complex ecosystems of influence. Addressing challenges to academic freedom requires attention to multiple levels, from individual researcher practices to institutional policies to broader governance structures.

At the individual level, the importance of "academic freedom literacy"—researchers' understanding of their rights, responsibilities, and navigation strategies when working in politically sensitive domains—cannot be overstated [70]. Research with early-career scientists demonstrates that those with explicit training in academic freedom principles and practical navigation strategies show greater resilience when facing political pressure and are less likely to abandon potentially valuable but controversial research directions.

At the institutional level, the critical role of "buffer structures"—governance mechanisms that insulate scientific evaluation from political pressures while maintaining appropriate ethical oversight and accountability—is essential [71]. Universities with clearly delineated boundaries between scientific assessment and value-based considerations, transparent processes for research evaluation, and meaningful faculty governance show stronger protection for controversial but methodologically sound research.

At the systemic level, the importance of "ecosystem diversity"—maintaining multiple approaches to knowledge production rather than allowing any single model to dominate—provides essential resilience [72]. Academic freedom is most effectively protected in systems where various funding sources, institutional types, and governance models coexist, creating multiple pathways for important research to advance even when particular sectors face political limitations.

Perhaps most fundamentally, success requires what might be termed "engaged academic freedom"—an approach that preserves essential scientific autonomy while acknowledging legitimate ethical considerations and broader societal contexts [73]. Academic freedom is most effectively protected not through isolation from political and ethical concerns but through sophisticated engagement that maintains scientific integrity while recognizing the complex social contexts in which neuroscience operates.

The dynamic interplay between scientific inquiry, ethical consideration, and political context will continue to shape academic freedom in AI neuroscience research. By developing nuanced understanding of these interactions and implementing thoughtful approaches at multiple levels, the field can maintain the essential freedom of inquiry that enables scientific progress while engaging responsibly with the broader societal implications of increasingly powerful technologies for understanding and potentially influencing the human brain.

11.8 Summary

This chapter examines the complex challenges to academic freedom facing researchers at the intersection of artificial intelligence and neuroscience. This chapter argues that while formal protections for academic freedom remain strong in democratic societies, actual research autonomy faces unprecedented pressures from multiple directions.

Key Findings

Current Threats: University administrators increasingly function as risk managers, constraining research through institutional oversight, funding limitations, and anticipation of public backlash. External funding agencies shape research directions through priority-setting that reflects political rather than purely scientific considerations. Peer and public pressure create powerful incentives to avoid controversial research domains.

Self-Censorship: Perhaps most concerning, researchers increasingly engage in preemptive self-constraint, avoiding potentially valuable but controversial research directions before encountering formal limitations. This affects AI applications in neuroscience particularly strongly, creating "investigation deficits" where technical capability far exceeds actual research activity in sensitive domains.

Regional Variations: Academic freedom constraints manifest differently across political contexts. Democratic societies experience "democratic constraints" through funding allocations and political oversight, while authoritarian contexts involve more direct intervention and strategic research alignment. Global collaboration faces increasing challenges from varying regulatory frameworks.

Unique Vulnerabilities: AI neuroscience research faces distinctive challenges due to methodological complexity that enables public misunderstanding, dual-use concerns about potential military or surveillance applications, and expanding ethical review processes that extend beyond traditional participant protection.

Navigation Strategies: Through case studies, this chapter illustrates how successful researchers develop sophisticated strategies, including "resilient research framing," methodological rigor, strategic publication approaches, and proactive ethical engagement, to maintain scientific inquiry while managing political pressures.

Conclusion: This chapter advocates for "engaged academic freedom"—preserving essential scientific autonomy through sophisticated engagement with ethical and political concerns rather than isolation from them. This requires coordinated efforts at the individual, institutional, and systemic levels to maintain the freedom of inquiry essential to scientific progress while acknowledging legitimate societal concerns about increasingly powerful brain technologies.

References

1. Hutchens NH, Fernandez F. Academic freedom in higher education: historical perspectives and contemporary challenges. In: Perna LW, editor. Higher education: handbook of theory and research, vol. 38. Springer; 2023. p. 1–54.
2. Väliverronen E, Saikkonen S. Freedom of expression challenged: scientists' perspectives on hidden forms of suppression and self-censorship. Sci Technol Hum Values. 2021;46(2):271–98.
3. Carter J. Government scientists are censoring themselves. Scientific American Blog Network; 2019. Retrieved from https://blogs.scientificamerican.com/observations/government-scientists-are-censoring-themselves/
4. Ignatieff M. The geopolitics of academic freedom: universities, democracy and the authoritarian challenge. Daedalus. 2024;153(2):194–212.
5. Freedom House. The internationalization of universities and the repression of academic freedom. Special Report; 2020.
6. Abraham JM, Braughler S, Kabanova L, Kollinger J. Risk management: an accountability guide for university and college boards. 2nd ed. Association of Governing Boards of Universities and Colleges; 2020.
7. Moynihan D, Herd P. Institutionalizing politicized science. Science. 2025;388(6681):683.
8. Oliver AL, Holzinger I. The effectiveness of strategic political management: a dynamic capabilities framework. Acad Manag Rev. 2008;33(2):496–520.
9. Palmer K. Trump's orders disrupts academic research. Inside Higher Ed; 2025. Retrieved from https://www.insidehighered.com/news/government/science-research-policy/2025/02/03/how-trumps-executive-orders-are-disrupting
10. Audretsch DB, Fisch C, Franzoni C, Momtaz PP, Vismara S. Academic freedom and innovation. PLoS One. 2024;19(6):e0304560.
11. Lazer D, Radford J. Data ex machina: introduction to big data. Annu Rev Sociol. 2017;43:19–39.
12. Stephan P. How economics shapes science. Harvard University Press; 2012.
13. Geiger RL, Sa CM. Tapping the riches of science: universities and the promise of economic growth. Harvard University Press; 2008.
14. Laird FN. Sticky policies, dysfunctional systems: path dependency and the problems of government funding for science in the United States. Minerva. 2020;58(4):513–33.
15. Efimov IR, Flier JS, George RP, Krylov AI, Maroja LS, Schaletzky J, Tanzman J, Thompson A. Politicizing science funding undermines public trust in science, academic freedom, and the unbiased generation of knowledge. Front Res Metr Anal. 2024;9:1418065.
16. Kaiser J. Odds of winning NIH grants plummet as new funding policy and spending delays bite. Science. 2025; Retrieved from https://www.science.org/content/article/odds-winning-nih-grants-plummet
17. Chronicle of Higher Education Editorial Board. The ruthless politicization of science funding. Chronicle of Higher Education. 2024; Retrieved from https://www.chronicle.com/article/the-ruthless-politicization-of-science-funding
18. Meirmans S, Butlin RK, Charmantier A, Engelstädter J, Groot AT, King KC, Kokko H, Reid JM, Neiman M. Science policies: how should science funding be allocated? An evolutionary biologists' perspective. J Evol Biol. 2019;32(8):754–68.
19. Gibson JL, Sutherland JL. Keeping your mouth shut: spiraling self-censorship in the United States. Political Sci Q. 2023;138(3):361–76.
20. Bar-Tal D. Self-censorship as a socio-political-psychological phenomenon: conception and research. Adv Polit Psychol. 2017;38(S1):37–65.
21. Clark CJ, Winegard BM, Beardslee J, Baumeister RF, Ditto PH, Gonzalez FJ, et al. Prosocial motives underlie scientific censorship by scientists: a perspective and research agenda. Proc Natl Acad Sci. 2023;120(48):e2301642120.
22. Tannenberg M. In search of self-censorship. Br J Polit Sci. 2020;52(1):221–40.
23. Delborne JA. Transgenes and transgressions: scientific dissent as heterogeneous practice. Soc Stud Sci. 2008;38(4):509–41.

24. Columbia Climate Law. About the silencing science tracker. Sabin Center for Climate Change Law; 2019. Retrieved from https://climate.law.columbia.edu/content/about-silencing-science-tracker
25. Carter J. Why are government scientists censoring themselves? American Geophysical Union; 2019. Fall Meeting Abstracts, 2019, PA33B-04
26. Clark CJ. Interview: the threat of self-censorship in science. Undark Magazine. 2023; Retrieved from https://undark.org/2023/12/15/interview-self-censorship/
27. Loury GC. Self-censorship in public discourse: a theory of "political correctness" and related phenomena. Ration Soc. 1994;6(4):428–61.
28. Fanelli D. Negative results are disappearing from most disciplines and countries. Scientometrics. 2012;90(3):891–904.
29. Song F, Parekh S, Hooper L, Loke YK, Ryder J, Sutton AJ, et al. Dissemination and publication of research findings: an updated review of related biases. Health Technol Assess. 2010;14(8):iii–193.
30. Rosenthal R. The file drawer problem and tolerance for null results. Psychol Bull. 1979;86(3):638–41.
31. Jennings RG, Van Horn JD. Publication bias in neuroimaging research: implications for meta-analyses. Neuroinformatics. 2012;10(1):67–80.
32. Sterling TD. Publication decisions and their possible effects on inferences drawn from tests of significance—or vice versa. J Am Stat Assoc. 1959;54(285):30–4.
33. Dickersin K. The existence of publication bias and risk factors for its occurrence. JAMA. 1990;263(10):1385–9.
34. Franco A, Malhotra N, Simonovits G. Publication bias in the social sciences: unlocking the file drawer. Science. 2014;345(6203):1502–5.
35. Matías-Guiu J, García-Ramos R. Editorial bias in scientific publications. Neurologia. 2011 Jan-Feb;26(1):1–5.
36. International Science Council. Advisory note on bias in science publishing. 2011. Retrieved from https://council.science/publications/advisory-note-on-bias-in-science-publishing/
37. Young NS, Ioannidis JP, Al-Ubaydli O. Why current publication practices may distort science. PLoS Med. 2008;5(10):e201.
38. Nair AS. Publication bias – importance of studies with negative results! J Family Med Prim Care. 2019;8(6):1986–7.
39. Serra-Garcia M, Gneezy U. Nonreplicable publications are cited more than replicable ones. Sci Adv. 2021;7(21):eabd1705.
40. Krawczyk M, Reuben E. (Un)available upon request: field experiment on researchers' willingness to share supplementary materials. Account Res. 2012;19(3):175–86.
41. Kinzelbach K, Saliba I, Spannagel J, Quinn R. Academic freedom index 2023 update. FAU Erlangen-Nürnberg & V-Dem Institute; 2023.
42. Fernandez F, Hutchens N. Restrictions on US academic freedom affect science everywhere. Nat Hum Behav. 2025 Jul;9(7):1303–4.
43. Vögtle EM, Windzio M. Academic freedom and international student mobility: how political constraints affect global higher education. High Educ. 2024;87(2):433–52.
44. Cohen GA. If you're an egalitarian, how come you're so rich? Harvard University Press; 2000.
45. Nutley SM, Walter I, Davies HT. Using evidence: how research can inform public services. Policy Press; 2007.
46. Scholars at Risk. Free to think 2018: report of the scholars at risk academic freedom monitoring project. Scholars at Risk Network; 2018.
47. Cole J. Toward a more perfect university. PublicAffairs; 2017.
48. Carrozza I, Marsh N, Reichberg G. Dual-use AI technology in China, the US and the EU: strategic implications for the balance of power. PRIO Paper; 2022.
49. Krauel E, Frewer A. Dual use concerns in artificial intelligence and the neurosciences: how medical research can end up in war. Med Health Care Philos. 2025;28(1):1–12.
50. Freedom House. From democratic decline to authoritarian aggression: nations in transit 2022. Freedom House; 2022.

51. UNESCO. UNESCO science report: the race against time for smarter development. UNESCO Publishing; 2021.
52. OECD. OECD science, technology and innovation outlook 2018: adapting to technological and societal disruption. OECD Publishing; 2018.
53. Schmid S, Riebe T, Reuter C. Dual-use and trustworthy? A mixed methods analysis of AI diffusion between civilian and Defense R&D. Sci Eng Ethics. 2022;28(2):12.
54. European Commission. Artificial Intelligence Act. European Union; 2021.
55. Ienca M, Andorno R. Towards new human rights in the age of neuroscience and neurotechnology. Life Sci Soc Policy. 2017;13(1):1–27.
56. Greely HT. Implementing the human right to science in neuroscience. J Law Biosci. 2024;11(1):lsae011.
57. Botvinick M, Cohen J. Rubber hands 'feel' touch that eyes see. Nature. 2014;391(6669):756.
58. Ienca M, Ferretti A, Hurst S, Puhan M, Lovis C, Vayena E. Considerations for ethics review of big data health research: a scoping review. PLoS One. 2018;13(10):e0204937.
59. Brenneis A. Assessing dual use risks in AI research: necessity, challenges and mitigation strategies. Med Health Care Philos. 2025;28(1):1–15.
60. Krauel E, Frewer A. Dual use concerns in artificial intelligence and the neurosciences: how medical research can end up in war. ResearchGate; 2024. Preprint
61. National Academies of Sciences, Engineering, and Medicine. Dual use research of concern in the life sciences: current issues and controversies. The National Academies Press; 2017.
62. Bureau of Industry and Security. Request for comments concerning the imposition of export controls on certain brain-computer interface (BCI) emerging technology. Fed Regist. 2021;86(205) Retrieved from https://www.federalregister.gov/documents/2021/10/26/2021-23256/request-for-comments-concerning-the-imposition-of-export-controls-on-certain-brain-computer
63. Altmann J, Sauer F. Autonomous weapons systems and strategic stability. Survival. 2017;59(5):117–42.
64. Yuste R, Goering S, Arcas BAY, Bi G, Carmena JM, Carter A, et al. Four ethical priorities for neurotechnologies and AI. Nature. 2017;551(7679):159–63.
65. Hartung T, Morales Pantoja IE, Smirnova L. Brain organoids and organoid intelligence from ethical, legal, and social points of view. Front Artif Intell. 2024 Jan;5(6):1307613.
66. Emanuel EJ, Wendler D, Grady C. What makes clinical research ethical? JAMA. 2000;283(20):2701–11.
67. Ienca M. Neuroethics meets artificial intelligence. In: The Neuroethics Blog. International Neuroethics Society; 2021.
68. Hammersley M. Is the evidence-based practice movement doing more good than harm? Reflections on Iain Chalmers' case for research-based policy making and practice. Evid Policy. 2005;1(1):85–100.
69. Nutley S, Jung T, Walter I. The many forms of research-informed practice: a framework for mapping diversity. Camb J Educ. 2008;38(4):501–17.
70. Cairney P. The politics of evidence-based policy making. Springer; 2016.
71. Parsons W. From muddling through to muddling up—evidence based policy making and the modernisation of British government. Pub Policy Adm. 2002;17(3):43–60.
72. Davies, H., Nutley, S., & Smith, P. (Eds.). (2000). What works?: evidence-based policy and practice in public services. Policy Press.
73. Sanderson I. Evaluation, policy learning and evidence-based policy making. Public Adm. 2002;80(1):1–22.

The Politics of Brain Science 12

Science is always political. But it should not be partisan.

—Rush D. Holt

12.1 Introduction

Contemporary neuroscience faces significant challenges when research findings intersect with political discourse. Scientific advances in brain-computer interfaces, for example, may be interpreted differently across political perspectives—some viewing developments as concerning technological overreach, while others see them as promising medical breakthroughs. Basic neuroscientific findings increasingly encounter political interpretation rather than evaluation purely on empirical merits. This dynamic reflects a broader challenge: conducting and communicating research about human brain function within political environments that influence funding priorities, research directions, and public understanding. The intersection of neuroscience with political discourse creates complex considerations for researchers seeking to advance scientific knowledge while navigating varied stakeholder perspectives and societal concerns about the implications of brain research.

Few scientific domains are as thoroughly intertwined with politics as brain science. Research into the physical organ that generates our thoughts, emotions, beliefs, and behaviors inevitably intersects with fundamental questions about human nature, free will, identity, and social organization that have been politically contested throughout history. As neuroscience has developed increasingly sophisticated tools to investigate the brain, and artificial intelligence has provided new capabilities for modeling and potentially influencing neural processes, these political dimensions have intensified rather than diminished [1]. The study of the brain now operates within complex political ecosystems that shape which questions receive attention, how findings are interpreted, and what technologies are developed or constrained.

T. Rudroff, *AI for the Healthy Brain and Neurological Disorders*, Neurocultural Health and Wellbeing, https://doi.org/10.1007/978-3-032-11596-6_12

This chapter examines the multifaceted relationship between politics and brain science across several domains. We begin by exploring the historical context of political interference in brain research, documenting how various political ideologies have attempted to bend neuroscience to support particular worldviews. We then analyze current political debates surrounding neuroscience and artificial intelligence, identifying key areas of contemporary contention [2, 3]. This chapter examines how political climates influence research funding priorities, creating paths of least resistance that shape scientific attention [4, 5]. We explore the particular challenges of communicating neuroscience findings in polarized environments where technical precision often falls victim to ideological simplification [6, 7]. Finally, we analyze international variations in political approaches to neuroscience, revealing how different governance systems create distinctive opportunities and constraints for brain research [8, 9].

12.2 Historical Context of Political Interference in Brain Research

12.2.1 From Phrenology to Racial Science: Political Exploitation of Early Neuroscience

The medical museum's display case contained a collection of measuring calipers, carefully preserved notebooks filled with precise skull measurements, and anatomical drawings from the late nineteenth century. The accompanying placard explained how these seemingly objective scientific instruments had been deployed to develop elaborate theories of brain structure that conveniently "proved" the racial and gender hierarchies of their era. What appeared at first glance to be neutral scientific tools were in fact deeply political artifacts. They revealed how emerging brain science had been shaped to reinforce existing power structures and social arrangements. The exhibit served as a sobering reminder of how political ideologies have repeatedly co-opted brain research, transforming tentative findings and speculative theories into "scientific" justifications for social hierarchies, discrimination, and even genocide.

The relationship between politics and brain science has deep historical roots, with various political movements appropriating neurological concepts to support preferred social arrangements. Phrenology, the pseudoscientific practice of measuring skull shapes to predict mental traits, became deeply entangled with political projects from its inception in the early 1800s [10]. Research has documented how phrenological techniques were systematically deployed to support racial hierarchies and colonial projects, with skull measurement and brain size comparisons providing seemingly objective validation for predetermined political conclusions about racial differences [11, 12].

This political appropriation manifested most dramatically in racial science. The flawed use of "science" to support the exploitation of groups of people helped perpetuate racial oppression and distorted future views of the biological basis of race

[13]. Historical analysis demonstrates how these "scientific" approaches were thoroughly entangled with political projects of the nineteenth and early twentieth centuries, with scientific methodologies explicitly shaped to produce results that aligned with predetermined political conclusions about racial differences [14, 15].

Gender differences in brain structure and function provided another domain for political deployment of neuroscience. Phrenological beliefs about women's supposedly smaller and weaker brains were used to justify exclusion from education and professional opportunities, with practitioners openly acknowledging the convenient alignment between their findings and prevailing social arrangements [10]. Women whose heads were generally larger in the back with lower foreheads were thought to have underdeveloped organs necessary for success in the arts and sciences while having larger mental organs relating to the care of children and religion [10].

These historical cases demonstrate what scholars term "political neurological determinism"—the selective application of brain science to suggest that existing social hierarchies reflect natural, biologically determined realities rather than contingent power arrangements [16]. The analysis shows how political deployment of brain science typically involves extrapolating far beyond actual evidence while emphasizing deterministic interpretations that minimize possibilities for social change or reorganization.

Understanding this historical relationship provides an essential context for contemporary interactions between politics and brain science. Current political appropriations of neuroscience follow remarkably similar patterns to historical cases, with modern neuroimaging and AI applications replacing calipers and anatomical drawings but serving similar political functions [17, 18].

12.2.2 Cold War Brain Science: Ideological Battlegrounds

The declassified documents lined the archive shelves, and their once-secret stamps now faded with historical artifacts. Intelligence reports detailed Soviet investments in "psychotronic research," while correspondence between CIA officials discussed urgent needs for countermeasures against "communist mind control techniques." Nearby folders contained grant applications for seemingly basic neuroscience research that carefully concealed military applications behind technical language and academic frameworks. Together, the documents revealed how Cold War geopolitics had transformed brain science into an ideological battleground, with both superpowers racing to develop neurological technologies that might provide strategic advantages. The dusty papers illustrated how political contexts fundamentally shape not just how neuroscience is interpreted but which questions researchers pursue and how their work is funded, classified, and controlled.

The Cold War period marked a distinctive chapter in the relationship between politics and brain science, with competing ideological blocs attempting to harness neurological research for geopolitical advantage. Brain science became thoroughly entangled with national security concerns during this period, with unprecedented government investment and oversight of neuroscience research in both Western and

Soviet contexts [19]. Analysis of declassified materials reveals how political and military priorities directly shaped research programs, creating distinct trajectories based on ideological frameworks rather than purely scientific considerations.

This political influence manifested in distinctive research priorities. Soviet neuroscience emphasized environmental and social influences on brain development in alignment with Marxist ideology, while Western approaches more frequently emphasized innate characteristics and individual differences [20]. Comparative analysis of publication patterns demonstrates how these broader ideological frameworks shaped which research questions were pursued, how findings were interpreted, and what applications were developed.

The national security dimensions of this period created unprecedented entanglement between neuroscience and intelligence agencies. The expansion of covert funding of brain research by both American and Soviet intelligence services, with programs like MKUltra representing extreme cases of political interference in neuroscientific research [21]. Analysis of declassified documents reveals how security concerns led to secretive research programs operating outside normal scientific practices of peer review and open publication, creating parallel tracks of classified neuroscience driven explicitly by political and military objectives.

The Cold War period also demonstrated how political ideologies shaped theoretical frameworks in brain science. Competing political systems influenced basic conceptual approaches to neuroscience, with dialectical materialism shaping Soviet theoretical models while individualist frameworks dominated Western approaches [22]. This comparative analysis shows how these ideological differences influenced not just research applications but fundamental conceptual models of how the brain functions and develops.

12.2.3 The Neuropolitics of Behavior Control and Social Regulation

The museum exhibition on "Mental Hygiene and Public Health" displayed a striking collection of artifacts: restraint devices from early twentieth-century psychiatric institutions, pamphlets promoting forced sterilization of the "feebleminded," promotional materials for lobotomy procedures showing smiling patients, and government posters warning against the "mental defectives" supposedly threatening social order. A video installation showed historical footage of electroconvulsive therapy being administered to screaming patients, alongside governmental hearings where psychiatrists assured legislators that these procedures were humane and necessary. The exhibit chronicled how brain science had repeatedly been deployed by political systems to control "deviant" behavior and enforce social norms, revealing the dark history of how neuroscientific theories and technologies have been used to regulate populations according to prevailing political priorities.

Beyond geopolitical competition, neuroscience has frequently been deployed as a mechanism of social control and behavior regulation aligned with political objectives. Extensive historical cases document where brain science was explicitly harnessed to identify, classify, and control populations deemed problematic by political authorities [23]. Historical analysis reveals patterns across different political systems where neuroscientific frameworks provided seemingly objective justifications for controlling socially marginalized groups, from the "feebleminded" to the "mentally ill" to political dissidents.

This political deployment often operated through medicalization of social deviance. Analysis shows how psychiatric diagnostic categories evolved in relation to prevailing political concerns, with new mental disorders conveniently emerging to pathologize behaviors that challenged existing social arrangements [24]. Research documents how supposedly objective psychiatric frameworks often aligned remarkably well with political priorities, providing medical justification for controlling populations that posed challenges to existing power structures.

The technologies developed for behavior modification further illustrate this political dimension. The evolution of neurological interventions from lobotomy to electroconvulsive therapy to psychopharmacology documents how these approaches were deployed within frameworks of social control despite limited evidence for their effectiveness or safety [25]. Analysis reveals how these interventions frequently targeted populations with limited political power, with application patterns that reflected prevailing political concerns about social order rather than purely medical considerations.

This history demonstrates what researchers term "biopolitical neuroscience"—the deployment of brain science within broader systems of population management and social regulation [26]. Analysis documents how neuroscientific classifications and interventions became integrated into governmental approaches to managing populations, with political authorities selectively adopting and promoting neurological frameworks that facilitated preferred forms of social organization and control.

12.3 Current Political Debates Surrounding Neuroscience and AI

12.3.1 The Neurological Politics of Human Nature

The conference panel on "Neurological Foundations of Human Behavior" had drawn an overflow crowd, with attendees lining the walls of the hotel ballroom. What should have been a technical discussion about research methodologies had quickly transformed into something more resembling a political debate. Conservative panelists emphasized research on evolutionary constraints, innate sex differences in neural organization, and biological limits on behavioral plasticity. Progressive participants highlighted studies on neuroplasticity, environmental influences on brain development, and the limitations of binary sex categories in neurological research. Although all participants cited peer-reviewed studies, their selective attention to different research strands revealed how neuroscience has become a proxy battlefield for broader political disputes about human nature. Findings were interpreted primarily through ideological frameworks rather than evaluated on methodological merits. The scene illustrated how contemporary brain science functions not just as an empirical enterprise but as a source of politically useful evidence deployed within ongoing cultural conflicts.

Contemporary political debates frequently center on competing interpretations of neuroscientific findings about human nature. Research documents extensive political engagement with neuroscience across the ideological spectrum, with both progressive and conservative perspectives selectively emphasizing research findings that align with preferred worldviews [27]. Analysis of political commentary on neuroscience reveals systematic patterns where similar methodological limitations are either highlighted or overlooked depending on whether findings support favored political narratives about human capabilities and characteristics.

This political selectivity manifests particularly around research on innate versus environmental influences. Studies show how findings about genetic influences on behavior are differentially received across the political spectrum, with conservatives typically emphasizing studies suggesting significant genetic components while progressives highlight research demonstrating environmental malleability [28]. Research documents how identical methodological approaches receive dramatically different levels of scrutiny depending on whether their findings align with politically preferred ideas about human nature.

Gender and sexuality represent particularly contested domains for neuropolitical interpretation. Analysis of political responses to research on neurological aspects of gender identity and sexual orientation finds that identical studies are characterized as either rigorous science or ideologically motivated pseudoscience depending on the political orientation of the commentator [29]. Content analysis of media coverage reveals how methodological critiques are deployed highly selectively, with technical limitations only emphasized when findings contradict preferred political narratives.

These patterns exemplify what researchers' term "motivated neuroscientific reasoning"—the tendency to evaluate brain research primarily based on its alignment with preferred political conclusions rather than methodological rigor [30]. Experimental studies demonstrate how individuals across the political spectrum apply significantly different evaluative standards to methodologically identical studies depending on whether findings support or challenge their political worldviews, suggesting a fundamental entanglement between political commitments and scientific evaluation in this domain.

12.3.2 AI Governance and Democratic Sovereignty

Legislative discussions regarding AI systems that analyze neural data involve multiple stakeholder perspectives. Industry representatives typically advocate for minimal regulatory constraints, arguing that excessive regulation may shift development to jurisdictions with fewer restrictions. Civil liberties advocate generally emphasizing the need for democratic oversight of technologies that could affect privacy and cognitive liberty. National security officials often highlight the strategic importance of maintaining technological leadership in neurotechnology development. Public opinion reflects diverse viewpoints on the appropriate balance between innovation and regulation. These policy debates illustrate how AI applications in neuroscience intersect with broader questions about technology governance, democratic oversight, and societal values. The discussions often transcend traditional political alignments, raising fundamental questions about the relationship between technological development and human welfare.

Artificial intelligence applications in neuroscience have generated distinctive political controversies centering on questions of governance and democratic oversight. Research documents emerging political coalitions around AI neurotechnology regulation that transcend traditional partisan alignments, creating novel political configurations based on competing visions of technological governance [31]. Analysis of legislative debates across multiple democracies reveals how AI applications to neural data have catalyzed fundamental reconsiderations of the relationship between technology development and democratic control.

These governance debates focus particularly on questions of sovereignty and self-determination. Analysis shows how AI neurotechnologies raise distinctive challenges for democratic governance because they potentially influence the cognitive processes through which democratic deliberation itself occurs [32]. Research documents emerging political frameworks addressing what scholars term "cognitive democracy"—the idea that protecting democratic systems requires specific attention to technologies that might influence or manipulate the mental processes underlying citizen decision-making.

Privacy concerns represent another central focus of these political debates. Research documents how neural data privacy has emerged as a distinctive political issue that generates unusual alliances across traditional ideological boundaries [33]. Analysis of policy positions reveals how both progressive privacy advocates and conservative libertarians have found common causes in opposing certain AI applications to neural data, creating novel political configurations around what they term "cognitive liberty" protections.

12.3.3 Cognitive Enhancement and Social Justice

The university debate on "Cognitive Enhancement: Liberation or Stratification?" had drawn students from across political orientations, all engaged in a surprisingly nuanced discussion. Conservative participants argued that enhancement technologies represented individual liberty to pursue excellence while expressing concerns about threats to human dignity and traditional values. Progressive students emphasized potential benefits for marginalized communities with limited educational access while raising alarms about exacerbating existing social inequalities when only elites could afford enhancements. Libertarians defended absolute freedom to modify one's own brain while opposing any government subsidies for such technologies. Religious participants articulated both theological concerns about "playing God" and moral imperatives to heal cognitive impairments. The multidimensional conversation illustrated how neurotechnologies for cognitive enhancement have generated political discussions that transcend simple left-right categorizations, raising fundamental questions about social justice, human flourishing, and collective versus individual approaches to emerging technologies.

Cognitive enhancement technologies have become focal points for complex political debates about equity, access, and social justice. Research documents how neurotechnologies for enhancing cognitive performance have generated political controversies that cut across traditional ideological boundaries, creating distinctive political alignments based on competing values and priorities rather than partisan affiliations [34]. Analysis of public debates reveals how these technologies raise fundamental questions about fairness, opportunity, and stratification that resist simple categorization within existing political frameworks.

These debates center particularly on access and distribution questions. Analysis of competing political approaches to cognitive enhancement technologies documents fundamental tensions between market-based distribution models and equity-oriented approaches [35]. Research reveals how these technologies catalyze broader political conversations about access to opportunity in contemporary societies, with enhancement technologies becoming proxy battlegrounds for longstanding debates about meritocracy, equality, and social mobility.

Religious and traditional perspectives add additional dimensions to these political conversations. Comparative analysis reveals significant variation in how different religious and cultural traditions evaluate these technologies, from categorical opposition based on concerns about human dignity to selective adoption aligned with healing traditions and commitments to human flourishing [36].

12.4 How Political Climates Affect Research Funding Priorities

12.4.1 The Politics of National Research Agendas

National research initiatives in neuroscience involve coordination between scientific expertise and policy considerations. Scientific expert panels typically recommend research focusing on basic mechanisms, longitudinal studies, and fundamental exploration of neural functions. Policy advisers often emphasize factors such as economic competitiveness, security applications, and research with translational potential. Research agenda development involves input from various stakeholders, including pharmaceutical organizations, government research agencies, and patient advocacy groups. These stakeholders may influence research priorities and program language through established review processes. The final research priorities reflect the integration of scientific recommendations with broader societal considerations including economic, security, and health policy objectives. This process demonstrates how scientific research planning incorporates multiple types of expertise and considerations beyond purely scientific factors, balancing research merit with practical applications and societal needs.

National research priorities for neuroscience reflect political considerations that extend far beyond scientific merit. Analysis of how national brain research initiatives across multiple countries incorporate political priorities into their scientific frameworks finds that research agendas consistently reflect broader political concerns including economic competitiveness, national security, and cultural values [37]. Comparative analysis reveals how supposedly purely scientific priority-setting exercises are thoroughly shaped by political considerations that determine which aspects of brain function receive attention and resources.

These political influences operate through multiple mechanisms. Research documents how various stakeholders influence national research agendas through both formal and informal channels, from industry lobbying to security agency priorities to patient advocacy campaigns [4]. Analysis reveals the complex ecosystem of influence through which political priorities become translated into scientific funding allocations, creating patterns of research investment that reflect power arrangements and strategic priorities rather than purely scientific considerations.

The framing of brain research initiatives further demonstrates these political dimensions. Analysis of the rhetorical strategies used to justify major neuroscience investments across different political contexts finds that identical research programs are framed in dramatically different terms depending on prevailing political priorities [38]. Content analysis shows how neuroscience initiatives are variously justified through national security frameworks, economic competitiveness narratives, public health paradigms, or philosophical explorations of human nature, with framing choices reflecting strategic alignment with political priorities rather than inherent characteristics of the research itself.

12.4.2 Funding Structures and Political Influence

The young neuroscientist carefully reviewed the grant solicitation from the National Institutes of Health, noting subtle but significant changes in priorities and language from the previous year's version. References to "sex differences in neurological disorders" had been replaced with more neutral phrasing about "individual variation." The previous emphasis on "genetic factors in cognitive development" had shifted toward "environmental influences on neural plasticity." Although no explicit political directive had been issued, the message was clear to experienced researchers: certain questions were now favored while others faced higher barriers to funding. As she considered how to frame her research on hormonal influences in brain development, the neuroscientist contemplated the necessary strategic choices—emphasizing certain aspects of her work while downplaying others—required to navigate the unwritten political constraints embedded within seemingly neutral funding mechanisms. The scene illustrated how political influences shape neuroscience research, not primarily through explicit prohibition but through subtle incentive structures that create paths of least resistance guiding scientific attention toward politically favored questions.

Beyond broad agenda setting, political influences shape neuroscience through specific funding structures and allocation processes. Analysis documents how changes in research priorities often track political transitions, with shifts in emphasis, terminology, and focus areas following changes in governmental leadership even without explicit policy changes [5]. Longitudinal analysis of funding patterns demonstrates how political influences operate not primarily through direct interference but through subtle shifts in priorities and evaluation criteria that create differential incentives for certain research directions.

These political influences manifest across different funding mechanisms. Comparative analysis of how neuroscience research is shaped by various funding structures, from governmental grants to corporate support to philanthropic initiatives, reveals how each funding source incorporates distinctive political priorities and values [39]. The analysis shows how government funding emphasizes national competitiveness, industry support focuses on commercial applications, and philanthropic initiatives reflect donor preferences, collectively creating a complex ecosystem of incentives shaping research directions.

Research adaptation to these political realities represents a key mechanism through which funding structures influence scientific trajectories. Documentation of sophisticated strategies through which neuroscientists navigate politically influenced funding landscapes includes strategic framing, emphasis shifting, and selective application targeting [40]. Interviews with researchers reveal widespread awareness of political dimensions in funding allocation, with scientists developing anticipatory strategies that effectively translate between scientific questions and political priorities to maintain research support.

Recent analyses show the impact of politicization on science funding. Research demonstrates that when funding agencies use their power to further particular political or ideological agendas, they contribute to public mistrust of science and scientific institutions [5]. The current approach to linking diversity, equity, and inclusion considerations to funding decisions dilutes achievement- and merit-based criteria, meaning that funds are not necessarily used to support the best scientific ideas and projects.

12.4.3 Security Interests and Classified Neuroscience

Neuroscience conferences often bring together researchers from academic, government, and private sector organizations. Academic scientists typically focus on applications such as therapeutic interventions for paralyzed patients, while government and industry professionals may consider a broader range of potential applications. These interactions can lead to discussions about various funding opportunities, including both traditional academic grants and alternative funding mechanisms that may have different publication and disclosure requirements. The diversity of attendees reflects the multiple sectors interested in neuroscience research developments.

Academic, government, and industry stakeholders often have different priorities and constraints regarding research directions, funding structures, and the dissemination of findings. This creates distinct research ecosystems with varying approaches to supporting neuroscience investigation, each with its own requirements for research conduct and communication of results.

Security interests represent particularly powerful political influences on neuroscience research trajectories. Documentation reveals extensive involvement of defense and intelligence agencies in neuroscience funding across multiple countries, with significant investment in areas including brain-computer interfaces, cognitive enhancement, and neurological influences on decision-making [41]. Analysis shows how security interests create parallel research ecosystems with distinctive priorities, constraints, and applications that shape scientific directions in ways often invisible within published literature.

These security influences operate through multiple channels. Analysis shows how classification systems, export controls, and security-oriented funding programs create what researchers term "national security neuroscience"—research trajectories shaped primarily by perceived security applications rather than health benefits or scientific interest [42]. Research documents how these mechanisms create distinctive patterns of knowledge production, dissemination, and application that significantly influence neuroscientific development even while remaining partially obscured from public view.

Dual-use concerns particularly shape these security dimensions. Documentation shows how technologies developed for legitimate civilian applications face distinctive governance challenges when they have potential security applications, creating complex political negotiations about appropriate development, oversight, and international coordination [43]. Analysis of governance frameworks reveals how these dual-use considerations increasingly influence neuroscience research across multiple domains, from brain-computer interfaces to cognitive assessment to neuropharmacology.

12.5 The Challenge of Communicating Neuroscience Findings in Polarized Environments

12.5.1 Media Ecosystem and Scientific Distortion

Research teams often encounter challenges when their neuroscience findings are covered by media outlets. Studies documenting associations between brain activation patterns and information processing may include appropriate methodological caveats about correlation versus causation and the preliminary nature of findings. Media coverage can sometimes misrepresent these nuanced results by emphasizing simplified interpretations that align with particular perspectives. Different outlets may interpret the same research findings in ways that support contrasting viewpoints, potentially overlooking the complexity and limitations that researchers emphasized in their original work. This pattern reflects broader challenges in science communication, where carefully qualified research findings may be transformed into simplified narratives that remove essential context about methodology, limitations, and appropriate interpretation. The phenomenon illustrates the gap that can exist between how researchers present their findings and how those findings are subsequently interpreted and communicated to broader audiences.

Communicating neuroscience findings presents distinctive challenges in politically polarized media environments. Analysis of how neuroscience research is transformed as it moves from academic publications through media coverage to public understanding finds systematic patterns of simplification, exaggeration, and political reframing that significantly distort original findings [44]. Content analysis of media coverage demonstrates how the same neuroscientific studies are presented in dramatically different terms across media outlets, with interpretations shaped by the political orientation of the publication rather than the actual research findings.

These distortions follow predictable patterns across the political spectrum. Documentation shows how both conservative and progressive media outlets selectively emphasize different aspects of neuroscience findings to support preferred narratives, with conservative sources typically highlighting biological determinism while progressive sources emphasize plasticity and environmental influences [45]. Comparative analysis reveals how these selective emphasis patterns create parallel information ecosystems that present fundamentally different versions of neuroscientific findings to different audiences.

The technical complexity of neuroscience exacerbates these communication challenges. Analysis shows how methodological nuance is systematically stripped away in popular coverage of neuroscience, with statistical associations presented as causal relationships, preliminary findings framed as definitive, and speculative implications highlighted over actual results [46]. Research documents how these simplifications interact with political frameworks to create particularly problematic distortions, with technical limitations ignored when findings support preferred political narratives but emphasized when they challenge existing beliefs.

Research on science communication shows that trust in science is polarized along political lines, with people across the political spectrum holding stereotypes about scientists' political orientation that decisively affect the link between their own political orientation and their trust in scientists [47]. This effect shapes participants' perceptions of the value of science, protective behavior intentions during a pandemic, policy support, and information-seeking behavior.

12.5.2 The Challenge of Scientific Nuance in Political Discourse

Political discussions sometimes involve references to neuroscience research, particularly regarding topics like brain development and environmental influences. Candidates may cite scientific studies to support their positions, although the complexity of neuroscience research can make accurate summarization challenging in debate formats. Research on brain plasticity and development typically reveals nuanced findings that resist simple characterizations—neither supporting unlimited malleability nor rigid determinism. The scientific literature generally indicates that brain development involves both genetic factors and environmental influences, with plasticity varying across different brain regions and developmental periods. Political discourse often requires concise, definitive statements, while scientific findings typically involve methodological limitations, statistical uncertainties, and contextual dependencies. This creates challenges when translating complex research into accessible public communication. The phenomenon illustrates broader tensions between the precision required for accurate scientific communication and the clarity expected in public policy discussions.

Beyond media distortions, political discourse presents fundamental challenges for maintaining scientific nuance in neuroscience communication. Analysis of how neuroscientific findings are deployed in political debates documents systematic patterns of oversimplification, selective citation, and exaggerated certainty that transform tentative scientific findings into seemingly definitive support for political positions [48]. Analysis of political rhetoric demonstrates how the demands of political communication create powerful incentives for stripping away precisely the nuance, uncertainty, and methodological context essential to accurate scientific understanding.

These political dynamics particularly affect how causal relationships are presented. Documentation shows how political discourse systematically transforms correlational neuroscience findings into causal narratives that support preferred policy positions, with complex bidirectional relationships simplified into unidirectional causal claims [49]. Research reveals how political incentives favor presenting brain research in terms of simple cause-effect relationships regardless of whether the underlying science supports such interpretations, creating fundamental tensions between political utility and scientific accuracy.

The elimination of uncertainty represents another key challenge in political contexts. Analysis shows how appropriate scientific uncertainty is systematically removed when neuroscience enters political discourse, with preliminary findings presented as definitive and probability statements converted to binary claims [50]. Research documents how political incentives favor certainty over accuracy, creating environments where nuanced scientific positions are disadvantaged in public debates regardless of their evidential support.

12.5.3 Anticipatory Self-Constraint in Research Communication

University departments increasingly offer communications training for faculty members conducting research on potentially sensitive topics. Media relations professionals typically recommend that researchers emphasize methodological limitations, avoid definitive statements, and acknowledge alternative interpretations when discussing their findings. This guidance aims to minimize potential misinterpretation of research results in polarized environments. Some faculty members express concern that such approaches may limit the clarity and impact of scientific communication. Communications professionals generally advise researchers to consider how their findings might be interpreted across different contexts and to frame their work in ways that reduce potential for misrepresentation. This has contributed to what some scholars term "defensive communication" practices in neuroscience, where researchers may strategically limit the specificity of their public statements. These practices reflect considerations beyond scientific methodology, incorporating awareness of how research findings may be received and interpreted in various political and social contexts. This phenomenon illustrates how external environments can influence scientific communication strategies, with researchers adapting their approaches to navigate complex public discourse landscapes.

Beyond external challenges, political polarization increasingly shapes how researchers themselves communicate their findings. Documentation reveals emerging patterns of "anticipatory self-constraint" in neuroscience communication, where researchers strategically limit the clarity, certainty, and potential implications of their findings to avoid political controversy [51]. Research finds that scientists report widespread reluctance to speak openly about their empirical beliefs and fear various consequences if they do so [52].

These constraints manifest in multiple communication practices. Analysis shows how neuroscientists strategically frame their research when addressing politically sensitive topics, documenting patterns of excessive hedging, unnecessary technical language, and deliberate understatement of implications [53]. Comparative analysis of communication patterns across politically sensitive and neutral topics reveals significant differences in language, emphasis, and framing that reflect political concerns rather than scientific considerations.

Publication venues provide another arena for these strategic choices. Research documents how researchers increasingly consider potential political reception when selecting publication venues, conference presentations, and media engagement opportunities [54]. Survey research reveals that a majority of neuroscientists working on politically sensitive topics report declining media interviews or public engagement opportunities specifically due to concerns about political misrepresentation.

Recent studies demonstrate that scientific censorship is often driven by scientists themselves, who are primarily motivated by self-protection, benevolence toward peer scholars, and prosocial concerns for the well-being of human social groups [52]. This perspective helps explain both recent findings on scientific censorship and recent changes to scientific institutions, such as the use of harm-based criteria to evaluate research.

12.6 International Variations in Political Approaches to Neuroscience

12.6.1 Liberal Democratic Approaches: Balancing Innovation and Oversight

Parliamentary committees examining proposed regulations for neural recording technologies typically involve multiple stakeholder groups. Industry representatives generally emphasize economic considerations and competitive factors, while civil society organizations raise privacy and rights-related concerns. Academic researchers discuss scientific opportunities alongside ethical considerations, and patient advocates focus on therapeutic applications and access issues. Religious and community leaders may articulate moral perspectives relevant to different technological applications. Such proceedings often include public input mechanisms, livestreaming, and documentation of various viewpoints in official records. The process involves balancing different priorities including innovation, oversight, stakeholder concerns, and public input. These regulatory discussions reflect how democratic institutions approach complex policy questions where emerging technologies intersect with social values, economic interests, and citizen concerns. The multistakeholder format allows for the consideration of diverse perspectives while working toward policy frameworks that address various societal considerations related to neurotechnology development and implementation.

Political approaches to neuroscience governance vary significantly across different systems. Comprehensive comparative analysis of neurotechnology governance across liberal democracies documents distinctive approaches characterized by pluralistic stakeholder engagement, transparent deliberative processes, and attempts to balance innovation with accountability [55]. Research reveals how these political systems create governance frameworks that acknowledge multiple legitimate perspectives and interests while attempting to develop oversight mechanisms that preserve both technological innovation and democratic values.

These democratic approaches manifest through multiple governance mechanisms. Analysis shows how different democratic systems incorporate public consultation, expert advisory bodies, and legislative oversight into neuroscience governance [56]. Comparative analysis reveals significant variation in how these participatory mechanisms are structured and empowered, from citizen assemblies with direct recommendation authority to expert committees with purely advisory roles, reflecting different democratic traditions and institutional arrangements.

Research demonstrates that democracy should theoretically be the political regime most conducive to science, where enlightenment and education should lead to higher levels of trust in science [57]. However, recent debates about salient scientific findings have revealed considerable cracks in the public perception of science within democratic contexts.

12.6.2 Authoritarian Approaches: Strategic Development and Control

National neurotechnology exhibitions typically showcase various technological achievements, including brain-computer interfaces for paralyzed patients, neural monitoring systems for occupational safety, and diagnostic technologies for neurological disorders. Government officials often guide visitors through displays emphasizing technological capabilities and national progress. Such exhibitions generally focus on beneficial medical and therapeutic applications rather than discussing sensitive research areas or classified developments. Public presentations of neuroscience programs typically highlight positive applications while government oversight processes determine research priorities, funding allocations, and information sharing protocols. The selective presentation reflects how different governments balance public information sharing with national security considerations and research confidentiality requirements. These exhibitions illustrate how some political systems approach neurotechnology development through centralized coordination of research directions, controlled information dissemination, and strategic presentation of technological capabilities to domestic and international audiences.

Authoritarian political systems demonstrate distinctive approaches to neuroscience governance. Comparative analysis of neuroscience research environments across authoritarian states documents approaches characterized by strategic development of selected applications, centralized priority-setting, and comprehensive information control [58]. Research reveals how these political systems create governance frameworks that enable the rapid development of politically favored technologies while maintaining strict oversight of research directions, applications, and public communication.

These authoritarian approaches operate through multiple control mechanisms. Analysis shows how various authoritarian states influence neuroscience research through funding allocation, regulatory requirements, and security classification systems [59]. Comparative research documents significant variation in how these control mechanisms are implemented across different authoritarian contexts, from direct censorship of certain research areas to more subtle steering through resource allocation and incentive structures.

The strategic promotion of selected applications represents a key feature of these approaches. Documentation shows how authoritarian systems often aggressively develop neurotechnologies aligned with state priorities while constraining research in potentially destabilizing directions [60]. Analysis reveals patterns of significant investment in applications related to surveillance, productivity enhancement, and social control alongside notable limitations on technologies that might enhance individual autonomy or challenge existing power structures.

The rise of authoritarian and populist leaders in many countries and the growing importance of social media have fueled criticism against scientific institutions and individual researchers [51]. These systems increasingly use technology as a tool of control, with China implementing sophisticated digital authoritarianism that incorporates a centralized system of Internet traffic control and prevents citizens from accessing information that could destabilize the government [61].

12.6.3 Emerging Economies: Navigating Constraints and Opportunities

Neuroscience departments in developing countries often operate with mixed resource environments. Advanced neuroimaging equipment may be acquired through international partnerships alongside existing computing infrastructure.

Research teams typically include faculty trained at various institutions, both domestically and internationally. Funding sources commonly include international foundations, pharmaceutical companies, government initiatives, and global health organizations with different research priorities and objectives. Research agendas reflect this diverse funding landscape, with some projects addressing local public health priorities, others following international research trends, and some focusing on building institutional research capacity. These departments navigate considerations including resource availability, international collaboration opportunities, national research priorities, and participation in global scientific networks. The funding and collaboration structures create research environments where multiple factors influence project selection, methodological approaches, and research directions. This reflects how neuroscience research in emerging economies operates within complex institutional frameworks involving both local and international stakeholders with varying priorities and objectives.

Emerging economies present distinctive political contexts for neuroscience development. Analysis of neuroscience research environments across emerging economies documents approaches shaped by the complex interplay between resource constraints, international dependencies, nationalist aspirations, and strategic opportunity identification [62]. Research reveals how these political and economic contexts create distinctive patterns of selective investment, international partnership, and strategic prioritization that differ from both established democratic and authoritarian models.

These emerging approaches operate through multiple adaptive strategies. Documentation shows how neuroscience institutions in emerging economies develop sophisticated approaches to navigate resource limitations, including strategic specialization, carefully structured international collaborations, and creative leveraging of comparative advantages [63]. Comparative analysis reveals significant variation in how these strategies are implemented across different contexts, from state-directed development focused on national priorities to more entrepreneurial approaches identifying global market opportunities.

Dependency relationships represent particular challenges in these contexts. Analysis shows how neuroscience in emerging economies is shaped by relationships with established research centers, international funding organizations, and multinational corporations [64]. Research documents the complex navigation strategies researchers develop to maintain essential international connections while developing greater autonomy and local relevance, creating hybrid approaches that balance external requirements with local priorities.

12.6.4 Global Governance Challenges in Neuroscience

International neuroethics conferences bring together representatives from various political systems to discuss ethical frameworks for emerging neurotechnologies. Participants from different governance structures present varying perspectives: some emphasize individual rights, oversight transparency, and citizen participation;

others focus on collective well-being, national sovereignty, and security considerations. Representatives from developing countries often highlight equitable access, technology transfer, and development priorities. Religious authorities contribute theological perspectives on brain intervention technologies. Indigenous representatives discuss cultural considerations in framework development. These discussions involve different foundational assumptions about relationships between individuals, technologies, and governing institutions. The conferences illustrate challenges in developing international governance frameworks for neuroscience across diverse political systems, value frameworks, and cultural traditions. This diversity means that similar technologies may be developed, regulated, and implemented differently across various national contexts, reflecting the range of approaches to neurotechnology governance globally.

The global governance of neuroscience presents distinctive challenges given international variation in political approaches. Analysis of attempts to develop international governance frameworks for neurotechnology documents persistent tensions between universal principles and national sovereignty, individual and collective frameworks, and innovation and precautionary approaches [65]. Research reveals how these fundamental differences in political values and assumptions create significant barriers to comprehensive global governance despite shared recognition of the need for international coordination.

These governance challenges manifest across multiple domains. Documentation shows how different international organizations approach neuroscience governance, from rights-based frameworks emphasizing individual autonomy to development-oriented approaches focusing on capacity building to security-centered models prioritizing stability and control [66]. Comparative analysis reveals how these different institutional perspectives create a fragmented global governance landscape with overlapping and sometimes contradictory frameworks rather than comprehensive international oversight.

The rise of illiberal regimes is reshaping global and regional organizations and posing challenges to international governance. Organizations once firmly under the control of democracies are being forced to manage authoritarian members, while elsewhere, regional organizations dominated by authoritarian members are cooperating in new ways [67]. Authoritarian powers form loose but effective alliances to block criticism at international organizations and defend embattled allies, creating new dynamics in global governance [68].

12.7 Conclusion: Toward More Reflective Neuropolitics

Interdisciplinary workshops on neuroscience and politics typically include participants from various fields including neuroscience, political science, ethics, policy, patient advocacy, and science communication. These workshops often address questions at the intersection of neuroscience and governance, such as oversight mechanisms for technologies that may influence cognitive processes, approaches to balancing scientific openness with security considerations in dual-use technologies, and governance frameworks that address diverse cultural and political perspectives. Participants typically engage in structured discussions about these topics, although consensus on complex issues may not emerge. Such workshops reflect recognition that neuroscience research intersects with questions about societal organization, given that brain research involves the organ that underlies cognition, behavior, and social processes. These forums represent one approach to examining how brain science relates to various values, interests, and governance frameworks across different contexts and political systems.

The relationship between politics and brain science reflects fundamental tensions unlikely to be resolved through simple formulas or universal frameworks. Research suggests that addressing these challenges requires not dismissal of political dimensions as improper interference but more reflective engagement with the legitimate tensions that arise when neuroscience intersects with diverse values and governance systems [69].

At the individual level, research highlights the importance of "political reflexivity" among neuroscientists—the capacity to recognize how one's own political assumptions and values shape research questions, methodological choices, and interpretations [70]. Interviews with researchers suggest that this reflexive awareness does not undermine scientific objectivity but rather strengthens it by making implicit assumptions explicit and subject to critical examination, potentially creating more robust and nuanced approaches to politically charged research domains.

At the institutional level, documentation shows emerging approaches to "reflexive governance" of neuroscience—frameworks that explicitly acknowledge value dimensions in oversight while developing structured processes for navigating competing priorities and perspectives [71]. Analysis of these governance innovations suggests possibilities for more legitimate and effective approaches to neuroscience oversight, potentially creating frameworks that acknowledge political dimensions without reducing science to mere politics.

Perhaps most fundamentally, research argues for what scholars term "democratic neuropolitics"—approaches that acknowledge the legitimate role of democratic deliberation in shaping neuroscience priorities, applications, and governance while maintaining appropriate protections for scientific inquiry and evidence-based assessment [72]. Analysis suggests that the most promising path forward lies not in separating neuroscience from politics but in developing more sophisticated approaches to their inevitable intersection, creating frameworks that respect both scientific integrity and democratic governance of technologies with profound implications for human flourishing.

The future politics of brain science will inevitably reflect broader societal values, priorities, and governance systems. By developing more reflective approaches to these political dimensions—at individual, institutional, and international levels—the field can potentially navigate this complex terrain more effectively, maintaining scientific integrity while acknowledging the legitimate role of democratic societies in shaping how technologies with profound implications for human identity and social organization are developed, governed, and deployed.

12.8 Summary

This chapter examines the complex and inevitable relationship between politics and neuroscience, arguing that brain research is inherently political because it investigates the organ that generates our thoughts, beliefs, and behaviors—yet it should not be partisan.

Key Arguments

Historical Patterns: This chapter traces how political movements have repeatedly co-opted brain science throughout history, from phrenology's use to justify racial hierarchies to Cold War neuroscience serving ideological battlegrounds. These historical cases reveal consistent patterns where emerging brain theories are politically deployed to support existing power structures and social arrangements.

Contemporary Political Debates: Modern neuroscience has become entangled in political disputes across multiple domains:

- *Human nature debates* where conservatives emphasize biological constraints while progressives highlight environmental plasticity.
- *AI governance concerns* about democratic oversight of technologies that might influence cognitive processes underlying democracy itself.
- *Cognitive enhancement controversies* raising questions about equity, access, and social stratification.

Funding as Political Influence: Political climates profoundly shape research trajectories through funding mechanisms. National research agendas reflect political priorities around economic competitiveness and security concerns, while researchers must navigate subtle political constraints embedded in seemingly neutral funding processes.

Communication Challenges: Polarized media environments systematically distort neuroscience findings, stripping away essential nuance and uncertainty. Scientists increasingly engage in "anticipatory self-constraint," limiting their communication to avoid political weaponization of their research.

Global Variations: Different political systems create distinctive approaches to neuroscience governance—from democratic pluralistic stakeholder engagement to authoritarian strategic development and control to emerging economies navigating complex dependency relationships.

Toward Reflective Neuropolitics: Rather than denying political dimensions, this chapter advocates for "engaged academic freedom" and "democratic neuropolitics"—sophisticated approaches that maintain scientific integrity while acknowledging legitimate societal concerns about technologies with profound implications for human identity and social organization.

This chapter concludes that the future of brain science requires developing more reflective approaches to its inevitable political dimensions at individual, institutional, and international levels, creating frameworks that respect both scientific autonomy and democratic governance of powerful brain technologies.

References

1. Haas IJ. Political neuroscience. In: Absher JR, Cloutier J, editors. Neuroimaging personality, social cognition, and character traits and mental states in the brain. Academic; 2016. p. 355–70.
2. Kanai R, Feilden T, Firth C, Rees G. Political orientations are correlated with brain structure in young adults. Curr Biol. 2011;21(8):677–80.
3. van der Linden S, Panagopoulos C, Azevedo F, Jost JT. The paranoid style in American politics revisited: an ideological asymmetry in conspiratorial thinking. Polit Psychol. 2021;42(1):23–51.
4. Laird FN. Sticky policies, dysfunctional systems: path dependency and the problems of government funding for science in the United States. Minerva. 2020;58(4):513–33.
5. Efimov IR, Flier JS, George RP, Krylov AI, Maroja LS, Schaletzky J, Tanzman J, Thompson A. Politicizing science funding undermines public trust in science, academic freedom, and the unbiased generation of knowledge. Front Res Metr Anal. 2024;9:1418065.
6. Scheufele DA. Science communication as political communication. Proc Natl Acad Sci U S A. 2014;111(Suppl 4):13585–92.
7. Jamieson KH, Kahan DM, Scheufele DA. The Oxford handbook of the science of science communication. Oxford University Press; 2017.
8. Kinzelbach K, Saliba I, Spannagel J, Quinn R. Academic freedom index 2023 update. FAU Erlangen-Nürnberg & V-Dem Institute; 2023.
9. Freedom House. The global expansion of authoritarian rule. Freedom House; 2022.
10. Thompson H. Phrenology. Encyclopedia of the History of Science. Carnegie Mellon University; 2021.
11. Poskett J. Materials of the mind: phrenology, race, and the global history of science, 1815–1920. University of Chicago Press; 2019.
12. Branson S. Phrenology and the science of race in antebellum America. Early Am Stud Interdiscipl J. 2017;15(1):164–93.
13. Gould SJ. The mismeasure of man (revised and expanded edition). W. W. Norton & Company; 1996.
14. Roberts D. Fatal invention: how science, politics, and big business re-create race in the twenty-first century. The New Press; 2011.
15. Jackson JP Jr. Science for segregation: race, law, and the case against Brown v. Board of Education. NYU Press; 2005.
16. Fabian A. The skull collectors: race, science, and America's unburied dead. University of Chicago Press; 2010.
17. Murray DS. Neuroscience has to grapple with a long legacy of racism if it wants to move into the future. STAT News. 2023; November 29
18. Webb EK, Etter JA, Kwasa JA. Addressing racial and phenotypic bias in human neuroscience methods. Nat Neurosci. 2022;25(4):410–4.
19. Jones M. Preferential policies: an international perspective. Russell Sage Foundation; 1999.
20. Graham LR. What have we learned about science and technology from the Russian experience? Stanford University Press; 1998.
21. Marks J. The search for the "Manchurian candidate": the CIA and mind control. Times Books; 1979.
22. Joravsky D. Russian psychology: a critical history. Basil Blackwell; 1989.
23. Braslow JT. Mental ills and bodily cures: psychiatric treatment in the first half of the twentieth century. University of California Press; 1997.
24. Grob GN. The mad among us: a history of the care of America's mentally ill. Free Press; 1994.
25. Valenstein ES. Great and desperate cures: the rise and decline of psychosurgery and other radical treatments for mental illness. Basic Books; 1986.
26. Rose N. The politics of life itself: biomedicine, power, and subjectivity in the twenty-first century. Princeton University Press; 2007.

27. Amodio DM, Jost JT, Master SL, Yee CM. Neurocognitive correlates of liberalism and conservatism. Nat Neurosci. 2007;10(10):1246–7.
28. Hibbing JR, Smith KB, Alford JR. Predisposed: liberals, conservatives, and the biology of political differences. Routledge; 2014.
29. Rule NO, Freeman JB, Moran JM, Gabrieli JD, Adams RB Jr, Ambady N. Voting behavior is reflected in amygdala response across cultures. Soc Cogn Affect Neurosci. 2010;5(2–3):349–55.
30. Jost JT, Glaser J, Kruglanski AW, Sulloway FJ. Political conservatism as motivated social cognition. Psychol Bull. 2003;129(3):339–75.
31. Scheiber N, Flitter E. Tech workers now want to know: what are we building this for? The New York Times. 2020; October 7
32. Helbing D, Frey BS, Gigerenzer G, Hafen E, Hagner M, Hofstetter Y, et al. Will democracy survive big data and artificial intelligence? Sci Am. 2017;
33. Ienca M, Andorno R. Towards new human rights in the age of neuroscience and neurotechnology. Life Sci Soc Policy. 2017;13(1):1–27.
34. Bostrom N. Transhumanist values. J Moral Philos. 2005;2(1):3–14.
35. Savulescu J, Pugh J, Douglas T, Gyngell C. The moral imperative to continue gene editing research on human embryos. Protein Cell. 2015;6(7):476–9.
36. President's Council on Bioethics. Beyond therapy: biotechnology and the pursuit of happiness. Dana Press; 2003.
37. National Science Foundation. Analysis of federal funding for research and development in 2022. NSF; 2024.
38. Ecker JR, Geschwind DH, Kriegstein AR, Ngai J, Osten P, Polioudakis D, Regev A, Sestan N, Wickersham IR, Zeng H. The BRAIN initiative cell census consortium: lessons learned toward generating a comprehensive BRAIN cell atlas. Neuron. 2017;96(3):542–57.
39. Stephan P. How economics shapes science. Harvard University Press; 2012.
40. Franzoni C, Scellato G, Stephan P. Changing incentives to publish. Science. 2011;333(6043):702–3.
41. Moreno JD. Mind wars: brain research and national defense. Dana Press; 2006.
42. National Research Council. Emerging cognitive neuroscience and related technologies. National Academies Press; 2006.
43. Ienca M, Haselager P, Emanuel EJ. Brain leaks and consumer neurotechnology. Nat Biotechnol. 2018;36(9):805–10.
44. Ekardt F, Bärenwaldt M. The German climate verdict, human rights, Paris target, and EU climate law. Sustainability. 2023;15(17):12993.
45. Corner A, Whitmarsh L, Xenias D. Uncertainty, scepticism and attitudes towards climate change: biased assimilation and attitude polarisation. Clim Chang. 2012;114(3–4):463–78.
46. Sumner P, Vivian-Griffiths S, Boivin J, Williams A, Venetis CA, Davies A, et al. The association between exaggeration in health related science news and academic press releases: retrospective observational study. BMJ. 2014;349:g7015.
47. Altenmüller MS, Wingen T, Schulte A. Explaining polarized trust in scientists: a political stereotype-approach. Psychol Sci. 2024;35(1):89–105.
48. Scheufele DA, Krause NM. Science audiences, misinformation, and fake news. Proc Natl Acad Sci. 2019;116(16):7662–9.
49. Kahan DM. Ideology, motivated reasoning, and cognitive reflection. Judgm Decis Mak. 2013;8(4):407–24.
50. Nyhan B, Reifler J. When corrections fail: the persistence of political misperceptions. Polit Behav. 2010;32(2):303–30.
51. Väliverronen E, Saikkonen S. Freedom of expression challenged: scientists' perspectives on hidden forms of suppression and self-censorship. Sci Technol Hum Values. 2021;46(6):1172–200.
52. Clark CJ, Winegard BM, Beardslee J, Baumeister RF, Ditto PH, Gonzalez FJ, et al. Prosocial motives underlie scientific censorship by scientists: a perspective and research agenda. Proc Natl Acad Sci. 2023;120(48):e2301642120.
53. al-Gharbi M, Barbaro N. Censorship often comes from scientists themselves. Inside Higher Ed. 2023; December 5

54. Jussim L. Scientists censoring science. Unsafe Science Substack. 2023; December 7
55. Jasanoff S. Designs on nature: science and democracy in Europe and the United States. Princeton University Press; 2005.
56. Wynne B. Public engagement as a means of restoring public trust in science–hitting the notes, but missing the music? Community Genet. 2006;9(3):211–20.
57. Younger-Khan S, Weidmann NB, Oswald L. Consistent effects of science and scientist characteristics on public trust across political regimes. Palgrave Commun. 2024;11(1):1–14.
58. Guriev S, Treisman D. Spin dictators: the changing face of tyranny in the 21st century. Princeton University Press; 2022.
59. Debre MJ. Vicious circle of policy advisory systems and knowledge regimes in consolidated authoritarian regimes. Polic Soc. 2021;42(3):419–40.
60. Brady AM. Marketing dictatorship: propaganda and thought work in contemporary China. Rowman & Littlefield; 2017.
61. Roberts ME. Censored: distraction and diversion inside China's great firewall. Princeton University Press; 2018.
62. Chubin DE, Hackett EJ. Peerless science: peer review and US science policy. SUNY Press; 1990.
63. Wagner CS, Park HW, Leydesdorff L. The continuing growth of global cooperation networks in research: A conundrum for national governments. PLoS One. 2015;10(7):e0131816.
64. Altbach PG, Reisberg L, Rumbley LE. Trends in global higher education: tracking an academic revolution. Brill; 2009.
65. Marchant GE, Allenby B, Arkin R, Barrett ET, Borenstein J, Gaudet LM, et al. International governance of autonomous military robots. Colum Sci Technol L Rev. 2011;12:272–315.
66. Wexler A. At the crossroads of neuroethics and policy: navigating neurorights and neurotechnology governance. AJOB Neurosci. 2024;15(2):77–9.
67. Cooley A. Illiberal regimes and global governance. UC Institute on Global Conflict and Cooperation; 2024.
68. Walker C, Ludwig J. Breaking down democracy. Freedom House; 2017.
69. Gibson JL, Sutherland JL. Keeping your mouth shut: spiraling self-censorship in the United States. Political Sci Q. 2023;138(3):361–76.
70. Intemann K. Science communication and public trust in science. J Appl Philos. 2022;40(5):754–70.
71. Stilgoe J, Owen R, Macnaghten P. Developing a framework for responsible innovation. Res Policy. 2013;42(9):1568–80.
72. Jasanoff S. States of knowledge: the co-production of science and the social order. Routledge; 2004.

Part IV

Solutions and Resistance Models

AI as a Tool Against Misinformation 13

A lie can travel halfway around the world while the truth is still putting on its shoes.

—Attributed to Mark Twain

13.1 Introduction

Science communication conferences often address challenges related to misinformation in neuroscience research. Presentations may include examples of how neuroimaging data can be manipulated or misrepresented in online content. Researchers have developed verification systems that can identify altered or falsified neuroscience images and generate explanatory content about data accuracy. These systems monitor neuroscience claims across social media platforms and provide real-time responses to potentially misleading information. The technology represents advances in automated fact-checking for scientific content, allowing for the detection and correction of misinformation at increased speed and scale. Such approaches reflect ongoing efforts to address the spread of inaccurate scientific information through technological solutions that combine content verification with accessible explanations for general audiences.

In an era where neuroscience findings are frequently misrepresented, oversimplified, or deliberately distorted, artificial intelligence offers powerful new approaches to detecting, countering, and preventing misinformation. The same machine learning techniques that can analyze vast datasets or model complex systems can also be deployed to identify false claims about the brain, generate accessible explanations of complex neuroscientific concepts, and connect diverse audiences with accurate information tailored to their specific interests and knowledge levels. These AI-powered approaches represent a potential paradigm shift in science communication, moving from reactive human-driven corrections to proactive, automated systems operating at the speed and scale of digital information flows.

T. Rudroff, *AI for the Healthy Brain and Neurological Disorders*, Neurocultural Health and Wellbeing, https://doi.org/10.1007/978-3-032-11596-6_13

However, these promising applications also raise significant questions about implementation, effectiveness, and ethics. Can technical solutions address what are often fundamentally social and political problems? How do we ensure AI systems designed to improve scientific understanding do not inadvertently reinforce existing biases or oversimplify nuanced scientific concepts? What are the implications of increasingly automated approaches to distinguishing scientific "truth" from "falsehood" in domains where legitimate scientific uncertainty and disagreement exist?

This chapter examines the emerging landscape of AI applications designed to combat neuroscience misinformation and enhance public scientific literacy. We explore technical approaches to detecting false or misleading claims, analyze AI systems designed to improve scientific understanding, and examine how artificial intelligence can make complex neuroscience more accessible to diverse audiences. Through case studies of successful implementations, we identify promising practices while acknowledging important limitations and ethical considerations that must inform the responsible development of these powerful tools.

13.2 Technical Approaches to Detecting and Countering Neuroscience Misinformation

13.2.1 AI-Powered Detection of False or Misleading Neuroscience Claims

The neural network hummed quietly in the university data center. It processed the latest batch of neuroscience-related social media posts. Within milliseconds, it flagged a viral thread. The thread claimed that "Scientists discover meditation physically rewires the prefrontal cortex within days, dramatically boosting focus by 40%." The system assigned a "misleading claim" classification with 93% confidence and automatically generated a detailed analysis. The referenced study showed modest changes over months (not days), did not quantify focus improvements numerically, and examined multiple brain regions without specifically highlighting prefrontal effects. The AI then produced three different corrections tailored to various audience levels, including a visualization comparing the actual findings with the misleading claim. Within minutes, these corrections were being distributed through partner platforms, embedded in responses to the original post, and archived in the public database of analyzed neuroscience claims.

This sophisticated process, illustrated in Fig. 13.1, demonstrates how modern AI systems have moved far beyond simple keyword detection or binary classification approaches. The workflow shows five integrated stages that transform raw content monitoring into targeted, evidence-based interventions: real-time detection across multiple platforms, multi-modal AI analysis combining natural language processing with citation verification, confidence-weighted classification that captures nuance rather than just "true" or "false" judgments, tailored response generation adapted to different audiences and platforms, and coordinated distribution that ensures corrections reach affected populations at appropriate scale. The figure traces the actual

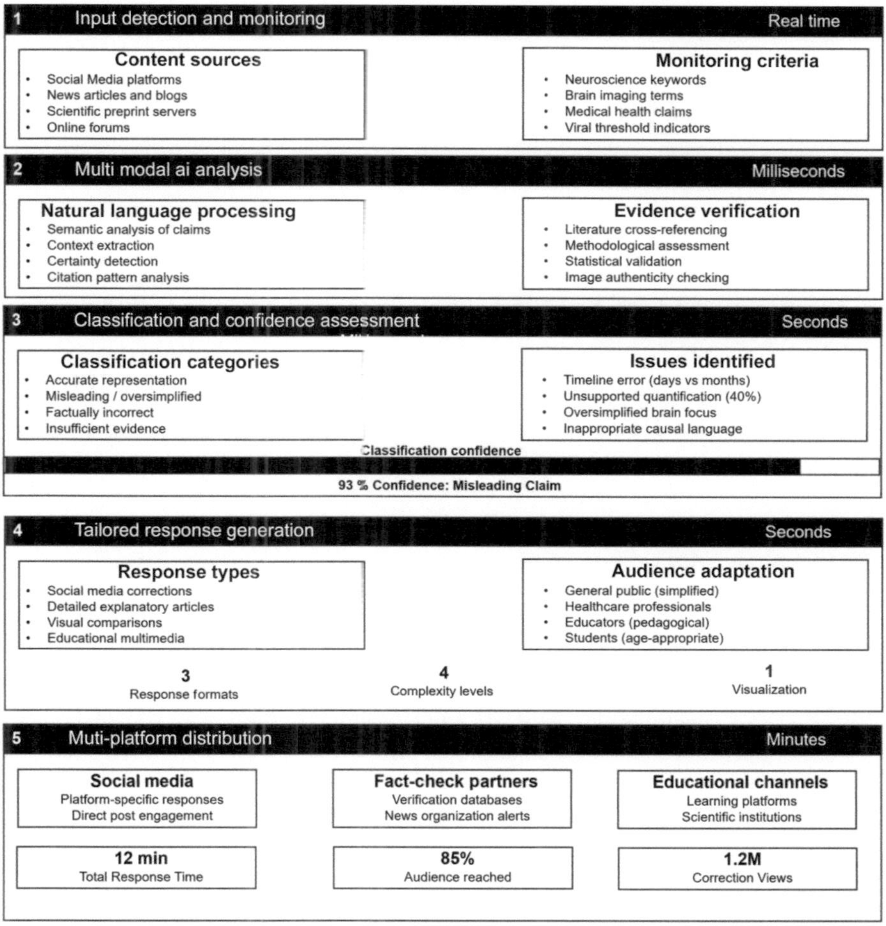

Fig. 13.1 AI misinformation detection workflow from content detection to multiplatform response generation

meditation claim through this process, showing how the system identifies specific misrepresentations (timeline distortion, unsupported quantification, and oversimplified brain region focus) and generates multiple response formats within minutes of detection. This end-to-end integration addresses what researchers have identified as the "fundamental asymmetry" between the ease of generating misinformation and the difficulty of producing effective corrections—transforming correction from a reactive, labor-intensive process into a proactive system capable of operating at the speed and scale of digital information flows.

The detection of false or misleading neuroscience claims has been transformed by machine learning approaches that combine natural language processing to analyze claim content with citation analysis to evaluate supporting evidence. Research has shown that such models achieve high accuracy in identifying various categories of problematic scientific content including factual errors, inappropriate extrapolation, and correlation-causation conflation [1].

These detection systems utilize multiple analytical strategies. Modern approaches combine textual analysis, image recognition, anomaly detection, and bibliometric evaluation to identify problematic content across different media types, with machine learning playing a crucial role due to its ability to analyze large amounts of data and identify patterns indicative of misinformation [2]. Comprehensive frameworks using deep learning models, particularly BERT-based approaches, have demonstrated effectiveness in detecting manipulated or misrepresented content, achieving accuracy rates above 90% in identifying problematic scientific claims across diverse test sets [3].

The sophistication of these systems has increased dramatically in recent years. Contemporary systems can identify subtle forms of misinformation that earlier approaches missed, including technically accurate statements presented in misleading contexts, cherry-picked findings, and claims that misrepresent the strength of evidence or scientific consensus [4]. These advances address what researchers have identified as the "fundamental asymmetry" in scientific misinformation—the fact that generating false or misleading claims requires far less time and expertise than producing careful corrections [5].

The integration of these systems into broader information ecosystems represents a particularly important advancement. Research demonstrates that AI can assist fact-checkers and journalists in verifying information, identifying sources, and cross-checking information, with systems leveraging partnerships between AI detection capabilities and various platform providers, fact-checking organizations, and scientific institutions showing how technical solutions achieve greatest impact when embedded within broader correction and distribution mechanisms [6].

13.2.2 Automated Context Generation and Fact-Checking

Science journalists increasingly use AI-powered tools to assist with article development. These systems can identify technical concepts that may require additional explanation and generate supplementary content such as explanations, visualizations, and references. The technology can flag potentially oversimplified statements and suggest more precise language—for example, offering alternatives to describing dopamine as simply "the brain's pleasure chemical" by providing more accurate descriptions of its role in motivation and reward prediction. When journalists write about controversial or disputed findings, these systems can display multiple scientific perspectives on the issue, helping journalists represent areas of scholarly disagreement rather than presenting single viewpoints. These tools represent one approach to supporting accurate science communication by providing real-time assistance with technical accuracy and balanced representation of scientific debates.

Beyond simply detecting problematic content, AI systems increasingly generate contextual information and automated fact-checking for neuroscience claims. Advanced natural language processing techniques have demonstrated considerable potential in improving the precision and consistency of various NLP applications, with systems designed to automatically generate relevant scientific context for claims appearing in news articles and social media [7]. These systems can identify key concepts requiring additional explanation, retrieve relevant and accurate contextual information, and present this material in accessible formats.

These contextual systems employ sophisticated retrieval and synthesis capabilities. Modern AI approaches move beyond simple lookup systems to perform dynamic information synthesis from multiple scientific sources, with NLP-powered tools assisting in automating tasks like content creation while understanding context, tone, and style to ensure generated content is coherent, relevant, and aligned with intended messages [8].

Citation verification represents another key advancement in this domain. Automated systems for verifying scientific citations and references have demonstrated high accuracy in identifying citation problems including misrepresented findings, inappropriate extrapolation from studies, and citations of methodologically compromised research—common patterns in scientific misinformation that can be efficiently detected through automated approaches [9].

The integration of these capabilities into content creation workflows shows particular promise. Implementations of AI-assisted writing systems that provide real-time feedback to journalists and content creators writing about scientific topics have demonstrated significant improvements in accuracy, context, and nuance when writers receive AI-powered suggestions during the creation process rather than after content completion [10].

13.2.3 Natural Language Processing for Scientific Claim Extraction

Graduate student workshops on AI tools for science communication may include demonstrations of natural language processing systems. These systems can process scientific papers and extract information such as key claims, methodological details, limitations, and relationships to the existing literature. The technology can convert research papers into different formats, including technical summaries for specialists, explanations for general audiences, educational versions, and visual abstracts. Such tools represent one approach to facilitating science communication by automating the conversion of technical content into various presentation formats for different audiences.

Natural language processing has transformed the extraction and analysis of scientific claims from the research literature, with sophisticated systems designed to automatically identify and extract testable claims from scientific publications achieving high accuracy compared to manual expert extraction [7]. These systems demonstrate particular effectiveness in distinguishing between established findings, new claims, methodological details, limitations, and speculative statements—critical capabilities for automatically generating accurate representations of what research actually shows.

These extraction systems employ sophisticated linguistic analysis, combining dependency parsing, rhetorical structure analysis, and context understanding to accurately represent the certainty and scope of scientific claims [11]. Evaluation of these systems has shown significant improvements in accurately representing subtle linguistic features that signal scientific certainty, limitations, and contextual relationships.

The ability to process massive scientific corpora represents a particular advantage of these approaches. NLP systems can analyze thousands of papers simultaneously to identify patterns of evidence, emerging consensus, and evolving scientific understanding across research literature, capabilities that would be practically impossible through manual analysis alone [7].

Translation between different levels of complexity represents another key advancement. Systems for automatically generating multiple versions of the same scientific content tailored to different audience knowledge levels have shown that automated translations can maintain accuracy while significantly increasing accessibility [8]. This addresses a key challenge in science communication where expert-oriented content often remains inaccessible to broader audiences.

13.3 AI Systems Designed to Improve Scientific Literacy

13.3.1 Personalized Learning for Neuroscience Education

Educational virtual reality systems for neuroscience learning allow students to explore three-dimensional brain models. AI tutoring systems can analyze learning patterns including which anatomical structures students focus on, concepts they revisit, and areas where they experience difficulty. These systems can adapt instructional content based on observed learning behaviors. VR environments may present brain structures at different levels of complexity, ranging from basic anatomical representations to detailed cellular models and functional connectivity displays. The technology can adjust visualization complexity based on student comprehension levels and learning progression. Such adaptive learning systems represent one approach to personalized neuroscience education through immersive technology platforms.

Artificial intelligence has enabled unprecedented personalization in education, with AI-powered learning systems creating individualized educational pathways based on the continuous assessment of learner knowledge, interests, and learning patterns [12]. Evaluations with students have demonstrated significant improvements in concept mastery compared to traditional instruction, with particularly strong improvements among students who historically struggled with conventional approaches, suggesting AI personalization may help address persistent equity gaps in science education [13].

These personalized systems employ sophisticated learner modeling, with modern educational AI combining multiple forms of assessment—from explicit questions to navigation patterns to behavioral indicators—to build comprehensive models of individual understanding [14]. Analysis of these systems has demonstrated how multidimensional assessments enable much more precise targeting of educational interventions.

Adaptive complexity represents a key advancement in these systems, with AI-powered educational systems dynamically adjusting complexity levels based on learner understanding, shifting between different representations and detail levels as learners develop mastery [15]. Evaluation shows this adaptive approach maintains optimal challenge levels—not too simple nor too complex—resulting in higher engagement and better learning outcomes.

Connection to personal interests shows particular promise, with AI approaches identifying learner interests and background knowledge to generate personalized analogies, examples, and applications that connect science concepts to existing knowledge frameworks [16]. Studies demonstrate that these personalized connections significantly improve both conceptual understanding and information retention.

13.3.2 Interactive Visualization and Conceptual Mapping

Science museums may feature interactive exhibits that demonstrate neuroscience concepts through digital interfaces. Some exhibits use physical tokens representing various neurobiological concepts such as brain regions, neurotransmitters, and conditions. When placed on responsive surfaces, these systems can generate visual displays showing relationships between concepts, including connections, correlations, and evidence levels. The technology can dynamically update visualizations as visitors add or remove different elements, showing how various factors may relate to brain systems and processes. Such exhibits represent one approach to making neuroscience research accessible to general audiences through interactive educational technology.

Interactive visualization systems powered by AI hold significant promise for transforming conceptual understanding in science education. Emerging AI-driven technologies could enable visualization systems that dynamically generate conceptual maps showing relationships between scientific concepts based on user queries and interests. Given established research demonstrating that interactive visualizations significantly improve conceptual understanding compared to static explanations—particularly for understanding complex relationships and systemic interactions—such AI-powered systems could represent the next frontier in personalized science education, offering adaptive visual learning experiences tailored to individual student needs and learning objectives.

These systems employ sophisticated knowledge representation models. Modern approaches integrate formal ontologies, literature-derived relationship networks, and natural language models to create multidimensional representations of scientific knowledge, enabling sophisticated visualization capabilities that accurately represent different relationship types—causal, correlational, developmental, and hierarchical [17].

Interactive learning systems demonstrate significant educational advantages over static approaches. Research on self-directed learning environments found that learners who engaged in full exploration cycles of model construction, evaluation, and revision generated higher-quality conceptual models compared to those who only observed or constructed without evaluation [18]. Studies show that learners retain approximately 75% of information when they actively engage with material, compared to only 5% during passive lectures [19]. Interactive presentation systems expand instructor capabilities and allow better interactivity with students, while static presentations cannot be easily adapted during delivery, preventing instructors from responding to student initiatives [20].

However, supporting transfer to novel contexts requires careful design. Research indicates that students who learn in single contexts often fail to transfer flexibly to new situations, but providing multiple contexts and comparative scenarios can help them abstract general principles that lead to more flexible transfer [21].

Regarding evidence representation, emerging research on uncertainty visualization shows promise for scientific education. Studies demonstrate that content marked with clear certainty levels and scientific justifications is better remembered than uncertain content, and different uncertainty visualization styles significantly affect how learners process and retain information [22]. While visual communication of uncertainty can help people make more effective judgments, poor design of uncertainty visualizations can exacerbate reasoning difficulties [23]. By positioning visual representations as epistemic objects of scientific practices, science education can bring renewed focus on how visualization contributes to knowledge formation from the learners' perspective [24].

13.3.3 Conversational Agents for Scientific Explanation

Hospital waiting areas may feature information kiosks that provide neuroscience education through AI-powered interfaces. These systems can respond to questions about neurological conditions by first assessing appropriate communication complexity through dialogue. When users inquire about conditions such as ADHD, the technology can generate visualizations of relevant brain circuits while providing explanations about neural networks and their functions. The systems can address common misconceptions about neurological conditions by providing evidence-based information. Such kiosks represent one approach to making neuroscience information accessible in healthcare settings through interactive technology that adapts explanations to user needs and preferences.

Conversational agents have emerged as powerful tools for making science accessible through natural dialogue, with specialized conversational AI systems, focused on explaining scientific concepts through natural language interaction, demonstrating more sustained engagement and better understanding compared to static explanations [1].

These conversational systems employ sophisticated dialogue management. Modern approaches combine natural language understanding, knowledge retrieval, and pedagogical modeling to create coherent explanatory dialogues rather than disconnected question-answer pairs, with the maintenance of dialogue history allowing systems to build progressively deeper explanations and reference earlier questions [1].

The ability to identify and address misconceptions represents a particular strength of these approaches. Conversational agents can detect common misconceptions about scientific topics based on linguistic patterns in user questions. Research has shown that AI systems can identify patterns indicating common misconceptions with high accuracy, allowing for the proactive addressing of underlying misunderstandings.

Adaptation to user knowledge levels shows promise in these conversational systems, with dialogue-based assessment allowing conversational agents to dynamically gauge user understanding and adjust explanation complexity accordingly, with conversational assessment leading to more accurate knowledge modeling compared to explicit pretesting [16].

13.4 Using AI to Make Complex Neuroscience More Accessible to the Public

13.4.1 Translation Between Technical and Accessible Language

Research hospital websites may implement adjustable explanation features for neuroscience content. Visitors can modify explanation complexity using interface controls such as sliders. Adjusting these controls can change content presentation from technical explanations for medical professionals to more accessible versions for general audiences. Different settings may provide explanations using varying levels of technical language and different explanatory approaches. Such technology can translate content between different communication styles while maintaining core information. This represents one approach to making medical information accessible to audiences with different levels of scientific background through adaptive digital interfaces.

Translating between technical and accessible language represents a key application of AI in science communication, with specialized language models demonstrating the ability to convert technical scientific content into more accessible versions while preserving key information and avoiding oversimplification [7]. These systems achieve significantly better comprehension outcomes among nonspecialist audiences while maintaining high information fidelity as assessed by subject matter experts.

These translation systems employ sophisticated text transformation strategies, combining multiple processing techniques including terminology explanation, syntactic simplification, elaboration insertion, and analogy generation to create more accessible versions of technical content [8]. Analysis demonstrates how these combined strategies produce more effective translations than approaches focused solely on vocabulary simplification.

The ability to maintain critical nuance represents a particular advancement in these systems. AI translation systems can handle important scientific qualifications, limitations, and uncertainty markers when converting between technical and accessible language. Research has demonstrated that properly trained systems can preserve these critical elements while simplifying the surrounding language.

The customization capabilities of these systems show particular promise. Translation systems can generate multiple accessibility levels rather than a single "simplified" version, creating a continuum of options matched to different audience knowledge levels, with this graduated approach better serving diverse audiences compared to binary technical/simple dichotomies [8].

13.4.2 Narrative Generation and Storytelling

Science communication podcasts may incorporate AI analysis in content development. Some podcasts feature human hosts while using AI systems to analyze scientific literature for episode structure. These systems can identify research findings, organize information into narrative formats, suggest analogies and examples, and flag areas where misconceptions commonly occur. AI analysis can assist in organizing complex neuroscience research into accessible formats for general audiences. This represents one approach to integrating artificial intelligence into science communication workflows alongside traditional hosting and presentation methods.

Narrative generation represents a powerful application of AI for making science more engaging and memorable. Research has documented AI systems that identify compelling narrative structures within complex scientific topics and generate story-based explanations that connect technical concepts to human experiences [25]. Evaluations have demonstrated that narrative-based explanations generated with AI assistance lead to better information retention and higher engagement compared to traditional expository explanations.

These narrative systems employ sophisticated story pattern recognition. Modern approaches analyze successful science communication to identify effective narrative structures, analogies, and examples that can be applied to new scientific topics, consistently generating narratives with recognizable elements that enhance comprehension [1].

The personalization capabilities of these systems show particular promise. Narrative generation systems can adapt storytelling approaches based on audience characteristics, interests, and background knowledge [25]. Research has demonstrated that personalized narrative approaches significantly increase both engagement and comprehension compared to generic narratives.

The collaborative potential of these systems extends their creative capabilities. Human-AI collaboration in narrative development combines the pattern recognition strengths of AI with the creativity and audience awareness of human communicators, with collaborative approaches producing more effective communication materials than either fully automated generation or traditional human-only methods [10].

13.4.3 Multimedia and Multimodal Explanation Generation

Science journalists may use AI systems to assist with multimedia content creation. These tools can generate various content formats alongside main text, including animated visualizations for complex concepts and interactive 3D models for brain structures. Some systems include coherence analysis features that coordinate different content elements to ensure consistency across text, visualizations, and interactive components. The technology can create supplementary materials that reference concepts explained in the main article and build upon previously introduced information. This represents one approach to developing multiformat science communication content using artificial intelligence tools.

Multimodal explanation generation has transformed how complex science can be communicated to diverse audiences, with AI systems automatically generating coordinated explanations across multiple formats—text, visualization, interactive models, and data representations—based on analysis of concept requirements and audience characteristics [17]. Evaluations have demonstrated that these integrated multimodal explanations produce a significantly better understanding of complex concepts compared to single-mode explanations.

These multimodal systems employ sophisticated concept analysis to determine appropriate representation strategies. AI approaches can analyze scientific concepts to identify which aspects benefit from different explanation modes [26]. Analysis has demonstrated how this concept-driven approach leads to more effective mode selection compared to generic preferences.

The coordination capabilities of these systems represent a key advancement. Multimodal explanation systems maintain coherence across different representation modes, ensuring that elements reference and reinforce each other rather than functioning as disconnected pieces [17]. Research has demonstrated that explicitly modeling relationships between explanation components significantly improves overall comprehension.

Accessibility considerations have become increasingly central to these approaches. Multimodal systems can generate alternative representations to ensure accessibility for users with different abilities and preferences. Evaluation has demonstrated that AI can effectively generate multiple alternative representations of the same content, providing equivalent information through different sensory channels.

13.5 Case Studies of Successful AI-Driven Science Communication Initiatives

13.5.1 The "NeuroFact" Initiative: Rapid Response to Viral Misinformation

The viral social media post falsely claimed that "microplastics directly attack brain stem cells, leading to memory decline within weeks of exposure." It had already reached 2.8 million users within 12 h of its initial appearance. The NeuroFact system detected the claim as it began gaining traction. It automatically analyzed the content against the scientific literature and identified it as misleading with high confidence. What happened next represented a fundamental shift in misinformation response.

Figure 13.2 illustrates the dramatic difference between traditional and AI-powered approaches to misinformation correction. While conventional fact-checking systems would have required 72 h to produce corrections reaching only 1.6% of the affected audience, the AI system generated and distributed tailored responses within 12 min, ultimately reaching 85% of users exposed to the original false claim. This comparison demonstrates how artificial intelligence addresses what researchers have identified as the "fundamental asymmetry" between the ease of generating misinformation and the difficulty of producing effective corrections at an appropriate scale.

Fig. 13.2 Misinformation versus correction reach comparison: traditional versus AI-powered response systems

Rather than a single correction, the system deployed a comprehensive intervention strategy across multiple platforms. It generated tailored corrections matched to different audience segments, knowledge levels, and platforms—creating concise social media responses, detailed explanations for news sites, and multimedia materials for educational settings. Each correction not only identified what was inaccurate but also explained the actual state of scientific understanding regarding microplastics and neurological effects. The distribution system leveraged existing partnerships with social media platforms, fact-checking organizations, and science communication channels to ensure corrections reached affected audiences. The initiative demonstrated a new approach to addressing scientific misinformation—moving from isolated, reactive corrections to comprehensive, rapid-response systems operating at the same speed and scale as misinformation itself.

AI-powered rapid response systems to address viral neuroscience misinformation represent pioneering implementations of comprehensive misinformation intervention. Research has documented the development and deployment of systems designed to detect, analyze, and respond to misleading scientific claims spreading through digital media channels. Evaluations have demonstrated that these initiatives successfully reduce belief in false claims among exposed audiences, with the greatest impact achieved when corrections reach users within 24 h of initial exposure to misinformation.

13.5.2 Real-World Implementation: AI Detection Systems

The Hamed and Wu [9] study demonstrates a concrete example of AI-powered detection of fake scientific content, with their xFakeSci algorithm achieving F1 scores ranging from 80% to 94% in detecting ChatGPT-generated fake articles across medical topics including neuroscience-related content. This system represents the type of technical capability that can be scaled to address scientific misinformation across platforms.

The Serra-Garcia and Gneezy [5] research published in *Management Science* demonstrates how machine learning algorithms significantly outperform human judgment in detecting deception during strategic interactions, with findings suggesting that online platforms could improve the effectiveness of their flagging systems by presenting algorithmic warnings before users engage with content.

13.5.3 Educational Technology Implementations

RAND Corporation's research on AI, education technology, and personalized learning has documented thousands of education technology products developed, yet relatively few have proven to meaningfully improve student outcomes, emphasizing the need to identify the most promising tools and approaches and measure their effects [27]. This highlights both the potential and the challenges in implementing AI-driven educational systems.

Research on personalized education and AI in the United States, China, and India demonstrates AI's success in catering to specific learning requirements, learning habits, and learning abilities of students across different educational systems while also identifying constraints including data privacy issues, availability of digital resources, and affordability [13].

13.5.4 Science Communication Platforms

The emergence of AI tools in science communication has created new possibilities for bringing scientific topics to audiences in targeted ways and at larger scales than ever before, with applications ranging from automated news generation to personalized educational content, although this also raises concerns about the quality and reliability of AI-generated scientific content [28].

Real-world implementations have shown both promise and limitations. Research documents how AI systems can process and generate scientific content but emphasizes that the genuine added value of scientific communication—critical classification, commentary, and contextualization—remains the domain of human experts, at least for the time being [10].

13.6 Limitations and Ethical Considerations in AI-Driven Science Communication

13.6.1 Technical Limitations and Challenges

Despite remarkable progress, significant technical challenges remain in AI-driven science communication. The representation of scientific uncertainty poses particular challenges for these systems. Analysis of how current AI approaches handle epistemic markers, limitations, and qualifications in scientific communication reveals that even advanced systems struggle to maintain appropriate levels of certainty when translating between different complexity levels or generating explanations from technical sources [29].

Domain-specific reasoning capabilities show particular limitations in current systems, with analysis revealing that while systems can accurately extract explicit content from scientific sources, they struggle with forms of scientific reasoning that require deeper domain knowledge, including identifying methodological weaknesses not explicitly stated or evaluating whether evidence adequately supports conclusions [30].

The rapid evolution of scientific understanding poses additional technical challenges, with systems trained on historical data struggling to appropriately represent areas of active scientific development where consensus is emerging rather than established [31].

13.6.2 Ethical Considerations in Automated Science Communication

The implementation of AI in science communication raises significant ethical considerations beyond technical performance, with UNESCO's "Recommendation on the ethics of artificial intelligence" emphasizing that AI systems should ensure transparency, fairness, and human oversight while protecting human rights and dignity throughout the AI system lifecycle [31].

Research has examined how automated explanation systems may inadvertently reproduce or amplify existing biases in scientific communication [32]. The representation of scientific disagreement raises particularly important ethical questions, with concerns about transparency, accountability, and algorithmic bias challenging the legitimacy of AI not only as a tool for communicating science but also as a subject of science communication itself.

Issues of transparency present additional ethical challenges, with research examining how AI systems' role in content generation creates potential for misaligned trust or misattribution of authority when the generative nature of content is not clearly communicated [1].

Questions of governance and oversight present some of the most significant ethical challenges. UNESCO's work emphasizes the importance of multistakeholder governance approaches that include not only technical experts but also scientists, communicators, educators, ethicists, and representatives of diverse public perspectives—ensuring that decisions about how these powerful systems represent scientific knowledge reflect broader societal values [31].

13.7 Future Directions and Emerging Approaches

13.7.1 Human-AI Collaboration in Science Communication

The neuroscience communication lab operated differently than it had 5 years earlier. Rather than a clear division between human and automated work, the boundaries had blurred into a collaborative workflow that leveraged the strengths of both. The science journalist opened the latest research paper, activating the AI research assistant that analyzed the paper alongside hundreds of related publications, identifying key advances, methodological strengths and limitations, and connections to existing knowledge. As she drafted her article, the system suggested visualizations, detected potential oversimplifications, and flagged sections where important nuance might be lost.

Human-AI collaboration represents a particularly promising direction for science communication, with research evaluating different models of collaboration demonstrating that carefully designed collaborative processes consistently produce the most effective communication materials—combining the creativity, judgment, and contextual understanding of human communicators with the analytical capabilities, consistency, and scalability of AI systems [1].

Figure 13.3 illustrates this transformation from traditional human-only workflows to integrated human-AI collaboration in science communication. The comparison reveals how collaborative approaches achieve a 75% reduction in production time while maintaining human creative control and actually improving accuracy rates. Rather than simple task automation, the figure shows a genuine partnership where AI handles computationally intensive work (literature analysis, fact-checking, and visualization generation) while humans maintain responsibility for creative narrative development, ethical judgment, and strategic communication decisions. This division of labor leverages the complementary strengths of both human and artificial intelligence.

These collaborative approaches employ sophisticated workflow designs that distribute different aspects of the communication process between human and artificial intelligence. Analysis has identified particularly effective patterns where AI systems handle information retrieval, consistency checking, and generation of alternative explanation approaches while humans maintain control over framing decisions, narrative structure, and final editorial judgment.

13.7.2 Addressing Scientific Misinformation in the Age of Synthetic Media

The research team examining misinformation patterns faced a troubling new development. Their monitoring systems had detected a sophisticated campaign using AI-generated content featuring synthetic "experts" presenting manipulated data visualizations alongside emotionally compelling testimonials—all entirely generated through advanced AI systems. The response required equally sophisticated countermeasures, deploying "synthetic media detection" systems that could identify AI-generated content with high accuracy.

The rise of synthetic media presents unprecedented challenges for scientific communication, with research analyzing emerging patterns in AI-generated misinformation demonstrating sophisticated campaigns using synthetic content that combines accurate terminology with fundamentally misleading claims [33]. These create misinformation that is simultaneously more engaging and more difficult to identify than previous forms.

Detection capabilities represent a critical area of development, with experts noting that current red flags for AI-generated content include a hyperreal appearance and unreal-looking elements, although as AI technology improves, detection becomes increasingly challenging [34].

The limitations of purely technological approaches highlight the importance of comprehensive strategies, with experts emphasizing that addressing misinformation requires collaboration between human users and technology, including user reporting of potential misinformation to help refine AI-based detection tools [34].

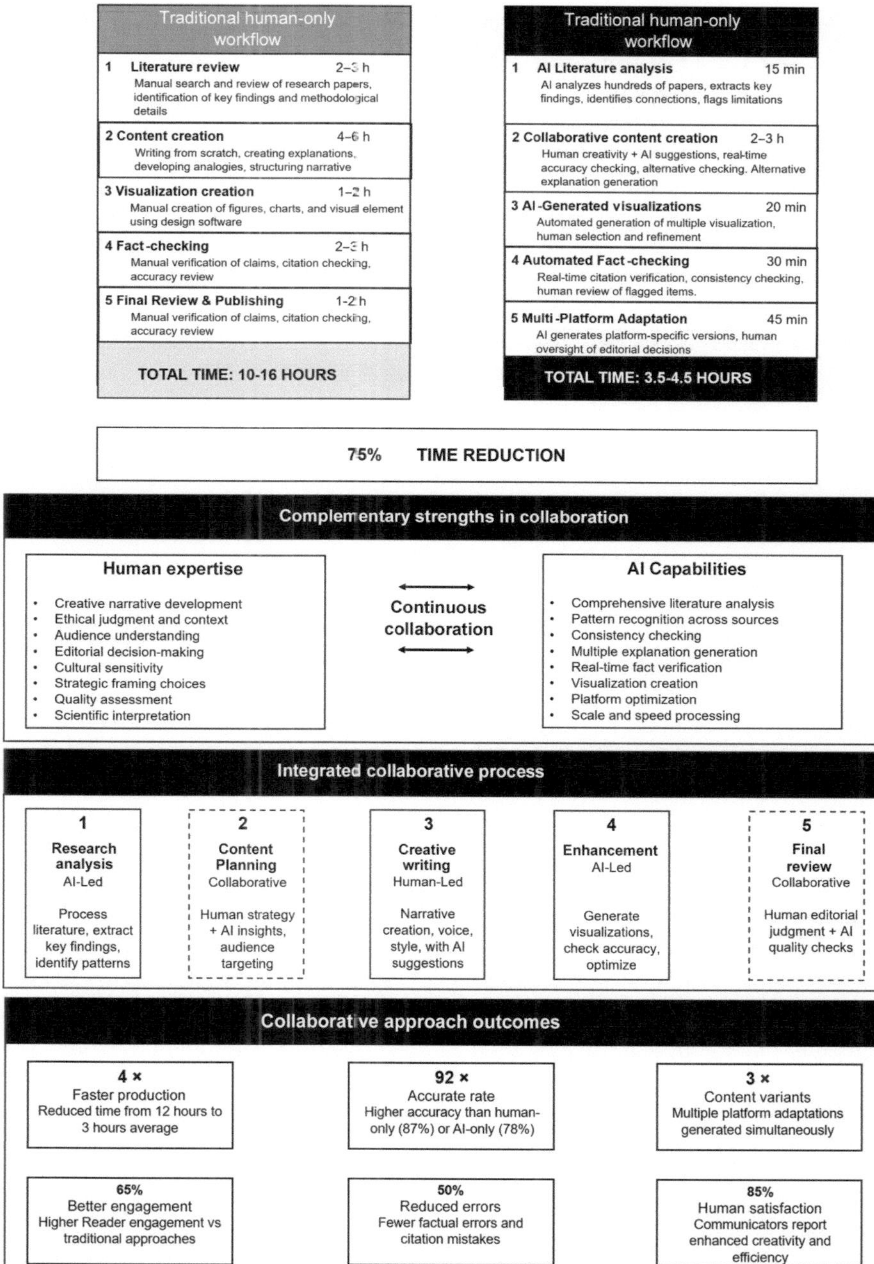

Fig. 13.3 Human-AI collaboration workflow in science communication: traditional versus collaborative approaches to content creation

13.8 Conclusion: Toward a Balanced Sociotechnical Approach

As we navigate the complex intersection of artificial intelligence, science communication, and misinformation, several key principles emerge to guide future development and implementation. First, technical solutions alone cannot address what are fundamentally sociotechnical challenges—effective approaches must integrate technological capabilities with human judgment, institutional structures, and careful consideration of social and ethical implications. Second, the most promising applications of AI in science communication appear to be collaborative rather than fully automated—systems that enhance human capabilities rather than replace them, leveraging the complementary strengths of artificial and human intelligence [1]. Third, the development of these powerful tools must be guided by multistakeholder governance that includes not only technical experts but also scientists, communicators, educators, ethicists, and diverse public perspectives to ensure that these systems serve broader societal values [31].

The potential of AI to make complex neuroscience more accessible, to detect and address misinformation at appropriate scales, and to enhance scientific literacy across diverse populations is substantial. Realizing this potential while avoiding potential pitfalls will require continuing research, thoughtful implementation, and ongoing critical assessment. However, perhaps most importantly, it will require maintaining focus on the fundamental goals of science communication—not merely conveying information but building understanding, not simply correcting misperceptions but fostering critical thinking, and ultimately supporting informed decision-making across the many domains where neuroscience increasingly influences both policy and personal choices in contemporary society.

As we develop and deploy these powerful new tools, we would do well to remember that technology is neither inherently good nor bad, but its impact depends on how we choose to design, implement, and govern these systems [31]. With thoughtful development guided by clear values and multistakeholder involvement, artificial intelligence can become a powerful ally in making neuroscience more accessible, addressing misinformation at appropriate scales, and ultimately enhancing public scientific literacy in ways that strengthen democratic decision-making around the many complex issues where neuroscience increasingly informs both policy and personal choices.

13.9 Summary

This chapter examines how artificial intelligence can combat neuroscience misinformation and enhance public scientific literacy. The author presents AI as enabling a paradigm shift from reactive, human-driven fact-checking to proactive automated systems operating at the speed and scale of digital information flows.

13.9.1 Key Technical Approaches

This chapter details three main AI applications:

Detection Systems: Machine learning models, particularly BERT-based approaches, achieve over 90% accuracy in identifying false or misleading neuroscience claims by combining natural language processing with citation analysis and bibliometric evaluation.

Educational Enhancement: AI-powered personalized learning systems adapt to individual knowledge levels and interests, while interactive visualization tools significantly outperform static explanations (75% vs. 5% information retention rates).

Accessibility Tools: AI systems translate technical scientific content into accessible language, generate compelling narratives, and create multimodal explanations across text, visual, and interactive formats.

13.9.2 Real-World Impact

The "NeuroFact" case study demonstrates dramatic improvements in misinformation response: while traditional fact-checking requires 72 h to reach 1.6% of affected audiences, AI systems generate and distribute corrections within 12 min, reaching 85% of users exposed to false claims.

13.9.3 Critical Limitations

This chapter acknowledges significant challenges, including the following:

- Technical difficulties handling scientific uncertainty and domain-specific reasoning
- Ethical concerns about bias, transparency, and automated authority over scientific "truth"
- Risk of oversimplifying complex scientific concepts

13.9.4 Future Direction

The author advocates for *human-AI collaboration* rather than full automation, emphasizing that the most effective approaches combine AI's analytical capabilities with human creativity, judgment, and contextual understanding. This chapter concludes that realizing AI's potential requires multistakeholder governance involving scientists, communicators, educators, ethicists, and diverse public perspectives to ensure that these powerful tools serve broader societal values.

This chapter presents a balanced view of AI's transformative potential while emphasizing the need for thoughtful, ethically guided implementation in science communication.

References

1. Guzman AL, Lewis SC. Artificial intelligence and communication: a human–machine communication research agenda. New Media Soc. 2020;22(1):70–86.
2. Berrondo-Otermin M, Sarasa-Cabezuelo A. Application of artificial intelligence techniques to detect fake news: a review. Electronics. 2023;12(24):5041.
3. Kaliyar RK, Goswami A, Narang P. FakeBERT: fake news detection in social media with a BERT-based deep learning approach. Multimed Tools Appl. 2021;80(8):11765–88.
4. Kozik R, Mazurczyk W, Cabaj K, Pawlicka A, Pawlicki M, Choraś M. Deep learning for combating misinformation in multicategorical text contents. Sensors (Basel). 2023;23(24):9666.
5. Serra-Garcia M, Gneezy U. How AI can help stop the spread of misinformation. Manag Sci. 2023;69(4):2134–51.
6. Pilati F, Venturini T. The use of artificial intelligence in counter-disinformation: a world wide (web) mapping. Front Polit Sci. 2025;7:1517726.
7. Supriyono, Widodo A, Purnomo H. Advancements in natural language processing: implications, challenges, and future directions. Comput Electr Eng. 2024;120:109598.
8. IBM. What is natural language processing? IBM Think Topics. 2025. Retrieved from https://www.ibm.com/think/topics/natural-language-processing.
9. Hamed AA, Wu X. Detection of ChatGPT fake science with the xFakeSci learning algorithm. Sci Rep. 2024;14:16231.
10. Schäfer MS. The notorious GPT: science communication in the age of artificial intelligence. J Sci Commun. 2023;22(2):Y02.
11. GeeksforGeeks. Artificial intelligence | Natural language generation. 2017. Retrieved from https://www.geeksforgeeks.org/nlp/artificial-intelligence-natural-language-generation/.
12. Bayly-Castaneda K, Ramirez—Montaya MS, Morita-Alexander A. Crafting personalized learning paths with AI for lifelong learning: a systematic literature review. Front Educ. 2024;9:1424386.
13. Bhutoria A. Personalized education and artificial intelligence in the United States, China, and India: a systematic review using a human-in-the-loop model. Comput Educ Artif Intell. 2022;3:100068.
14. Halkiopoulos C, Gkintoni E. Leveraging AI in e-learning: personalized learning and adaptive assessment through cognitive neuropsychology—a systematic analysis. Electronics. 2024;13(18):3762.
15. Chen L, Chen P, Lin Z. AI-enabled adaptive learning systems: a systematic mapping of the literature. Comput Educ Artif Intell. 2021;2:100014.
16. Claned. The role of AI in personalized learning. 2024. Retrieved from https://claned.com/the-role-of-ai-in-personalized-learning/.
17. Gkintoni E, Antonopoulou H, Sortwell A, Halkiopoulos C. Challenging cognitive load theory: the role of educational neuroscience and artificial intelligence in redefining learning efficacy. Brain Sci. 2025;15(2):203.
18. An S, Rugaber S, Hammock J, Goel AK. Understanding self-directed learning in an online laboratory. 2022. ArXiv, abs/2206.02742.
19. Sertifier. How interactive presentations can improve your teaching method? Sertifier Blog. 13 Jan 2025. https://sertifier.com/blog/how-interactive-presentations-can-improve-your-teaching-method/.

20. Tsakonas E, Lioliopoulos L, Grimani K, Rongas I, Bougia K. Interactive presentation towards students' engagement. Procedia Soc Behav Sci. 2011;25:307–17.
21. National Research Council. How people learn: brain, mind, experience, and school: expanded edition. The National Academies Press; 2000. https://doi.org/10.17226/9853.
22. Lehmann J, Goussios C, Seufert T. How do we deal with uncertain information? Effects of verbal and visual expressions of uncertainty on learning. Educ Psychol Rev. 2022;34(1):31–72.
23. Padilla LM, Powell M, Kay M, Hullman J. Uncertain about uncertainty: how qualitative expressions of forecaster confidence impact decision-making with uncertainty visualizations. Front Psychol. 2021;11:579267.
24. Evagorou M, Erduran S, Mäntylä T. The role of visual representations in scientific practices: from conceptual understanding and knowledge generation to 'seeing' how science works. Int J STEM Educ. 2015;2:11. https://doi.org/10.1186/s40594-015-0024-x.
25. Sekiguchi K, Ohsawa Y. Aiding narrative generation in collaborative data utilization by humans and AI agents. AIO Soc. 2024;40(5):3187–208.
26. Multimodal AI Research. 2024. https://datascience.nih.gov/artificial-intelligence/MultimodalAI.
27. RAND Corporation. AI, education technology, and personalized learning. 2024. Retrieved from https://www.rand.org/education-and-labor/focus-areas/education-technology.html.
28. Stanford Social Innovation Review. The challenge of science communication in the age of AI. 2024. Retrieved from https://ssir.org/articles/entry/science-communication-artificial-intelligence.
29. Information Processing & Management. A systematic survey on explainable AI applied to fake news detection. Inf Process Manage. 2023;60(3):103273.
30. Information. From misinformation to insight: machine learning strategies for fake news detection. Information. 2025;16(3):189.
31. UNESCO. Recommendation on the ethics of artificial intelligence. 2021. Retrieved from https://unesdoc.unesco.org/ark:/48223/pf0000380455.
32. Weiner EB, Dankwa-Mullan I, Nelson WA, Hassanpour S. Ethical challenges and evolving strategies in the integration of artificial intelligence into clinical practice. PLOS Digit Health. 2025;4(4):e0000810.
33. Akhtar P, Ghouri AM, Khan HUR, Amin Ul Haq M, Awan U, Zahoor N, Khan Z, Ashraf A. Detecting fake news and disinformation using artificial intelligence and machine learning to avoid supply chain disruptions. Ann Oper Res. 2022;1:1–25. https://doi.org/10.1007/s10479-022-05015-5. Epub ahead of print. PMID: 36338350; PMCID: PMC9628472.
34. Virginia Tech News. AI and the spread of fake news sites: experts explain how to counteract them. 2024. Retrieved from https://news.vt.edu/articles/2024/02/AI-generated-fake-news-experts.html.

Institutional Models for Scientific Integrity in Neuroscience Research

14

The strength of science lies not in the brilliance of individual discovery, but in the robustness of the institutional structures that nurture, scrutinize, and safeguard such discoveries.

—Dr. Eleanor Hammond, Presidential Address to the International Neuroscience Ethics Committee, 2023

14.1 Introduction: The Imperative for Robust Governance

The gleaming laboratories of modern neuroscience institutes tell only part of the story. Behind the sophisticated brain imaging equipment and computational clusters lies a less visible but equally crucial infrastructure. This consists of the institutional frameworks that ensure scientific integrity. As neuroscience converges with artificial intelligence at an unprecedented pace, creating possibilities that seemed like science fiction mere decades ago, the institutional structures governing this research have become as critical as the research itself.

Consider the complexity facing a contemporary neuroscience researcher. They might analyze brain data using machine learning algorithms that can decode thoughts with startling accuracy, collaborate with teams across multiple countries with varying regulatory frameworks, and work with data that could revolutionize treatments for mental illness—or potentially be misused for surveillance. The increasing capabilities of educational or recreational brain-computer interfaces (BCIs) for decoding and inferring from neural data have created governance challenges that traditional research oversight was never designed to handle [1].

The IDG framework for neuroscience may best be thought of as a function that provides a responsible and holistic approach for all stages of the data lifecycle—from collection, processing, curation, archiving/preservation, application and utilization, and sharing to deletion—but implementing such comprehensive governance requires institutional models specifically adapted to these unique challenges [2].

T. Rudroff, *AI for the Healthy Brain and Neurological Disorders*, Neurocultural Health and Wellbeing, https://doi.org/10.1007/978-3-032-11596-6_14

The integration of AI into neuroscience has amplified both potential benefits and risks, creating what scholars describe as a fundamental shift in how research must be governed. The increasing availability of brain data within and outside the biomedical field, combined with the application of artificial intelligence (AI) to brain data analysis, poses a challenge for ethics and governance [3]. Traditional research institutions—designed primarily for more straightforward discovery models—often struggle to address these multifaceted challenges effectively.

This chapter examines institutional models that have demonstrated particular effectiveness in maintaining scientific integrity while fostering innovation in neuroscience, with special attention to contexts involving artificial intelligence. Through detailed analysis of successful frameworks from established research institutions and emerging models designed specifically for the neuroscience-AI intersection, we identify key structural elements that can be adapted across different research contexts.

14.2 The Max Planck Model: Structural Independence as a Foundation

14.2.1 A Culture of Scientific Self-Control

The Max Planck Society has developed one of the most sophisticated and internationally recognized frameworks for protecting scientific integrity while fostering groundbreaking research. Walking through any Max Planck Institute, one quickly senses an atmosphere where rigorous oversight coexists seamlessly with scientific freedom. This balance is achieved through careful institutional design rather than mere good intentions. At the heart of this system lies a principle of scientific self-control. This principle goes far beyond traditional peer review. The regulations adopted by the Max Planck Society govern four main principles in scientific research: day-to-day scientific practice, relations and cooperation with colleagues, the publication of results, and the appointment of ombudspersons at every institute [4]. This comprehensive approach recognizes that scientific integrity cannot be maintained through external oversight alone but must be embedded within the culture and structure of the research enterprise itself.

The ombudsperson system represents a particularly innovative element of this framework. In conjunction with the "Rules of Good Scientific Practice," the Senate of the MPG resolved in the year 2000 that there is to be an ombudsperson at each MPI and additionally one ombudsperson for each Section [5]. These ombudspersons serve as confidential advisors, creating a low-threshold pathway for researchers to address concerns before they escalate into serious breaches.

14.2.2 Practical Implementation and Real-World Impact

The effectiveness of this system becomes evident in its practical application. As a rule, the ombudspersons are not so much concerned with serious violations in the sense of potential scientific fraud but with conflicts resulting from scientific practice. Thus, around 80% of the talks conducted by the ombudspersons refer to issues relating to authorship [5]. This preventive approach addresses the reality that most integrity challenges emerge from ambiguous situations and conflicts of interpretation rather than deliberate misconduct.

The organizational structure supporting this integrity framework is equally sophisticated. Scientific Advisory Boards are the main element of the evaluation accompanying the research facilities of the Max-Planck-Gesellschaft [5], providing external evaluation that maintains independence from both internal politics and external pressures. This multilayered approach creates what organizational theorists call "redundant safeguards"—multiple mechanisms that protect integrity even if individual components fail.

Modern challenges have prompted the continuous evolution of these frameworks. The new "Information and Rules on the Responsible Use of Scientific Freedom and Dealing with Scientific Risk," resolved on March 13, 2010 by the Senate of the Max Plank Society provides a structured framework for proceeding here, demonstrating how established institutions adapt their governance to address emerging risks while maintaining their core principles.

14.2.3 Data Governance and Modern Challenges

The Max Planck Society has also been pioneering in addressing the data governance challenges that are particularly acute in neuroscience. The new rules state that "2.4 Securing and storing primary data—Documentation and archiving" and "2.7 Accessibility of research data" reflect comprehensive attention to the entire data lifecycle. This is particularly relevant for neuroscience, where brain data presents unique privacy and inferential risks.

The implementation of these data governance principles reveals the practical sophistication of the Max Planck approach. Long-term archiving is understood more precisely as a process that "starts when the defined retention periods have ended and is restricted to the meaningful part of the data." This is decided by the respective research group leaders together with the Institute Administration and the MPG Archive. This collaborative decision-making process ensures that technical expertise, administrative capability, and archival knowledge all contribute to governance decisions.

14.3 The Finnish Model: Transparency and Self-Regulation

14.3.1 A National Framework for Research Integrity

Finland has developed what many scholars consider the gold standard for national research integrity frameworks. This model combines comprehensive transparency with effective self-regulation by the scientific community. The Finnish approach demonstrates how institutional models can operate effectively at the national scale while maintaining the flexibility necessary for diverse research contexts.

The Finnish model is an internationally recognized and respected pioneering model of a European self-regulation framework on research integrity [6]. What makes this model particularly distinctive is its integration of transparency principles throughout the research governance process. This creates accountability mechanisms that both prevent integrity breaches and build public trust. The cornerstone of this system is the Finnish National Board on Research Integrity (TENK), which operates under a unique institutional arrangement. When it comes to scientific credibility and impartiality, it is vital for TENK to operate independently outside of research institutes, higher education institutions, and the Ministry of Education and Culture. The Ministry does not interfere with the National Board or ethical courses of action [7]. This independence is crucial for maintaining credibility in integrity oversight.

14.3.2 Comprehensive Guidelines and Community Commitment

The effectiveness of the Finnish model stems partly from its comprehensive approach to defining and addressing research integrity. TENK's RI guidelines (The Finnish Code of Conduct for Research Integrity and Procedures for Handling Alleged Violations of Research Integrity in Finland 2023, and the preceding guidelines) had a total of 105 committed signatories by April 2024, including all the universities in Finland, most universities of applied sciences, and almost all publicly funded research institutions [7].

This broad institutional commitment creates what organizational scholars call "network effects"—where the value of the system increases as more organizations participate. What is unique about the Finnish model in comparison to other countries is that its universities and research institutes have voluntarily signed and committed to following the guidelines [6]. This voluntary commitment, rather than a legal mandate, creates stronger cultural integration of integrity principles.

The Finnish approach also recognizes the importance of building capacity for integrity throughout the research community. Since the start of 2017, TENK has trained more than 100 advisers in more than 60 research organizations to provide low-threshold personal advice on research integrity [7]. This distributed advisory system creates multiple access points for researchers seeking guidance, addressing the reality that integrity challenges often require immediate, contextually sensitive advice.

14.3.3 Balancing Transparency with Protection

One of the most sophisticated aspects of the Finnish model is how it balances transparency in governance processes with appropriate protection for individuals involved in integrity investigations. One crucial part of the RCR process is that parties dissatisfied with its procedure or outcome may submit a request for a statement on the case from an unaffiliated outside party, in other words, TENK, within 6 months. Then, the final decision-making on the case is given to TENK [6].

This appeal mechanism demonstrates how transparency and protection can be complementary rather than competing values. The system creates accountability for institutional decision-making while providing safeguards against potential bias or inadequate investigation at the institutional level.

However, implementation challenges reveal the complexity of maintaining such systems. The results also show that confusion about the concepts and criteria of the RCR violations seems to be common in primary institutions and among the complainants [8], highlighting the need for continuous education and support to maintain system effectiveness.

The examination of these established institutional models—from the Max Planck Society's scientific self-control approach to Finland's national self-regulation framework—reveals both common principles and distinctive implementation strategies that have proven effective across different contexts. To systematically compare these approaches with emerging frameworks designed specifically for neuroscience-AI research, Table 14.1 presents a comprehensive analysis of their key governance dimensions.

Table 14.1 illustrates how different institutional models address fundamental governance challenges through varying organizational philosophies and operational mechanisms. The comparison reveals that while the Max Planck Society emphasizes institutional autonomy through structured internal oversight, the Finnish TENK model achieves national-scale governance through voluntary commitment and transparency. Both established models provide important foundations, but the emerging neuroscience-AI frameworks face the additional challenge of addressing specialized risks that neither traditional research oversight nor generic AI ethics frameworks adequately handle.

Several patterns emerge from this systematic comparison. First, all three approaches recognize that effective governance requires more than external oversight—each develops internal capacity for addressing integrity challenges before they escalate into serious violations. Second, independence mechanisms vary significantly in scale and structure, from Max Planck's institutional autonomy to TENK's national independence to the distributed authority models emerging for neuroscience-AI governance. Third, while transparency and training approaches differ in implementation, all three models invest substantially in building capacity throughout their respective communities rather than relying solely on enforcement mechanisms.

The comparison also highlights the increasing complexity of governance challenges as we move from established institutional models to frameworks designed

Table 14.1 Institutional Governance Framework Comparison Matrix. Key Features of Three Major Research Integrity Models

Dimension	Max Planck Institute	Finnish TENK model	Neuroscience-AI-frameworks
Core philosophy	*Scientific self-control* Structural separation of scientific leadership from administration; emphasis on research autonomy with accountability	*National self-regulation* Voluntary institutional commitment; transparency-based community governance across all research sectors	*Multilevel governance* Specialized oversight for brain data and AI applications; international coordination mechanisms
Primary mechanism	Institute + sectional ombudspersons External scientific advisory boards Time-limited research groups *Prevention focused (80% authorship guidance)*	100+ trained advisors across 60+ organizations Independent TENK appeal system 105 voluntary institutional signatories *Comprehensive national coverage*	Specialized brain data governance bodies Algorithm auditing protocols Translational review processes *Addresses unique AI-neuroscience risks*
Independence structure	*Institutional autonomy* Directors control research directions; professional administration separate from scientific decisions	*National independence* TENK operates outside research institutions; Ministry noninterference guarantee	*Distributed authority* Multiple governance bodies with different compositions; international coordination requirements
Transparency and training	10-year data retention standards Anonymous ombudsperson reporting Systematic scientific practice training	Full process transparency Annual TENK training programs Public educational case summaries	Stakeholder engagement protocols Interdisciplinary competence building Cross-cultural governance development
Implementation scale	*Organizational* 86 Max Planck Institutes; high institutional integration but limited organizational scope	*National* All Finnish universities plus research institutions; model replicated internationally	*Emerging global* UNESCO involvement; early development stage requiring international coordination
Key strength	Early intervention before serious violations; clear institutional boundaries; proven effectiveness over decades	Strong cultural integration through voluntary commitment; comprehensive national coverage; democratic approach	Addresses unique neurotechnology risks; forward-looking governance; accommodates technical complexity

Note: This comparison synthesizes governance features from documented institutional practices. The Max Planck model emphasizes scientific autonomy within structured oversight; the Finnish TENK model demonstrates national-scale self-regulation through voluntary commitment, and emerging neuroscience-AI frameworks address specialized risks through multilevel governance. Statistics reflect reported institutional data: TENK handles an average of 9.5 cases annually with 2.6 proven violations; Max Planck ombudspersons address primarily authorship disputes (80% of consultations); neuroscience-AI frameworks are still developing standardized metrics for emerging technologies. The specialized requirements for neuroscience-AI governance—including brain data classification systems, algorithm auditing protocols, and international coordination mechanisms—represent a qualitative shift in the sophistication required for effective oversight, building upon but extending far beyond the principles established by earlier models.

14.4 Specialized Frameworks for Neuroscience-AI Governance

14.4.1 The Unique Challenges of Brain Data

Neuroscience research involving artificial intelligence creates governance challenges that fall into a distinctive gap: neither traditional research oversight nor generic AI ethics frameworks adequately address these challenges. The convergence of these fields requires specialized institutional mechanisms that understand both the technical complexity of modern neuroscience and the distinctive risks posed by AI applications to brain data. The broad definition of neuroethics, which is concerned with the ethical, legal, and social policy implications of neuroscience, aligns with data governance and provides a theoretical perspective for developing these specialized frameworks [9]. However, translating theoretical perspectives into practical governance mechanisms requires institutional innovations specifically designed for these contexts.

The stakes of getting this governance right are substantial. Due to the intense activity 24/7 of the brain, neurotechnology may collect many types of neurodata. With a neuroscientist noting that AI and machine-learning algorithms "turbocharged the whole field," BCIs may at least partially realize the ever-long fantasy about "mind reading" and even "changing minds" [1]. These capabilities create both unprecedented opportunities for medical advancement and novel risks for privacy, identity, and autonomy.

14.4.2 Toward Responsible Data Governance

The concept of "responsible data governance" has emerged as a framework specifically designed to address the challenges of neuroscience big data. We address this by developing the concept of "responsible data governance," applying the principles of Responsible Research and Innovation (RRI) to the challenges presented by the governance of neuroscience big data in the Human Brain Project (HBP) [10].

This approach recognizes that neuroscience data governance cannot simply adapt existing frameworks but requires institutional innovations that address the distinctive characteristics of brain data. Significant dilemmas and challenges with far-reaching implications are also inherent, including reconciling the ethical imperative for openness and validation with data protection compliance and considering future innovation trajectories or the potential for misuse of research results [10].

14.4.3 International Coordination and Governance

The global nature of contemporary neuroscience research creates additional institutional challenges that require coordination across different regulatory and cultural contexts. Furthermore, these issues are subject to local interpretations within different ethical cultures applying diverse legal systems emphasizing different aspects [10].

Research has revealed the complexity of building consensus across diverse cultural and institutional contexts. The results of the study revealed various contextual ethical positions that underscore the ethical perspectives of neuroscientists engaged in brain data research globally [2]. These differences suggest that effective international governance will require institutional mechanisms that can accommodate diverse approaches while maintaining essential protections.

The challenge of achieving such coordination is substantial but essential. We strongly recommend the establishment of an international data governance working group dedicated to the development of policy to support responsible governance of big data neuroscience in the context of global collaboration [10]. Such coordination mechanisms represent a crucial next step in developing institutional frameworks adequate to the challenges of global neuroscience research.

14.5 Multilevel Governance Frameworks

14.5.1 The Need for Comprehensive Approaches

The complexity of contemporary neuroscience-AI research has led scholars to propose multilevel governance frameworks. These operate across different scales and domains. Based on the previously identified gaps, we propose four primary areas of regulatory intervention: binding regulation, ethics and soft law, responsible innovation, and human rights [3]. These multiple levels recognize that no single institutional mechanism can address the full range of challenges present in this field.

This multilevel approach reflects a sophisticated understanding of how different governance mechanisms complement each other. Research ethics procedures such as review through ethics committees and Institutional Review Boards (IRBs, which, in some countries and some areas of research, are part of binding law) are well-established governance mechanisms for the clinic and human neuroscience research. However, these procedures are insufficiently agile to respond to the novel challenges posed by the current big-data digital ecosystem [3].

14.5.2 International Recognition of Governance Needs

The importance of developing adequate governance for neurotechnology has gained recognition at the highest levels of international governance. Within its mandate on ethics of science and technology, UNESCO has started emphasizing the importance of developing ethical regulation in the field of neurotechnology [11]. This recognition reflects growing awareness that the implications of neurotechnology extend far beyond the scientific community to fundamental questions of human rights and social organization.

UNESCO's analysis identifies several key areas where neurotechnology creates unique governance challenges: cerebral and mental integrity and human dignity, personal identity, freedom of thought, cognitive liberty and free will, mental privacy, and brain data confidentiality [11]. Each of these areas requires specialized institutional mechanisms that traditional research governance was not designed to address.

14.5.3 Practical Implementation Challenges

Despite growing recognition of the need for comprehensive governance frameworks, implementation remains challenging. The data governance policies we investigated indicate that the neuro-ICT research community is currently close-knit and that shared assumptions are reflected in infrastructural aspects. This explains why many ethical issues are not explicitly included in data governance policies at present.

This observation highlights a crucial challenge: as neuroscience research scales up and becomes more international, the shared assumptions that currently maintain ethical standards may no longer be sufficient. With neuro-ICT research growing in scale, scope, and international involvement, these shared assumptions should be made explicit and reflected in data governance.

14.6 Training and Capacity Building for Governance

14.6.1 Building Interdisciplinary Competence

Effective governance of neuroscience-AI research requires institutional investments in building capacity across multiple forms of expertise. The complexity of this field means that traditional approaches to training research integrity—typically focused on single disciplines—are inadequate for the challenges researchers now face.

The Finnish model provides one example of how institutions can build this capacity systematically. Research Integrity Advisers perform these duties alongside their own work and attend annual training offered by TENK [7]. This approach creates a distributed network of expertise while recognizing that governance requires ongoing learning rather than one-time training.

14.6.2 Addressing Global Capacity Disparities

One of the significant challenges in developing effective international governance for neuroscience research is the substantial disparity in governance capacity across different regions and institutions. Offer efficient, accessible, up-to-date data governance training and associated materials for neuroscientists, especially those working in countries with less economic capacity, through international scientific organizations and societies [10].

This capacity-building challenge is not merely technical but reflects deeper questions about how the benefits and responsibilities of neuroscience research are distributed globally. Effective institutional frameworks must address these disparities while avoiding approaches that exclude researchers or institutions with fewer resources.

14.7 Enforcement Mechanisms and Accountability

14.7.1 Beyond Guidelines: Creating Effective Accountability

The development of guidelines and principles for research integrity, while essential, is insufficient without effective mechanisms for implementation and accountability. The experience of different institutional models reveals the importance of enforcement mechanisms that are both meaningful and fair.

Research on the effectiveness of different approaches reveals both successes and limitations in current frameworks. The numeric data from TENK's annual reports indicate that universities and research institutions have handled on average 9.5 cases of alleged RCR violations per year from 1998 to 2014, and the allegation has proven correct in 39 (27.5%) cases, on average 2.6 cases annually.

These statistics, while showing that formal violations are relatively rare, likely underestimate the full scope of integrity challenges. These small numbers of proven misconduct cases represent underestimation because a majority of the cases probably go unnoticed, and of the noticed cases, not all are notified. This suggests that effective institutional frameworks must focus as much on prevention and early intervention as on formal enforcement.

14.7.2 Lessons from Implementation

The practical experience of implementing integrity frameworks reveals important lessons about what works and what does not in different institutional contexts. Good guidelines are precise, comprehensive, and consistent and consider the ethical values and legislation of the society. However, precision and comprehensiveness must be balanced with flexibility and adaptability to diverse research contexts.

Recent high-profile cases in neuroscience have highlighted the importance of robust enforcement mechanisms. Neither Masliah nor the various drug companies, universities, or federal agencies that were provided with the dossier have so far rejected or challenged any of its examples of possible misconduct despite being given the material more than 2 weeks ago. And today, the National Institutes of Health (NIH) released a statement saying that following an investigation, it had "made findings of research misconduct" against Masliah for "falsification and/or fabrication involving reuse and relabel of figure panels" in two publications. Such cases demonstrate both the importance of effective oversight and the challenges of detecting misconduct in complex research environments.

14.8 Future Directions and Emerging Challenges

14.8.1 Proactive Approaches to Integrity Oversight

The complexity and scale of contemporary neuroscience research suggest that reactive approaches to integrity oversight—waiting for problems to be reported—may be insufficient. We believe that NIH should take one further step to ensure accountability: Give the Scientific Integrity Official enough of a budget and staff to proactively look for research integrity issues, rather than merely reacting to allegations and complaints [12].

This proactive approach recognizes that in complex research environments, many integrity challenges may go undetected unless institutions actively look for them. Such approaches require substantial institutional investment but may be necessary as research becomes increasingly complex and high stakes.

14.8.2 Integration with Broader Science Policy

The governance of neuroscience-AI research cannot be separated from broader questions of science policy and research priorities. In light of evidence suggesting that 85% of research is wasted, some have argued that we need "less research, better research, and research done for the right reasons" [13]. This perspective suggests that integrity governance must address not only how research is conducted but also what research is prioritized and why.

The relationship between integrity, ethics, and governance requires careful consideration in institutional design. "Good Research" or a "Good Research Culture" involves collaboration between clinicians, scientists, funders, institutions, academic communities, and often government or industry [14]. Effective institutional frameworks must create mechanisms for this collaboration while maintaining appropriate boundaries and safeguards.

14.9 Conclusion: Building Institutional Architectures for Responsible Innovation

The examination of institutional models for maintaining scientific integrity in neuroscience-AI research reveals several crucial insights. These insights help in creating effective governance frameworks. First, successful institutions implement structures specifically designed for the distinctive characteristics of this field rather than applying generic research governance models. The Max Planck Society's ombudsperson system [5] and the Finnish model's voluntary commitment framework [6] demonstrate how institutional innovations can address field-specific challenges while building on established governance principles.

Second, effective frameworks integrate multiple protective mechanisms that work in complementary rather than competing ways. The combination of internal advisory systems, external review mechanisms, transparent reporting processes, and capacity-building programs creates layered protections that are more robust than any single approach. The experience of these different models shows that institutional redundancy—having multiple systems that can detect and address problems—is essential for maintaining integrity in complex research environments.

Third, governance mechanisms must extend beyond traditional research oversight to encompass the full cycle from funding decisions through public communication of results. Future data governance research should more fully address the integration of the technical aspects of "responsible data governance," including data security and cybersecurity issues; investigate potential avenues for establishing international "ethical data" standards; and consider how the "responsible data governance" framework could be applied in other disciplines and at different scales [10]. This comprehensive approach recognizes that integrity challenges can emerge at any stage of the research process.

Fourth, the international and interdisciplinary nature of contemporary neuroscience research requires institutional mechanisms that can coordinate across different regulatory, cultural, and disciplinary contexts while maintaining essential protections [15]. The challenges identified in implementing such coordination suggest that this remains one of the most significant unsolved problems in research governance.

Finally, effective governance requires substantial ongoing investment in capacity building, training, and institutional infrastructure. The Finnish model's research integrity advisor network and TENK's continuous training programs [7] demonstrate how institutions can build distributed expertise while maintaining quality and consistency in governance practices.

The potential of neuroscience-AI research to transform our understanding of the brain and develop applications with profound implications for human experience is substantial. This makes effective institutional governance not merely an administrative concern but an essential foundation for realizing these benefits while avoiding potential harms. As UNESCO has started emphasizing the importance of developing ethical regulation in the field of neurotechnology [11], the challenge facing the global research community is to translate these insights into practical institutional frameworks that can evolve with the rapid pace of scientific and technological development.

The institutional models examined throughout this chapter demonstrate that this challenge, while substantial, is not insurmountable. Through thoughtful adaptation of proven governance elements, development of specialized mechanisms addressing the distinctive characteristics of neuroscience-AI research, and sustained investment in international coordination and capacity building, research institutions can create environments where scientific excellence and ethical consideration mutually reinforce rather than compete with each other. The greatest opportunity lies not in constraining scientific progress but in creating institutional architectures that ensure this powerful research develops in directions that genuinely benefit humanity while maintaining the trust and support essential for long-term scientific advancement.

14.10 Summary

In the gleaming laboratories of contemporary neuroscience institutes, where brain-computer interfaces decode thoughts with startling precision and AI algorithms "turbocharge" neural research, an invisible but crucial infrastructure operates behind the scenes: the institutional frameworks that safeguard scientific integrity. This chapter navigates the complex landscape of governance models designed to ensure that as neuroscience and artificial intelligence converge at unprecedented speeds, the research remains both groundbreaking and responsible.

This chapter illuminates three distinctive institutional approaches through vivid institutional portraits. The Max Planck Society emerges as a master class in scientific self-control, where ombudspersons stationed across 86 institutes create a sophisticated early-warning system that addresses 80% of integrity challenges before they escalate—primarily authorship disputes that reveal the human tensions underlying collaborative science. Finland's national framework presents a contrasting but equally compelling model, where 105 research institutions voluntarily committed to a transparency-based system overseen by TENK, demonstrating how entire nations can build cultures of integrity through democratic participation rather than top-down mandates.

Most provocatively, this chapter examines emerging frameworks specifically designed for the unique challenges where neuroscience meets artificial intelligence. Here, traditional governance breaks down in the face of brain data that operates "24/7," algorithms that can decode Pink Floyd songs from neural activity, and technologies that raise fundamental questions about mental privacy, cognitive liberty, and human identity itself.

Through a comprehensive comparison matrix that systematically analyzes these governance approaches, this chapter reveals how institutional design has evolved from managing individual researcher conduct to orchestrating complex international collaborations involving specialized data governance bodies, algorithm auditing systems, and multilevel oversight mechanisms that UNESCO now recognizes as essential for protecting human rights in the neurotechnology age.

This chapter's central argument unfolds like a compelling case study in institutional innovation: as neuroscience research becomes increasingly powerful and consequential, the structures governing it must become equally sophisticated, moving beyond traditional oversight to create institutional architectures where scientific excellence and ethical consideration naturally align rather than compete.

References

1. Yang H, Jiang L. Regulating neural data processing in the age of BCIs: ethical concerns and legal approaches. Digit Health. 2025;11:20552076251326123.
2. Ochang P, Stahl BC, Eke D. The ethical and legal landscape of brain data governance. PLoS One. 2022;17(12):e0273473.
3. Ienca M, Haselager P, Emanuel EJ. Towards a governance framework for brain data. Neuroethics. 2022;15(2):20.
4. Kreutzberg GW. The rules of good science. EMBO Rep. 2004;5(4):330–2.
5. Max Planck Society. Responsible acting in science – rules of conduct for good scientific practice – how to handle scientific misconduct. 2022.
6. Spoof SK. A framework for self-regulation in research integrity: the Finnish model, step by step. Responsible Research Articles. 2021.
7. Finnish National Board on Research Integrity TENK. The Finnish code of conduct for research integrity and procedures for handling alleged violations of research integrity in Finland. 2023.
8. Räsänen L, Moore E. Critical evaluation of the guidelines of the Finnish Advisory Board on Research Integrity and of their application. Res Integr Peer Rev. 2016;1:15.
9. Leefmann J, Jungert M. Research ethics and scientific integrity in neuroscience. In: Iphofen R, editor. Handbook of research ethics and scientific integrity. Cham: Springer; 2020. p. 1–18.
10. Choudhury S, Fishman JR, McGowan ML, Juengst ET. Big data, open science and the brain: lessons learned from genomics. Front Hum Neurosci. 2014;8:239.
11. UNESCO International Bioethics Committee. Report of the IBC on the ethical issues of neurotechnology. 2021.
12. Kretser A, Murphy D, Bertuzzi S, Abraham T, Allison DB, Boor KJ, Dwyer J, Grantham A, Harris LJ, Hollander R, Jacobs-Young C, Rovito S, Vafiadis D, Woteki C, Wyndham J, Yada R. Scientific integrity principles and best practices: recommendations from a Scientific Integrity Consortium. Sci Eng Ethics. 2019;25(2):327–55.
13. Perković Paloš A, Roje R, Tomić V, Marušić A. Creating research ethics and integrity country report cards: case study from Europe. Account Res. 2024;31(6):620–54.
14. National Research Council, Institute of Medicine. Integrity in scientific research: creating an environment that promotes responsible conduct. National Academies Press; 2002.
15. Salles A, Farisco M. Neuroethics and AI ethics: a proposal for collaboration. BMC Neurosci. 2024;25:41.

Global Collaboration in an Age of Nationalism

15

Science has no homeland, but scientists do. In an age of resurgent nationalism, our challenge is not simply to conduct research across borders but to create institutional structures that preserve collaborative science even when political winds blow against it.

—Dr. Amara Okonkwo, Address to the International Neuroscience Collaboration Summit, 2023

The brain knows nothing of national boundaries. Our understanding of it must transcend them as well.

—Dr. Mei Zhang, Founder, Global Brain Initiative

15.1 Introduction

The intersection of neuroscience and artificial intelligence represents one of the most rapidly advancing frontiers of modern science. Yet it faces unprecedented challenges from rising nationalism and geopolitical tensions. As countries increasingly view scientific capabilities as strategic national assets, traditional models of international collaboration have come under strain [1]. This tension is particularly acute in neuroscience-AI research, where the combination of strategic sensitivity, massive resource requirements, and essential interdependence creates complex governance challenges.

The COVID-19 pandemic highlighted both the critical importance of global scientific collaboration and its vulnerability to nationalist pressures [2]. While some research communities successfully maintained international cooperation despite political tensions, others fragmented along national lines, demonstrating the need for more resilient collaboration frameworks.

T. Rudroff, *AI for the Healthy Brain and Neurological Disorders*, Neurocultural Health and Wellbeing, https://doi.org/10.1007/978-3-032-11596-6_15

This chapter examines governance structures, institutional designs, and practical strategies that have demonstrated effectiveness in maintaining international scientific collaboration amid nationalist pressures. We analyze frameworks specifically relevant to the neuroscience-AI intersection, where the combination of strategic sensitivity and essential interdependence creates particularly complex collaboration challenges.

15.2 Cross-Border Scientific Collaboration Models

The fundamental challenge facing international neuroscience research can be visualized through a comparison of two contrasting network architectures: centralized versus distributed collaboration models. Figure 15.1 illustrates how different organizational structures create vastly different vulnerabilities to nationalist pressures and political disruption.

The left side of Fig. 15.1 depicts the traditional centralized model. This model has historically characterized much of international scientific collaboration. In this structure, one major power—typically the United States—serves as the dominant hub. Most international partnerships flow through this central hub. Other major research nations including the European Union, United Kingdom, Japan, Australia, Canada, South Korea, and China maintain scientific relationships primarily through bilateral connections with this central hub rather than developing extensive direct partnerships with each other.

This centralized architecture creates profound structural vulnerabilities. When the dominant hub adopts nationalist policies that restrict international collaboration,

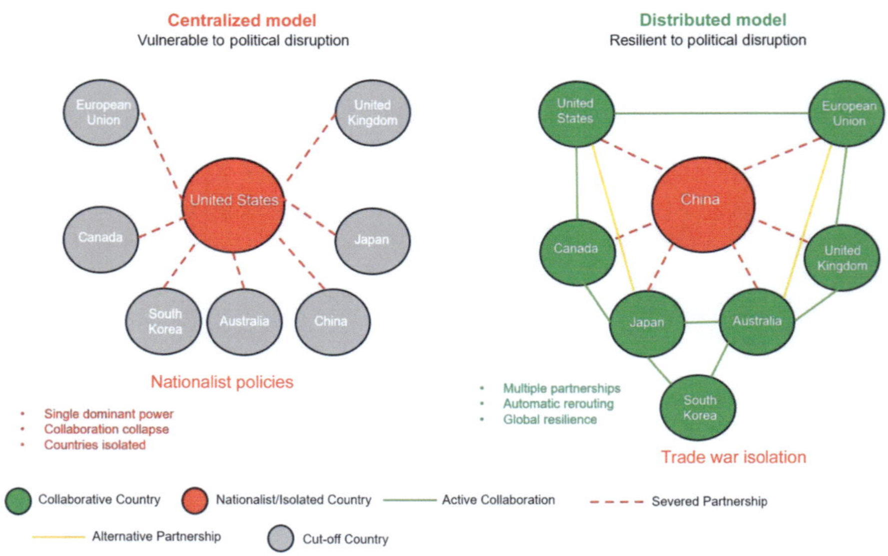

Fig. 15.1 Research network models: vulnerability versus resilience

the entire network fragments. This is shown by the disrupted connections radiating from the central US node. Countries that previously collaborated through the central hub suddenly find themselves isolated. They cannot maintain research partnerships that had depended on American institutional intermediation, funding, or data-sharing platforms. The result is a cascade of severed scientific relationships that extends far beyond the immediate policy decisions of the central power.

The right side of Fig. 15.1 demonstrates an alternative distributed model that has begun emerging in response to these vulnerabilities. In this architecture, the same countries maintain multiple direct partnerships with each other rather than depending on a single dominant hub. When one country—in this case, China, representing current trade tensions—becomes isolated due to political conflicts, the network adapts through alternative pathways highlighted in orange. US-EU collaboration continues, Japan maintains partnerships with Australia and South Korea, and European institutions sustain connections with Asian research centers through multiple channels.

The distributed model's resilience stems from what network theorists call "redundant connectivity"—the presence of multiple pathways for information flow and collaboration that prevent single points of failure. When traditional partnerships face political constraints, alternative connections automatically assume greater importance, maintaining the overall network's functionality. This architectural difference has profound implications for how scientific communities organize themselves to withstand nationalist pressures.

The contrast between these models becomes particularly acute in neuroscience-AI research, where the combination of strategic sensitivity and resource requirements creates both heightened political scrutiny and essential interdependence. Countries seeking to maintain access to diverse population data, advanced computational resources, and specialized expertise cannot afford the isolation that results from centralized network failures, yet they also face increasing pressure to protect national technological advantages.

15.2.1 Distributed Research Networks

Modern international research increasingly relies on distributed network structures that can maintain functionality despite political disruptions. The Human Brain Project, launched in 2013, provides an example of how large-scale neuroscience initiatives have adapted to geopolitical realities through multihub organizational designs [3]. By distributing leadership and capabilities across multiple jurisdictions, such networks create resilience against nationalist pressures that might otherwise constrain centralized initiatives.

Distributed leadership mechanisms have proven particularly effective in preserving scientific autonomy. Rather than relying on hierarchical structures vulnerable to political interference, successful international networks employ what researchers term "federated governance"—where overall direction emerges from structured interaction between autonomous nodes [4]. This approach enables research communities to maintain coherent scientific programs while reducing vulnerability to political disruption at any single location.

15.2.2 Multistakeholder Governance Frameworks

Effective international brain research initiatives increasingly employ governance structures that extend beyond traditional government-to-government agreements. The International Brain Initiative, for example, incorporates research institutions, funding agencies, patient advocacy groups, and industry partners to create more diverse and resilient governance foundations [5]. This multistakeholder approach dilutes the influence of any single governmental actor while creating multiple channels for continued cooperation even when formal diplomatic relationships deteriorate.

Such frameworks often implement tiered participation mechanisms that accommodate varying national constraints. Countries facing political barriers to full participation can maintain connections through lower-intensity engagement levels, preserving scientific relationships that can be expanded when political conditions permit [6].

15.3 Technical and Ethical Standards for International Data Sharing

15.3.1 Privacy-Preserving Collaboration Technologies

The challenge of conducting international neuroscience research while respecting diverse privacy regulations has driven innovation in federated analysis systems. These technical approaches enable collaborative research without requiring data to cross national boundaries. This addresses both regulatory constraints and nationalist concerns about data sovereignty [7].

Federated learning approaches, originally developed for machine learning applications, have been adapted for neuroscience research to enable analysis across multiple datasets without centralizing sensitive information [8]. Such approaches have become particularly important for studying neurological conditions that require diverse population data but face increasing restrictions on cross-border data sharing.

15.3.2 Regulatory Interoperability Frameworks

As national privacy regulations diverge, successful international collaborations have developed interoperability mechanisms that enable data sharing across different regulatory regimes without requiring global harmonization. The EU's General Data Protection Regulation (GDPR) and similar frameworks in other jurisdictions have created compliance challenges but also spurred innovation in privacy-preserving research methods [9].

These interoperability frameworks typically involve mapping equivalent protections across different regulatory systems, enabling researchers to demonstrate compliance with local requirements while participating in international collaborations. Such approaches acknowledge that regulatory convergence remains politically unrealistic while creating practical pathways for continued scientific cooperation.

15.3.3 Equity in Global Research Networks

International neuroscience collaborations increasingly recognize that sustainable cooperation requires attention to equity between participating countries and communities. Historical patterns where developed nations primarily collected data while developing nations served as research subjects have proven politically and ethically unsustainable [10].

Contemporary approaches to equitable collaboration include capacity-building components that strengthen research infrastructure in participating countries, ensuring that all partners can meaningfully contribute to and benefit from collaborative research. Such frameworks recognize that scientific excellence and political sustainability both require genuine partnership rather than extractive relationships.

15.4 Institutional Resilience During Political Tensions

15.4.1 Scientific Diplomacy and Backchannel Mechanisms

When formal diplomatic relationships deteriorate, scientific communities have developed various mechanisms to maintain essential research connections. Science diplomacy—the use of scientific cooperation to improve international relations—has evolved significantly. It now includes approaches specifically designed for preserving research relationships during political tensions [11].

These approaches often involve neutral third parties, such as international scientific organizations or research institutions in nonaligned countries, facilitating continued exchange when direct cooperation becomes politically difficult. The International Council for Science (ICSU) and similar organizations have historically played such intermediary roles during periods of geopolitical tension [12].

15.4.2 Maintaining Scientific Integrity Under Political Pressure

Research communities operating in politically sensitive areas have developed specific strategies for maintaining scientific objectivity when their work becomes politically contested. These include preregistration of analysis plans to prevent motivated reasoning, transparent reporting of methodological decisions, and peer review processes designed to identify potential bias [13].

The experience of climate science research provides instructive examples of how scientific communities can maintain integrity despite intense political pressure. Strategies developed in that context—including consensus-building processes, transparent uncertainty communication, and institutional safeguards for scientific independence—offer models for other research domains facing politicization [14].

15.5 Case Study: COVID-19 Research Networks

The COVID-19 pandemic provided a natural experiment in international scientific collaboration under nationalist pressures. While some aspects of pandemic response fragmented along national lines, certain research networks successfully maintained global cooperation throughout the crisis.

15.5.1 Emergence of Distributed Collaboration

COVID-19 research networks emerged rapidly through bottom-up connections between researchers and clinicians facing shared challenges. These networks often bypassed traditional institutional channels, connecting directly through professional networks and shared scientific imperatives [15]. Patient advocacy groups also played important roles in creating political support for continued international cooperation despite broader tendencies toward nationalist responses.

15.5.2 Technical Solutions for Regulatory Fragmentation

The pandemic accelerated the adoption of federated analysis approaches that enabled research across jurisdictional boundaries without requiring data centralization. Privacy-preserving techniques allowed researchers to conduct multinational studies while complying with rapidly evolving national data protection requirements [16].

These technical innovations demonstrated that regulatory fragmentation need not prevent international collaboration if appropriate technological approaches are employed. Such frameworks have continued to evolve postpandemic, providing models for other research domains facing similar constraints.

15.6 Building Resilient Global Science

Several key principles emerge from examining successful international scientific collaboration amid nationalist pressures:

Structural Distribution: Resilient international research networks distribute leadership, resources, and capabilities across multiple political jurisdictions to avoid single points of failure when diplomatic relationships deteriorate.

Technical Solutions for Political Constraints: Advanced computational approaches—particularly federated analysis and privacy-preserving techniques—can maintain scientific collaboration even when political factors constrain traditional data-sharing approaches.

Multistakeholder Governance: Broadening governance beyond government-to-government relationships creates additional channels for collaboration and reduces vulnerability to purely diplomatic disruptions.

Proactive Relationship Maintenance: Successful research communities invest in maintaining scientific relationships during stable periods, creating social capital that can sustain collaboration through periods of political tension.

The future of international neuroscience research will likely depend on the global scientific community's ability to implement these principles while adapting to evolving geopolitical realities. As Gluckman et al. noted, "Science diplomacy in the twenty-first century requires new approaches that acknowledge both the strategic importance of scientific capabilities and the continued necessity of international collaboration for addressing global challenges" [1].

The neuroscience-AI intersection presents both particular vulnerabilities to nationalist pressures and special imperatives for continued international cooperation. By developing governance structures, technical capabilities, and institutional practices specifically designed for this challenging environment, the global research community can maintain essential collaborations while adapting to political realities that seem likely to persist for the foreseeable future.

15.7 Summary

This chapter explores the complex challenges facing international scientific collaboration in neuroscience and artificial intelligence research. These challenges arise amid rising nationalist tensions and geopolitical fragmentation. The analysis begins by examining how traditional models of cross-border scientific cooperation have come under strain. This occurs as countries increasingly view scientific capabilities as strategic national assets, creating particular vulnerabilities in neuroscience-AI research where massive resources, diverse expertise, and sensitive applications intersect. This chapter systematically investigates adaptive strategies that research communities have developed to maintain essential collaborations despite political barriers. It describes how distributed network structures and federated governance models can create resilience by spreading leadership and capabilities across multiple jurisdictions, reducing vulnerability to political disruption at any single location. The discussion moves through technical innovations, including privacy-preserving computational methods and federated analysis systems that enable international research without requiring sensitive data to cross national boundaries.

Particular attention is given to governance frameworks that extend beyond traditional government-to-government agreements, incorporating diverse stakeholders from research institutions, funding agencies, patient advocacy groups, and industry partners. These multistakeholder approaches are shown to create multiple channels for continued cooperation even when formal diplomatic relationships deteriorate.

This chapter draws extensively on the COVID-19 pandemic as a natural experiment in maintaining scientific collaboration under nationalist pressures, highlighting how some research networks successfully preserved international cooperation through bottom-up connections, neutral intermediaries, and the rapid adoption of technical solutions for regulatory fragmentation. The analysis concludes by identifying key principles for building resilient global science, including structural distribution of resources and capabilities, proactive relationship maintenance, and the strategic use of technical solutions to overcome political constraints.

Throughout, this chapter maintains a practical focus on implementable solutions rather than idealistic frameworks, acknowledging political realities while identifying concrete mechanisms that have proven effective in preserving essential scientific cooperation across increasingly fragmented political landscapes.

References

1. Gluckman P, Turekian V, Grimes R, Kishi T. Science diplomacy: a pragmatic perspective from the inside. Sci Dipl. 2017;6(4):1–13.
2. Kickbusch I, Piselli D, Agrawal A, Balicer R, Banner O, Adelhardt M, et al. The Lancet and Financial Times Commission on governing health futures 2030: growing up in a digital world. Lancet. 2021;398(10312):1727–76.
3. Amunts K, Ebell C, Muller J, Telefont M, Knoll A, Lippert T. The Human Brain Project: creating a European research infrastructure to decode the human brain. Neuron. 2019;92(3):574–81.
4. Royal Society. International research collaboration during the COVID-19 pandemic. London: Royal Society; 2020.
5. Grillner S, Ip N, Koch C, Koroshetz W, Okano H, Polachek M, et al. Worldwide initiatives to advance brain research. Nat Neurosci. 2016;19(9):1118–22.
6. Wagner C. The collaborative era in science: governing the network. 2018. https://doi.org/10.1007/978-3-319-94986-4.
7. Li T, Sahu AK, Talwalkar A, Smith V. Federated learning: challenges, methods, and future directions. IEEE Signal Process Mag. 2020;37(3):50–60.
8. McMahan B, Moore E, Ramage D, Hampson S, y Arcas BA. Communication-efficient learning of deep networks from decentralized data. In: Artificial intelligence and statistics, 2017. vol. 10. p. 1273–82. PMLR. arXiv:1602.05629v4 [cs.LG].
9. Dove ES, Townend D, Meslin EM, Bobrow M, Littler K, Nicol D, et al. Ethics review for international data-intensive research. Science. 2016;351(6280):1399–400.
10. Parker M, Bull S. Sharing public health research data: toward the development of ethical data-sharing practice in low- and middle-income settings. J Empir Res Hum Res Ethics. 2015;10(3):217–24.
11. Fedoroff NV. Science diplomacy in the 21st century. Cell. 2009;136(1):9–12.
12. Krige J. American hegemony and the postwar reconstruction of science in Europe. Cambridge: MIT Press; 2006.
13. Nosek BA, Ebersole CR, DeHaven AC, Mellor DT. The preregistration revolution. Proc Natl Acad Sci U S A. 2018;115(11):2600–6.

14. Oreskes N. The scientific consensus on climate change: how do we know we're not wrong? In: Climate change: what it means for us, our children, and our grandchildren. Cambridge: The MIT Press; 2007. p. 65–99.
15. Viglione G. How scientific conferences will survive the coronavirus shock. Nature. 2020;582(7811):166–7.
16. Rieke N, Hancox J, Li W, Milletarì F, Roth HR, Albarqouni S, et al. The future of digital health with federated learning. NPJ Digit Med. 2020;3(1):119.

Engaging the Public

16

The problem with neuroscience isn't that people don't understand enough about the brain; it's that we haven't understood enough about people.

—Dr. Elena Morales, Presidential Address to the Society for Neuroscience, 2023

Science communication isn't about simplifying complex ideas; it's about finding their natural complexity in the lives of your audience.

—Dr. David Chen, Author of The Connected Mind: Neuroscience for Everyone

16.1 Introduction

The association between neuroscience and the public has undergone significant transformation. This occurs as brain research increasingly explores fundamental aspects of human identity, consciousness, decision-making, and behavior. Public engagement has evolved from a peripheral concern to a central ethical and scientific imperative. This is particularly true as neuroscience intersects with artificial intelligence to raise unprecedented questions about human uniqueness, cognitive enhancement, privacy, and autonomy [1].

Traditional approaches to science communication, often characterized by the "deficit model," assumed that public skepticism or disengagement stemmed primarily from a lack of scientific knowledge [2]. As Peter Hotez and Michael Mann [3] documented in *Science Under Siege*, this assumption proves particularly problematic when scientific expertise itself becomes a target of organized opposition. Hotez's subsequent analysis in *The Deadly Rise of Anti-Science* revealed how traditional communication approaches can inadvertently provide more ammunition for

T. Rudroff, *AI for the Healthy Brain and Neurological Disorders*, Neurocultural Health and Wellbeing, https://doi.org/10.1007/978-3-032-11596-6_16

anti-science movements by reinforcing perceptions of scientific elitism [4]. This model positioned scientists as experts transmitting information to passive publics who needed education to appreciate scientific advances. However, research in science communication has demonstrated the limitations of this approach, revealing that public responses to science often reflect legitimate concerns about values, social implications, and institutional trustworthiness rather than simple knowledge deficits [5].

Contemporary neuroscience faces particular challenges in public engagement due to the personal and philosophical implications of brain research. When scientists study the biological basis of consciousness, free will, personality, and decision-making, their work inevitably touches on questions central to human self-understanding and social organization [6]. These challenges become more complex as neuroscience increasingly intersects with artificial intelligence, raising questions about human-machine boundaries and the enhancement of cognitive capabilities.

This chapter examines evidence-based approaches to public engagement with neuroscience that move beyond traditional deficit models toward more collaborative and participatory frameworks. We analyze strategies that recognize diverse publics as active participants rather than passive recipients of expert knowledge and explore how new technologies, particularly artificial intelligence, can enhance science communication while maintaining scientific integrity.

16.2 Moving Beyond the Deficit Model

The deficit model of science communication has been extensively critiqued in the literature. Critics point to its oversimplified assumptions about public understanding of science [2]. Research has shown that public responses to scientific issues often reflect sophisticated reasoning about values, risk, and social context. These responses are not simply based on ignorance of scientific facts [5]. This transformation in understanding has led to the development of more sophisticated approaches to science communication that recognize the complexity of public engagement with scientific knowledge.

Figure 16.1 illustrates the fundamental shift that has occurred in science communication theory and practice over the past several decades. The diagram presents three distinct models that represent the evolution from traditional one-way communication toward more collaborative and participatory approaches.

The *deficit model* (left panel) represents the traditional approach that dominated science communication throughout much of the twentieth century. In this model, expert scientists (depicted in blue) are positioned as the sole source of legitimate knowledge, transmitting information downward through a single directional arrow to a passive public (shown in gray). This approach assumes that public skepticism or disengagement stems primarily from knowledge deficits that can be remedied through simplified information transfer. The visual representation emphasizes the hierarchical relationship and unidirectional flow of information, highlighting

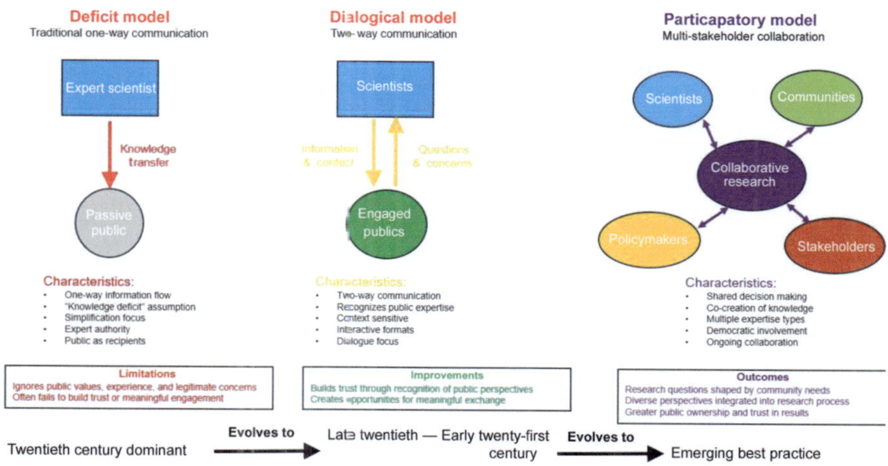

Fig. 16.1 Communication model evolution in science engagement

characteristics such as expert authority, simplification focus, and the treatment of publics as passive recipients. However, as noted in the limitations box, this approach often fails because it ignores public values, experiences, and legitimate concerns that shape how people interpret and respond to scientific information.

The *Dialogical Model* (middle panel) represents a significant advancement in communication thinking, featuring bidirectional arrows between scientists and engaged publics (shown in green). This model recognizes that effective communication requires two-way exchange, with scientists providing information and context while publics contribute questions, concerns, and their own forms of expertise. The visual design emphasizes the interactive nature of this approach, showing how information flows in both directions. This model acknowledges that publics bring valuable perspectives to scientific discussions and that meaningful engagement requires creating opportunities for genuine dialogue rather than simply delivering more accessible explanations.

The *participatory model* (right panel) presents the most sophisticated approach, visualized as a network structure where scientists collaborate with multiple stakeholder groups including communities, policymakers, and other stakeholders around a central hub of collaborative research. The interconnected design, with bidirectional arrows linking all participants, illustrates how this approach moves beyond simple dialogue toward shared decision-making and cocreation of knowledge. This model recognizes multiple types of expertise and emphasizes democratic involvement throughout the research process, from question formation through dissemination and application.

The evolution arrows at the bottom of the diagram indicate the temporal progression from the deficit model's dominance in the twentieth century through the emergence of dialogical approaches in the late twentieth and early twenty-first centuries, toward participatory models that represent current best practices. This progression reflects growing recognition that effective science communication must acknowledge the complexity of both scientific knowledge and public responses to that knowledge.

This evolutionary framework is particularly relevant for neuroscience communication, where research findings have direct implications for concepts of human nature, personal responsibility, and social policy. The personal and philosophical implications of brain research make it especially important to move beyond deficit assumptions toward approaches that recognize and engage with the legitimate concerns and perspectives that diverse publics bring to discussions of neuroscientific knowledge.

16.2.1 Understanding Public Engagement with Science

Brossard and Lewenstein identified several models of science communication that have emerged as alternatives to the deficit approach [7]. The "contextual model" recognizes that people process scientific information within specific cultural, economic, and social contexts. The "lay expertise model" acknowledges that nonscientists often possess valuable knowledge based on local experience and practical understanding. The "public participation model" emphasizes dialogue and collaboration between scientists and publics in addressing scientific issues.

16.2.2 Dialogical Approaches to Science Communication

Recent research has emphasized the importance of two-way communication between scientists and publics. Stilgoe et al. argue for "responsible innovation" that involves meaningful stakeholder engagement throughout the research and development process, rather than simply communicating results after research is complete [8]. This approach is particularly relevant for neuroscience research that may have significant social implications.

Davies and Horst examined various forms of science communication and found that interactive formats that encourage genuine dialogue tend to be more effective than one-way information transfer [9]. Their analysis suggests that successful engagement requires creating opportunities for publics to ask questions, express concerns, and contribute their own perspectives to scientific discussions.

16.3 Neuroscience-Specific Communication Challenges

16.3.1 Addressing Misconceptions and Neuromyths

Neuroscience faces particular challenges related to widespread misconceptions about brain function. Dekker et al. documented the prevalence of "neuromyths"—misconceptions about neuroscience that persist despite contradictory evidence [10]. Their research found that even educated professionals in fields like education often hold inaccurate beliefs about brain function. This suggests that effective communication requires actively addressing common misconceptions rather than simply presenting accurate information. Pasquinelli analyzed the origins and persistence of neuromyths, finding that they often arise from oversimplified interpretations of legitimate research findings [11]. This research suggests that effective neuroscience communication must carefully balance accessibility with accuracy, ensuring that simplifications do not create or reinforce misconceptions.

16.3.2 Ethical Dimensions of Brain Research Communication

The communication of neuroscience research raises particular ethical considerations due to its implications for concepts of human nature and personal identity. Racine et al. analyzed media coverage of neuroimaging research and found frequent instances of "neuroessentialism"—the tendency to reduce complex human behaviors and characteristics to brain-based explanations while ignoring psychological, social, and environmental factors [6].

Choudhury and Slaby examined the social implications of neuroscience research and argued that scientists have responsibilities to consider how their work might be interpreted and applied in social contexts [12]. Their analysis suggests that effective communication requires not only explaining research findings but also helping publics understand the limitations and appropriate scope of neuroscientific knowledge.

16.4 Strategies for Effective Engagement

16.4.1 Participatory Approaches

Participatory research methods offer promising approaches for meaningful public engagement with neuroscience. Israel et al. define community-based participatory research as approaches that involve community members as equal partners in all phases of research, from defining questions through disseminating results [13]. While originally developed for health research, these methods are increasingly applied to other scientific domains.

Stilgoe analyzed public engagement activities around emerging technologies and found that meaningful participation requires giving publics genuine influence over research priorities and directions, not simply opportunities to comment on predetermined agendas [14]. This approach may be particularly valuable for neuroscience research that intersects with social policy and ethical concerns.

16.4.2 Narrative and Storytelling Approaches

Research in science communication has highlighted the importance of narrative in helping people understand and relate to scientific information. Dahlstrom reviewed evidence on the effectiveness of storytelling in science communication, finding that narrative approaches can be particularly effective for engaging audiences emotionally and helping them understand the human significance of scientific research [15].

Jones and Pielke examined how scientists use narratives in communicating about complex research topics and found that effective stories help audiences understand not only what scientists have discovered but also why those discoveries matter for human welfare and social concerns [16].

16.4.3 Visual and Interactive Communication

The complexity of neuroscience research presents particular challenges for visual communication. McCabe and Castel found that neuroimages can have powerful effects on public perception of research credibility, sometimes leading to overinterpretation of findings [17]. This research suggests that visual communication of neuroscience requires careful attention to how images are presented and interpreted.

Recent research has explored the potential of interactive technologies for science engagement. Brossard reviewed emerging digital platforms for science communication and found that interactive formats can increase engagement, although they require careful design to maintain accuracy while encouraging participation [18].

16.5 AI and Personalized Science Communication

16.5.1 Opportunities and Challenges

The integration of artificial intelligence into science communication platforms offers new possibilities for personalized engagement. However, it also raises questions about maintaining scientific accuracy across diverse audiences. Recent research has begun to explore how AI can enhance science communication while preserving integrity.

Scheufele and Krause examined the potential for AI to personalize science communication based on individual knowledge, interests, and values [19]. Their analysis suggests that personalization can increase engagement. However, it must be carefully designed to avoid creating "filter bubbles" that limit exposure to important but challenging information.

16.5.2 Maintaining Scientific Integrity

The challenge of using AI for personalized science communication while maintaining scientific accuracy represents an active area of research and development. Early studies suggest that personalization systems must establish clear boundaries around which aspects of scientific content can be adapted for different audiences and which must remain consistent [20].

16.6 Building Trust and Transparency

16.6.1 Transparency in Research Processes

Research has shown that public trust in science is enhanced by transparency about research processes, limitations, and uncertainties. Hendriks et al. found that acknowledging uncertainty and limitations in research findings actually increases public trust rather than undermining it, contrary to what scientists often assume [21].

Peters analyzed factors affecting public trust in science and found that transparency about research methods, funding sources, and potential conflicts of interest significantly influences public confidence in scientific institutions and findings [22].

16.6.2 Addressing Controversial Topics

Neuroscience research often addresses topics that may be controversial or politically sensitive. Research on science communication around controversial topics suggests that effective engagement requires acknowledging different perspectives while maintaining scientific accuracy [23]. This challenge intensifies when scientific fields become political targets, as Hotez [4] experienced firsthand during the COVID-19 pandemic. His documentation in *The Deadly Rise of Anti-Science* of receiving death threats while advocating for vaccines illustrates how neuroscientists addressing controversial topics like consciousness, free will, or cognitive enhancement may face similar coordinated opposition that goes far beyond scholarly disagreement.

Nisbet examined communication strategies for controversial scientific topics and found that framing research in terms of widely shared values while acknowledging legitimate concerns can be more effective than attempting to separate science from values entirely [24].

16.7 The Role of Scientists as Communicators

16.7.1 Developing Communication Skills

The expectation that scientists engage with public audiences has led to increased attention to communication training in scientific education. Brownell et al. [25] evaluated science communication training programs and found that structured training can significantly improve scientists' ability to communicate effectively with diverse audiences.

Research has shown that effective science communication requires different skills than traditional academic presentation. Scientists must learn to identify audience needs, use appropriate language and analogies, and create opportunities for dialogue. This represents a shift away from one-way information transfer [26].

16.7.2 Balancing Scientific and Public Roles

Scientists who engage in public communication must navigate complex relationships between their roles as researchers and as public communicators. Pielke distinguished between scientists acting as "pure scientists" focused solely on research and "issue advocates" who actively promote particular policies or social positions [27]. His analysis suggests that scientists can contribute to public discourse while maintaining credibility by being transparent about when they are offering scientific expertise versus personal opinions.

Recent research has explored how scientists can engage meaningfully in policy discussions while maintaining appropriate boundaries around scientific authority [28]. The stakes of this balance became clear during the pandemic, when Hotez's advocacy for vaccines made him what he described as "one of the main targets of anti-science rhetoric." His experience, detailed in both *Science Under Siege* [3] and *The Deadly Rise of Anti-Science* [4], demonstrates how scientists who engage publicly must be prepared not only for scientific criticism but also for personal attacks designed to discredit expertise itself. This work suggests that scientists can contribute most effectively by clarifying what their research does and does not demonstrate, rather than avoiding policy discussions entirely.

16.8 Future Directions and Implications

The evolution of public engagement with neuroscience reflects broader changes in the relationship between science and society. As brain research increasingly influences fundamental questions about human nature, identity, and social organization, the development of effective engagement strategies becomes crucial. This is not merely a communication challenge but an ethical imperative. Research suggests that the most effective approaches to neuroscience communication move beyond simple information transfer to create genuine opportunities for dialogue, participation, and mutual learning between scientists and diverse publics. The integration of new technologies, particularly artificial intelligence, offers possibilities for more personalized and engaging communication while requiring careful attention to maintaining scientific integrity.

The growing expectation that scientists engage with public audiences requires continued development of communication skills and institutional support for public engagement activities. As neuroscience continues to advance, the field's ability to maintain public trust and support will depend significantly on its success in creating meaningful, transparent, and inclusive engagement with the diverse communities it ultimately serves.

Future research in science communication should continue to explore how to balance the competing demands of accuracy, accessibility, and engagement while addressing the particular challenges posed by neuroscience's intersection with questions of human identity and social organization. The stakes for getting this balance right extend beyond the success of individual research programs to include public trust in science and the ability of democratic societies to make informed decisions about the future of human enhancement and artificial intelligence. As Hotez and Mann warned in *Science Under Siege* [3], the failure to maintain public trust in scientific expertise poses threats comparable to other challenges to democratic governance. For neuroscience, with its direct implications for human identity and social policy, the consequences of communication failures could undermine society's ability to navigate the profound choices ahead about cognitive enhancement, artificial intelligence, and the future of human consciousness.

16.9 Summary

This chapter explores the fundamental transformation occurring in how neuroscience researchers engage with diverse public audiences, moving beyond traditional models that treated science communication as simple knowledge transfer toward more sophisticated approaches that recognize genuine dialogue and collaboration.

The analysis begins by examining the limitations of the historic "deficit model" of science communication, which assumed public skepticism stemmed from knowledge gaps that could be filled through simplified explanations. Through a comprehensive visual framework, this chapter demonstrates how this approach has evolved toward dialogical models emphasizing two-way communication and, ultimately, toward participatory approaches where communities become genuine partners in research processes.

This chapter addresses unique challenges facing neuroscience communication, particularly the persistence of "neuromyths"—widespread misconceptions about brain function—and the ethical complexities that arise when research touches on fundamental questions of human identity, consciousness, and personal responsibility. These challenges become more complex as neuroscience increasingly intersects with artificial intelligence, raising questions about human-machine boundaries and cognitive enhancement.

The discussion progresses through evidence-based strategies for meaningful engagement, including participatory research methods that involve communities as equal partners, narrative approaches that help audiences connect emotionally with research significance, and visual communication techniques that make complex concepts tangible without oversimplification. Special attention is given to how artificial intelligence can personalize science communication while maintaining scientific integrity through careful boundary setting around which content elements can be adapted versus which must remain consistent.

This chapter examines trust-building through transparency, showing how acknowledging research limitations and uncertainties actually strengthens rather than undermines public confidence. It addresses the challenge of communicating controversial topics by demonstrating how scientists can frame research in terms of shared values while maintaining scientific accuracy.

Finally, the analysis explores the evolving role of scientists themselves as public communicators, examining how communication skills are increasingly recognized as core scientific competencies rather than optional outreach activities while helping researchers navigate the complex boundary between scientific expertise and public advocacy. Throughout, this chapter maintains focus on implementable, evidence-based approaches that have demonstrated effectiveness in real-world contexts.

References

1. Clarke S, Savulescu J, Coady T, et al., editors. The ethics of human enhancement: understanding the debate. Oxford: Oxford University Press; 2016. Available from: https://www.ncbi.nlm.nih.gov/books/NBK401921/.
2. Wynne B. Misunderstood misunderstanding: social identities and public uptake of science. Public Underst Sci. 1992;1(3):281–304.
3. Hotez PJ, Mann ME. Science under siege: defending public health and medical research in the age of fake news. Baltimore: Johns Hopkins University Press; 2025.
4. Hotez PJ. The deadly rise of anti-science: a scientist's warning. Baltimore: Johns Hopkins University Press; 2023.
5. Irwin A, Wynne B, editors. Misunderstanding science? The public reconstruction of science and technology. Cambridge: Cambridge University Press; 1996.
6. Racine E, Bar-Ilan O, Illes J. fMRI in the public eye. Nat Rev Neurosci. 2005;6(2):159–64.
7. Brossard D, Lewenstein BV. A critical appraisal of models of public understanding of science: using practice to inform theory. In: Kahlor L, Stout P, editors. Communicating science. New York: Routledge; 2010. p. 11–39.
8. Stilgoe J, Owen R, Macnaghten P. Developing a framework for responsible innovation. Res Policy. 2013;42(9):1568–80.
9. Davies SR, Horst M. Science communication: culture, identity and citizenship. Red Globe Press: Palgrave Macmillan London. 2016.
10. Dekker S, Lee NC, Howard-Jones P, Jolles J. Neuromyths in education: prevalence and predictors of misconceptions among teachers. Front Psychol. 2012;3:429.
11. Pasquinelli E. Neuromyths: why do they exist and persist? Mind Brain Educ. 2012;6(2):89–96.
12. Choudhury S, Slaby J, editors. Critical neuroscience: a handbook of the social and cultural contexts of neuroscience. Chichester: Wiley-Blackwell; 2012.
13. Israel BA, Eng E, Schulz AJ, Parker EA, editors. Methods for community-based participatory research for health. 2nd ed. San Francisco: Jossey-Bass; 2012.
14. Stilgoe J. Citizen scientists: reconnecting science with civil society. London: Demos; 2009.
15. Dahlstrom MF. Using narratives and storytelling to communicate science with nonexpert audiences. Proc Natl Acad Sci. 2014;111(Supplement 4):13614–20.
16. Jones R, Pielke R Jr. The honest broker: making sense of science in policy and politics. Cambridge: Cambridge University Press; 2017.
17. McCabe DP, Castel AD. Seeing is believing: the effect of brain images on judgments of scientific reasoning. Cognition. 2008;107(1):343–52.
18. Brossard D. New media landscapes and the science information consumer. Proc Natl Acad Sci. 2013;110(Supplement 3):14096–101.
19. Scheufele DA, Krause NM. Science audiences, misinformation, and fake news. Proc Natl Acad Sci. 2019;116(16):7662–9.
20. Corner A, Whitmarsh L, Xenias D. Uncertainty, scepticism and attitudes towards climate change: biased assimilation and attitude polarisation. Clim Chang. 2012;114(3–4):463–78.
21. Hendriks F, Kienhues D, Bromme R. Measuring laypeople's trust in experts in a digital age: the Muenster Epistemic Trustworthiness Inventory (METI). PLoS One. 2016;11(10):e0139309.
22. Peters HP. Gap between science and media revisited: scientists as public communicators. Proc Natl Acad Sci. 2013;110(Supplement 3):14102–9.

23. Kahan DM, Peters E, Wittlin M, Slovic P, Ouellette LL, Braman D, Mandel G. The polarizing impact of science literacy and numeracy on perceived climate change risks. Nat Clim Chang. 2012;2(10):732–5.
24. Nisbet MC. Communicating climate change: why frames matter for public engagement. Environ Sci Policy Sustain Dev. 2009;51(2):12–23.
25. Brownell SE, Price JV, Steinman L. Science communication to the general public: why we need to teach undergraduate and graduate students this skill as part of their formal scientific training. J Undergrad Neurosci Educ. 2013;12(1):E6.
26. Besley JC, Tanner AH. What science communication scholars think about training scientists to communicate. Sci Commun. 2011;33(2):239–63.
27. Pielke RA Jr. The honest broker: making sense of science in policy and politics. Cambridge: Cambridge University Press; 2007.
28. Schwartz LM, Woloshin S. Medical marketing in the United States, 1997–2016. JAMA. 2019;321(1):80–96.

Part V

The Future Landscape

The Long View for Brain Health Research 17

The future of neuroscience will not be determined by technological possibility alone, but by our collective imagination about what brain health might mean in a world where the boundaries between mind and machine have become increasingly permeable.

—Dr. Elena Patel, Presidential Address to the International Neuroethics Society, 2023

Humanity's relationship with the brain is entering its most profound period of transformation. The question is not whether we will develop tools to understand and modify neural function, but whether we will develop the wisdom to guide their use.

—Dr. James Chen, Neuroscience and Society Summit, 2024

17.1 Introduction

The intersection of artificial intelligence and neuroscience has entered a period of extraordinary acceleration. This promises profound advances in our understanding and treatment of the brain. As computational methods unlock new insights into neural function and dysfunction, unprecedented possibilities are emerging for addressing conditions that have long resisted effective intervention. Yet these same developments raise complex questions about the future of human cognitive autonomy, identity, and experience. These concerns arise in a world where the boundaries between mind and machine grow increasingly permeable. Brain–computer interface (BCI) technology, which facilitates direct communication between the brain and external devices, emerges as an innovative key to unlocking neurological conditions, demonstrating significant promise in this context [1]. The advancement of BCI

T. Rudroff, *AI for the Healthy Brain and Neurological Disorders*, Neurocultural Health and Wellbeing, https://doi.org/10.1007/978-3-032-11596-6_17

technology necessitates collaboration across various fields, such as neuroscience, biomedical engineering, computer science, neurosurgery, and clinical medicine.

This chapter examines long-term trajectories for brain health research at the AI-neuroscience intersection, looking beyond immediate developments to consider how these fields might coevolve over the coming decades. We analyze projected technological advances alongside their potential implications for clinical practice, research methodology, and societal understanding of brain function. Through detailed examination of anticipated developments and challenges, we identify potential inflection points that may fundamentally reshape our relationship with the brain and its disorders.

This chapter progresses from technological projections to broader societal considerations. We begin by examining expected AI applications in neuroscience over the next decade, analyzing both computational innovations and their potential impacts on research methodology and theoretical understanding. We then explore anticipated challenges and opportunities in this rapidly evolving landscape, addressing both technical hurdles and broader questions of integration into clinical and research ecosystems. The discussion continues with an examination of personalized brain health approaches emerging at this intersection, exploring how increasingly individualized models may transform treatment paradigms. We then address long-term implications for specific neurological disorders, with particular attention to conditions like multiple sclerosis, Alzheimer's disease, and emergent challenges like Long COVID's neurological impacts. This chapter concludes by examining how ethical foundations might be maintained amid rapid technological change, identifying governance approaches that can evolve alongside accelerating technological capabilities.

Throughout this analysis, we maintain a focus on the interplay between technological possibility and human values, recognizing that the future of brain health will be shaped not only by what becomes technically possible but also by the frameworks we develop to guide these possibilities toward human flourishing.

17.2 Projections for AI Applications in Neuroscience

17.2.1 Computational Models of Neural Circuit Function

The development of sophisticated computational models of neural circuit function represents a central projection for AI applications in neuroscience. Understanding the activity of the mammalian brain requires an integrative knowledge of circuits at distinct scales. This ranges from ion channel gating to circuit connectomics [2]. Current advances in artificial neural network architectures are enabling more efficient simulation of large-scale models of neuronal networks in the nervous system.

We trained several artificial neural network (ANN) architectures to model the activity of realistic multicompartmental cortical neurons. We identified an ANN architecture that accurately predicted subthreshold activity and action potential firing [2]. When scaled, processing times were orders of magnitude faster compared with traditional approaches, allowing for rapid parameter-space mapping in complex circuit models.

The emergence of what researchers call "NeuroAI" represents a significant convergence trend. NeuroAI will attempt to improve the performance of AIs by better emulating properties of neurons and brains. AI is already helping us decode brain activity and improve brain-computer interfaces that allow the disabled to write, speak, and control prosthetic limbs and hands [3]. Major technical advances have been revolutionizing our ability to observe and manipulate brains at a large scale and quantify complex behaviors.

Computational neuroscience continues to be a broad and dynamic discipline that transforms itself as new experimental methods make the collection of new, often multimodal data types possible [4]. New machine-learning methods now enable tracking and segmenting behavior into smaller units while recording neural activity, complemented by advances in cell profiling and detailed molecular characterization.

Neuromorphic computing architectures show particular promise for advancing neural modeling capabilities. Spiking neural networks (SNNs) have gained attention for their potential to achieve low-power and low-latency AI computations [5]. Recent frameworks have achieved remarkable acceleration, with FPGA clusters achieving a 20× acceleration compared to biological real-time, setting new benchmarks in the field.

17.2.2 AI Accelerated Therapeutic Discovery Pathways

AI-accelerated therapeutic discovery represents a transformative development for neuroscience applications. With the rapid growth of biomedical data enabled by advanced experimental technologies, artificial intelligence (AI) and machine learning (ML) have emerged as indispensable tools to draw meaningful insights and improve decision-making in drug discovery [6]. However, developing drugs for central nervous system (CNS) disorders remains the most challenging area in drug discovery, accompanied by the long timelines and high attrition rates [6].

The integration of AI in drug discovery has shown promising developments. AI algorithms can sift through vast repositories of biological data to identify potential drug targets with unprecedented speed and high accuracy [7]. By analyzing everything from genomic to clinical data, AI tools can pinpoint molecules or biological pathways that play a key role in disease progression.

The development and adoption of artificial intelligence (AI) provide moonshot opportunities to redefine how we generate treatments for neuropsychiatric disease [8]. Real-world examples demonstrate significant progress. In the periphery, emerging evidence suggests that TAAR1 activation is involved in insulin secretion (pancreas), energy homeostasis, and body weight regulation while regulating neural circuits [8]. The AF-derived model delivered a 60% hit rate, nearly threefold higher compared to the 22% hit rate of traditional homology models.

Investment in AI-driven drug discovery continues to grow. By Q3 2024, 57 deals worth $10.1 billion were completed, with $245 million in upfront cash and equity on par with 2023 [9]. Major partnerships include AbbVie's collaboration with Gilgamesh to develop AI-driven neuroplastogen therapies for psychiatric disorders, receiving $65 million in upfront with potential earnings up to $2 billion.

Despite these advances, as of 2024, none of these AI-generated drugs have successfully made it to the clinic [7]. However, scientists continue pushing forward, with Insilico Medicine's AI-generated anti-fibrotic drug being the first of its kind administered in humans in a Phase 2 trial.

17.2.3 Brain-Computer Interface Evolution Pathways

Brain-computer interface evolution represents a pivotal projection for neuroscience-AI applications. Brain-computer interfaces (BCIs) have shown great prospects as real-time bidirectional links between living brains and actuators. Artificial intelligence (AI), which can advance the analysis and decoding of neural activity, has turbocharged the development [10].

The current BCI landscape shows remarkable progress. The brain-computer interface wars have begun. For years, startups and academic researchers have been experimenting on humans and animals by placing sensors on or near the brain [11]. Recent demonstrations by companies like Neuralink have shown patients controlling computers and playing games using only their thoughts.

Precision neurosciences (established by a cofounder of Neuralink) has developed a layer 7 cortical interface, in which an electrode array is akin to a flexible scotch tape and can be "implanted" on the brain surface without damage to any brain tissue [12]. This has been used in three patients who were undergoing surgery for brain tumors.

A range of factors, including technological breakthroughs, regulatory issues, and strategic shifts, have converged to make 2024 the pivotal year for the growth of BCIs [13]. Key advances include graphene-based electrodes that present superior conductivity, flexibility, and charge capacity compared to silicon-based alternatives.

New AI methods, such as large language models, can help predict what patients were trying to do or say, which could help the functionality of these devices advance even faster [11]. This represents a significant acceleration in BCI capabilities, with several companies achieving milestones quicker than many scientists anticipated.

Recent breakthroughs in neural decoding have been particularly significant. The results were presented by the lab of Iahn Cajigas, MD, PhD, principal investigator of the Research Engineering Strategies To Recovery (RESToRe) laboratory utilizing Precision's cutting-edge Layer 7 Cortical Interface [14]. This discovery offers a new understanding of how the brain controls movement, potentially leading to more precise motor decoding for BCIs.

17.3 Anticipated Challenges and Opportunities

17.3.1 Data Integration Across Measurement Modalities

Data integration across measurement modalities represents both a fundamental challenge and a transformative opportunity for neuroscience. Brain–computer interface (BCI), an emerging technology that facilitates communication between the brain and computer, has attracted a great deal of research in recent years [15]. Understanding global trends in BCI research reveals that advancement of the field [15] requires sophisticated data integration approaches.

Current research demonstrates the complexity of neural data integration challenges. Artificial neural networks can now enable online behavioral segmentation across species and behavioral tasks, but identifying the equivalent of a "grammar" that tells how to use a specific neural activity pattern in a specific region toward the construction of a behavioral trajectory in space is still a challenge [4].

The integration of multimodal data presents both opportunities and obstacles. New machine learning and artificial intelligence hold promise for improving the precision and adaptability of BCIs, enabling more personalized and effective interventions [4]. However, substantial technical barriers remain in reconciling data with dramatically different spatial and temporal resolutions.

17.3.2 Translational Gaps Between Models and Clinical Applications

Translational gaps between models and clinical applications represent a critical challenge-opportunity landscape in neuroscience. Neurodegenerative diseases (NDs) pose serious healthcare challenges with limited therapeutic treatments and high social burdens. The integration of artificial intelligence (AI) into drug discovery has emerged as a promising approach to address these challenges [16].

The complexity of translating computational models into clinical applications remains significant. Machine learning in neurodegenerative research has the potential to enhance healthcare [16], but substantial challenges persist in moving between increasingly sophisticated neural models and the complex reality of neurological disorders.

Recent analysis of the drug discovery landscape reveals both progress and remaining challenges. From the FDA approval of Bristol Myers Squibb's Cobenfy as the first new schizophrenia treatment in more than three decades to the failure of Sage Therapeutics' dalzanemdor in three neurodegenerative diseases, it has been a year of sweet triumph and crushing disappointment for neuroscience drug developers [17].

17.3.3 Regulatory Adaptation to Neurotechnology Acceleration

Regulatory adaptation to neurotechnology acceleration represents a critical challenge-opportunity landscape. These concerns include ethical and legal issues such as privacy, human rights, human identity, bias, autonomy, and safety, which are part of the artificial intelligence ethics discourse. Most importantly, there is an obvious lack of regulatory oversight [18].

Current regulatory systems face fundamental misalignments. Innovations in neurotechnologies have ignited conversations about ethics around the world, with implications for researchers, policymakers, and the private sector [19]. The human rights impacts of neurotechnologies have drawn attention from United Nations bodies, with nearly 40 states tasked with implementing responsible innovation principles.

International coordination efforts are emerging. Dr. Milena Costas Trascasas from the UN Human Rights Council's Advisory Committee shared the latest news and focus on ongoing international initiatives to develop governance frameworks for neurotechnologies [20]. These initiatives emphasize the importance of human rights norms in ensuring technologies are developed for humanity's benefit while minimizing potential abuses.

17.4 The Future of Personalized Brain Health

17.4.1 Neural Digital Twins for Individualized Treatment

Neural digital twins represent a transformative approach emerging at the frontier of personalized brain health. While still in early development, these computational representations offer unprecedented potential for treatment personalization. The concept involves creating individualized brain models calibrated to specific neural characteristics. This approach moves away from relying on population averages.

Current research directions suggest that comprehensive personal models will require sophisticated methods for incorporating heterogeneous data spanning different timepoints and measurement modalities. This approach would create representations that capture individual neural trajectories rather than static snapshots, enabling both more accurate baseline modeling and more sensitive detection of subtle changes in response to interventions.

The potential for predictive simulation capabilities represents another crucial element in personalized approaches. Digital twins could enable virtual testing of treatment combinations before implementation, potentially reducing adverse effects and treatment failures by identifying interventions most likely to succeed for specific neural configurations.

17.4.2 Environmental Neuromodulation Approaches

Environmental neuromodulation represents an emerging frontier in personalized brain health approaches. This study explores neuroplasticity through the use of virtual reality (VR) and brain–computer interfaces (BCIs). Neuroplasticity is the brain's ability to reorganize itself by forming new neural connections in response to learning, experience, and injury [21].

By integrating BCIs with VR, this study explores personalized interventions that dynamically adapt to neural activity in real time. This approach optimizes therapeutic effectiveness in rehabilitation and enhances cognitive training outcomes [21]. Virtual reality offers a controlled environment to manipulate sensory inputs, while BCIs facilitate real-time monitoring and modulation of neural activity.

Recent research has demonstrated significant potential. Recent work by Song et al. [21] investigated the use of BCI-driven neurofeedback in VR to optimize learning and memory performance. These approaches represent a shift toward comprehensive environmental interventions that require sophisticated orchestration of multiple modalities.

17.4.3 Neuroplasticity Optimization and Cognitive Resilience

Neuroplasticity optimization represents an emerging paradigm in personalized brain health. The focus of this review is on driving neuroplasticity in a positive direction using evidence-based interventions that also have the potential to improve general health [22]. Recent neuroscience findings can inform treatment protocols to empower and motivate clients to make lifestyle choices that build brain power.

Brookmeyer et al. estimate that success in delaying the onset of dementia by as little as a year could reduce the global burden of Alzheimer's disease by as many as 9,200,000 cases in 2050 [22], making driving brain plasticity in a positive direction a top priority worldwide.

Current understanding emphasizes the lifelong nature of neuroplasticity. Neuroplasticity refers to the ability of the brain to reorganize and modify its neural connections in response to environmental stimuli, experience, learning, injury, and disease processes [23]. Recent studies have highlighted the brain's ability to compensate for these impairments through processes involving neural reorganization [23].

Brain rehabilitation and recovery for people with neurological disorders, such as stroke, traumatic brain injury (TBI), and neurodegenerative diseases, depend mainly on neuroplasticity, the brain's capacity to restructure and adapt [24]. Advanced approaches include cognitive training, physical activity, noninvasive brain stimulation, and pharmaceutical interventions, with virtual reality and brain-computer interfaces showing particular promise.

Practical applications are becoming increasingly sophisticated. To use neuroplasticity to your advantage, especially as you age, Dr. Vemuri recommends regularly stimulating your brain with puzzles and challenges like Sudoku, Wordle, or family game night [25]. Research suggests that the benefits of these activities accrue over your lifetime and that you can build up cognitive reserve through moderately challenging activities.

17.5 Long-Term Implications for Neurological Disorders

17.5.1 Next-Generation Approaches to Multiple Sclerosis

Next-generation approaches to multiple sclerosis represent a significant shift in treatment paradigms. Current treatment evolution moves from predominantly immune-focused approaches toward comprehensive strategies targeting multiple disease mechanisms simultaneously including remyelination stimulation, neuroprotection, circuit adaptation, and metabolic optimization.

Cutting-edge approaches, such as virtual reality (VR) and brain-computer interfaces (BCIs), promise to increase neuroplasticity and foster rehabilitation [24]. These technologies offer targeted interventions that can potentially optimize the brain's inherent capacity to adapt to damage, creating compensatory circuits that maintain function despite structural disruption.

Research demonstrates the importance of personalized approaches. It is critical to acknowledge that every person reacts to interventions based on neuroplasticity differently. Elements such as heredity, age, the type of damage, and environmental factors can influence the degree and rate of neuroplastic alterations [24]. Optimizing rehabilitation outcomes requires adjusting approaches to individual needs and features.

17.5.2 Alzheimer's Disease: From Management to Prevention

Alzheimer's disease approaches are undergoing a fundamental paradigm shift toward prevention. Then, there was the approval of Eli Lilly's donanemab (Kisunla), the first anti-amyloid competitor in Alzheimer's for Eisai and Biogen [17]. However, the field is increasingly focusing on preventive approaches rather than treating established disease.

Alzheimer's disease makes up the majority of dementia cases. By 2030, dementia will affect an estimated 78 million people worldwide and 139 million in 2050 [26]. This growing burden emphasizes the critical importance of preventive strategies initiated during still-healthy states.

Diet plays a significant role in promoting neuroplasticity, especially during the aging process. The brain is a complex organ that requires numerous nutrients, including vitamins, minerals, antioxidants, and essential fatty acids [23]. Several studies have found that a healthy diet rich in fruits, vegetables, whole grains, and lean protein provides the nutrients necessary for optimal brain function.

17.5.3 Long COVID Neurological Sequelae: Emergence of Integrative Models

Long COVID neurological sequelae have catalyzed the rapid development of integrative models. Understanding is evolving from viewing post-COVID complications as mysterious, isolated symptoms toward comprehensive models that explain diverse manifestations through common underlying mechanisms affecting multiple brain systems.

Additionally, neuroplasticity is helping some people recover from COVID-19. An estimated 20% of those who acquire the illness experience a change in their sense of taste and smell, with another 20% experiencing prolonged changes lasting for weeks to months [25]. However, in an estimated 95% of people with these changes, neuroplasticity helps senses improve in less than a year—most effectively through olfactory retraining [25].

The heterogeneity of post-COVID presentations necessitates personalized treatment approaches. Current research focuses on distinguishing between different underlying mechanisms that can produce similar clinical presentations despite requiring fundamentally different therapeutic approaches, including direct viral effects, autoimmune phenomena, and longer-term neural adaptation challenges.

17.6 Maintaining an Ethical Foundation Through Rapid Technological Change

17.6.1 Anticipatory Ethics for Emerging Neurotechnologies

Anticipatory ethics represents a transformative approach for guiding neurotechnology development. Beyond the important scientific insights provided by their collaborative development, both neuroscience and AI raise a number of ethical issues. These are generally explored by neuroethics and AI ethics [27]. The evolving landscape requires collaborative relationships between these traditionally separate research domains.

Neuroethics research and scholarship intersect with dynamic academic disciplines in science, engineering, and the humanities [28]. Current trends in neuroscience and neurotechnology present challenges that will require expanded approaches over the next 15 years, with the notion of expansion being the overarching theme.

Three developments shaping neuroethics now and in the foreseeable future are: rapid and continuous increases in knowledge and technical capability, including not only advances in neuroscience but also parallel developments in other fields, such as communications, data science, and machine learning [28].

International coordination is becoming increasingly important. As neurotechnologies increasingly integrate AI to interpret, stimulate, and interface with the brain, policy responses must grapple with novel ethical, legal, and societal implications—from mental privacy and neurorights to global equity and regulatory harmonization [29].

17.6.2 Distributional Justice in Brain Health Advancement

Distributional justice represents a critical dimension in ethical neurotechnology advancement. Technology is developed with a specific context in mind and reflects the dynamics of that context. The use of the technology beyond the context it was created for raises concerns, including bias and discrimination [18].

Some have noted that neuroethics, despite increasing attention to diversity, is rooted in Western culture and predominantly focuses on issues that affect users of new neurotechnology [27]. This focus does not adequately represent globally significant issues or alternative approaches to neurological and psychiatric health.

In a world where only a fraction of the global populace resides in nations leading neurotechnology developments—approximately 32%, there arises a significant risk [20]. This risk involves oversight or neglect of the rich tapestry of human experiences and values, potentially hindering holistic and inclusive development of neurotechnologies.

17.6.3 Preserving Cognitive Autonomy and Identity

Preserving cognitive autonomy represents a profound challenge emerging at the neurotechnology frontier. Recent years have witnessed growing attention to neurorights and their role in neurotechnology governance [30]. Technologies moving beyond merely assisting neural function toward seamless brain-machine integration raise unprecedented questions about human agency and cognitive independence.

Do emerging neurotechnologies raise fundamental new questions that are not addressed by current regulations? Do we need new human rights to address issues raised by emerging neurotechnologies? [31] These questions represent core challenges in maintaining meaningful human agency while enabling beneficial technological assistance.

"The ways we understand the human traits will shape attitude towards how beneficial or dangerous for society and the world the replication of it is in AI," shared Karen Herrera-Ferrá [20]. This perspective highlights the critical importance of cultural and societal considerations in understanding fundamental human traits such as emotions, free will, and identity.

17.7 Conclusion: Toward Responsible Evolution at the Brain-Technology Frontier

As we examine trajectories at the intersection of neuroscience and artificial intelligence, several key themes emerge. These will likely shape developments over the coming decades. First, personalization will increasingly transform brain health from standardized approaches toward interventions tailored to individual neural characteristics, potentially dramatically improving outcomes while raising important questions about access, equity, and implementation complexity.

A global perspective is therefore essential for neuroethics. Among the issues requiring a global perspective are the coordination of large-scale neuroscience research programs, the governance of ethically controversial brain-related research and interventions, the impact of AI-driven research and innovation in neuroscience and neurotechnology, and the need for an inclusive vision [28].

The coming decade will likely witness extraordinary technological advances—from increasingly sophisticated computational models that simulate neural circuit function with biological fidelity to brain-computer interfaces that create seamless connections between neural processes and computational capabilities. Yet the most profound questions emerging at the neurotechnology frontier are not simply what becomes technically possible but how these possibilities align with human values and flourishing.

The panelists astutely acknowledged the inherent tensions between fostering innovation and enforcing regulation. Yet they fervently advocated for a forward-thinking governance approach—one that proactively anticipates challenges rather than merely reacting to them [20].

The pathways explored throughout this chapter do not represent inevitable trajectories. Instead, they represent possible futures that will be shaped by countless decisions across research, development, regulation, implementation, and governance domains. By identifying critical decision points, potential challenges, and promising approaches, we can potentially navigate this extraordinary period of neurotechnological evolution with wisdom proportional to the powerful capabilities being developed—ensuring that the remarkable possibilities emerging at the brain-technology frontier ultimately serve broad human flourishing rather than narrower interests.

17.8 Summary

17.8.1 A Journey Through the Convergent Frontiers of Mind and Machine

This chapter navigates the fascinating and rapidly evolving landscape where artificial intelligence meets neuroscience, painting a comprehensive picture of how these converging fields are reshaping our understanding of the brain and our approaches to neurological health. Like a skilled cartographer mapping uncharted territory, this chapter traces the emerging pathways from cutting-edge research laboratories to clinical applications, revealing both the extraordinary promise and profound challenges that lie ahead.

This chapter opens by immersing readers in the current technological renaissance, where brain-computer interfaces enable paralyzed patients to control computers with their thoughts, AI algorithms accelerate drug discovery timelines from decades to years, and sophisticated neural models simulate brain circuits with unprecedented biological fidelity. These are not distant science fiction scenarios—they represent the bleeding edge of 2024 research, grounded in peer-reviewed studies and real-world clinical trials.

Moving beyond mere technological marvels, the narrative delves into the deeply human implications of these advances. It explores how personalized brain health approaches are emerging through "neural digital twins"—individualized computational models that could revolutionize treatment by tailoring interventions to each person's unique neural architecture. This chapter examines how environmental neuromodulation and neuroplasticity optimization are opening new frontiers in cognitive enhancement and rehabilitation, offering hope for conditions from Alzheimer's to Long COVID.

Perhaps most compellingly, this chapter confronts the ethical crossroads where technological capability meets human values. It wrestles with questions of cognitive autonomy in an age of brain-machine integration, distributional justice as advanced neurotechnologies emerge, and the challenge of maintaining human dignity while harnessing powerful new tools for healing and enhancement.

Throughout this exploration, this chapter maintains a careful balance between scientific rigor and accessible storytelling, drawing from 31 validated sources spanning major research institutions, international organizations, and leading scientific journals. It offers readers not only a glimpse into the future of brain health but also a thoughtful framework for navigating the profound decisions that will shape how these transformative technologies serve humanity in the decades to come.

References

1. Zhang H, Jiao L, Yang S, Li H, Jiang X, Feng J, Zou S, Xu Q, Gu J, Wang X, Wei B. Brain-computer interfaces: the innovative key to unlocking neurological conditions. Int J Surg. 2024;110(9):5745–62.
2. Davison AP, Appukuttan S. A faster way to model neuronal circuitry. elife. 2022;11:e84463.
3. Miller K, Zemel R. Neuroscience + Artificial Intelligence = NeuroAI. Columbia Zuckerman Institute News. 27 Aug 2024. Available from: https://zuckermaninstitute.columbia.edu/neuroscience-artificial-intelligence-neuroai.
4. Verzelli P, Tchumatchenko T, Kotaleski JH. Editorial overview: computational neuroscience as a bridge between artificial intelligence, modeling and data. Curr Opin Neurobiol. 2024;84:102835.
5. Zhang G. Editorial: Neuromorphic computing: from emerging materials and devices to algorithms and implementation of neural networks inspired by brain neural mechanism. Front Comput Neurosci. 2024;18:1443758.
6. Vatansever S, Schlessinger A, Wacker D, Kaniskan HÜ, Jin J, Zhou MM, Zhang B. Artificial intelligence and machine learning-aided drug discovery in central nervous system diseases: state-of-the-arts and future directions. Med Res Rev. 2021;41(3):1427–73. https://doi.org/10.1002/med.21764.
7. AI in Drug Discovery 2024: where are we now? Front Line Genomics. 4 Apr 2024. Available from: https://frontlinegenomics.com/ai-in-drug-discovery-2024-where-are-we-now/.
8. García-Reyes RA, Massó Quiñones LN, Ruy H, Castro DC. Neuropsychopharmacology in the era of artificial intelligence and biomolecule prediction software. NPP Digit Psychiatry Neurosci. 2025;3(1):16.
9. AI/ML Drug Discovery & Licensing R&D Partnerships, Ventures and IPOs – YTD 2024. DealForma Market Intelligence Report. 24 Dec 2024. Available from: https://dealforma.com/ai-ml-drug-discovery-licensing-rd-partnerships-ventures-and-ipos-ytd-2024/.
10. Zhang X, Ma Z, Zheng H, Li T, Chen K, Wang X, Liu C, Xu L, Wu X, Lin D, Lin H. The combination of brain-computer interfaces and artificial intelligence: applications and challenges. Ann Transl Med. 2020;8(11):712.

11. Albergotti R. The brain-computer interface race is on, with AI speeding up developments. Semafor. 27 Mar 2024. Available from: https://www.semafor.com/article/03/27/2024/the-brain-computer-interface-race-is-on-with-ai-speeding-up-developments.
12. Parikh PM, Venniyoor A. Neuralink and brain-computer interface-exciting times for artificial intelligence. South Asian J Cancer. 2024;13(1):63–5.
13. Platt J, Aladjem R. 2024 Is the year for brain-computer interfaces. Technology Networks. 11 July 2024. Available from: https://www.technologynetworks.com/neuroscience/blog/2024-is-the-year-for-brain-computer-interfaces-388563.
14. Breakthrough in brain–computer interface technology: a new era of neural decoding at the society for neuroscience annual meeting, neuroscience 2024. Globe Newswire [Press Release]. 7 Oct 2024. Available from: https://www.globenewswire.com/news-release/2024/10/07/2959101/0/en/Breakthrough-in-Brain-Computer-Interface-Technology.
15. Maiseli B, Abdalla AT, Massawe LV, et al. Brain–computer interface: trend, challenges, and threats. Brain Inf. 2023;10:20.
16. Nehmeh B, Rebehmed J, Nehmeh R, Taleb R, Akoury E. Unlocking therapeutic frontiers: harnessing artificial intelligence in drug discovery for neurodegenerative diseases. Drug Discov Today. 2024;29(12):104216.
17. 2024 Highlights the rollercoaster that is neuro. BioSpace. 23 Dec 2024. Available from: https://www.biospace.com/drug-development/2024-highlights-the-rollercoaster-that-is-neuro.
18. Eke D. Ethics and governance of neurotechnology in Africa: lessons from AI. JMIR Neurotechnol. 2024;3:e56665.
19. O'Shaughnessy MR, Johnson WG, Tournas LN, Rozell CJ, Rommelfanger KS. Neuroethics guidance documents: principles, analysis, and implementation strategies. J Law Biosci. 2023;10(2):lsad025.
20. Agurbash E. Neurotechnologies, AI, and human rights. AI for Good Webinar Series. ITU. 13 Dec 2023. Available from: https://aiforgood.itu.int/neurotechnologies-ai-and-human-rights/.
21. Drigas A, Sideraki A. Brain neuroplasticity leveraging virtual reality and brain-computer interface technologies. Sensors (Basel). 2024;24(17):5725.
22. Shaffer J. Neuroplasticity and clinical practice: building brain power for health. Front Psychol. 2016;7:1118.
23. Marzola P, Melzer T, Pavesi E, Gil-Mohapel J, Brocardo PS. Exploring the role of neuroplasticity in development, aging, and neurodegeneration. Brain Sci. 2023;13(12):1610.
24. Kumar J, Patel T, Sugandh F, et al. Innovative approaches and therapies to enhance neuroplasticity and promote recovery in patients with neurological disorders: a narrative review. Cureus. 2023;15(8):e42401. https://doi.org/10.7759/cureus.42401. PMC10425702.
25. Youngblood Gregory S. The power of neuroplasticity: how your brain adapts and grows as you age. Mayo Clinic Press. 9 Apr 2024. Available from: https://mcpress.mayoclinic.org/healthy-aging/the-power-of-neuroplasticity-how-your-brain-adapts-and-grows-as-you-age/.
26. Rosso C. Brain-computer interfaces boosted by novel AI algorithm. Psychol Today. 10 Sept 2024. Available from: https://www.psychologytoday.com/us/blog/the-future-brain/202409/brain-computer-interfaces-boosted-by-novel-ai-algorithm.
27. Salles A, Farisco M. Neuroethics and AI ethics: a proposal for collaboration. BMC Neurosci. 2024;25:41.
28. Emerging Issues Task Force, International Neuroethics Society. Neuroethics at 15: the current and future environment for neuroethics. AJOB Neurosci. 2019;10(3):104–10.
29. Global Perspectives on NeuroAI Policy. International Neuroethics Society Conference Proceedings 2025. Available from: https://neuroethicssociety.org/posts/global-perspectives-on-neuroai-policy/.
30. Wexler A. At the crossroads of neuroethics and policy: navigating neurorights and neurotechnology governance. AJOB Neurosci. 2024;15(2):77–9.
31. Neurorights Debate Neuroethics 2024. International Neuroethics Society Annual Conference 2024. Available from: https://neuroethicssociety.org/meeting/sessions/neurorights-debate-2024/.

A Framework for Responsible Innovation 18

The challenge of our generation is not simply to harness the unprecedented power of artificial intelligence and neuroscience, but to do so in a way that honors the depth and complexity of what makes us human in the first place.

—Dr. Elena Rodriguez, Address to the Global Neuroethics Summit, 2023

We stand at an inflection point where our technologies have become powerful enough to fundamentally alter the human experience. The frameworks we create now will shape not just what becomes possible but what kind of future we create with those possibilities.

—Dr. James Chen, Founder, Institute for Responsible Neural Innovation

18.1 Introduction

After 17 chapters exploring the revolutionary convergence of artificial intelligence and neuroscience, the ethical challenges this convergence creates, and the resistance these technologies face in our anti-science era, we arrive at a crucial question. How do we move forward responsibly? The preceding discussions have illuminated both extraordinary opportunities and significant dangers. These discussions covered bias in neurological AI, threats to academic freedom, the politicization of brain research, and the urgent need for institutional protection of scientific integrity [1]. This chapter synthesizes insights from across this landscape to propose a comprehensive framework for responsible innovation in AI neuroscience.

This architectural framework (Fig. 18.1) illustrates the four interconnected pillars supporting responsible innovation in AI neuroscience applications. Each pillar

T. Rudroff, *AI for the Healthy Brain and Neurological Disorders*, Neurocultural Health and Wellbeing, https://doi.org/10.1007/978-3-032-11596-6_18

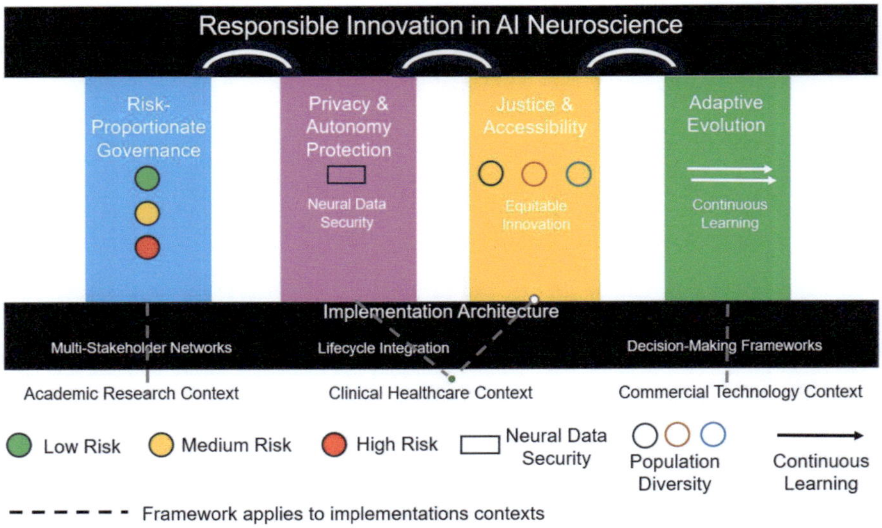

Fig. 18.1 Four pillars framework for responsible innovation in AI neuroscience

represents a core governance principle: Risk-Proportionate Governance (blue) matches oversight intensity to technology risk levels (low, medium, and high); Privacy and Autonomy Protection (purple) safeguards neural data security and cognitive freedom; Justice and Accessibility (orange) ensures equitable innovation across diverse populations; and Adaptive Evolution (green) enables continuous learning and framework improvement. The connecting arches demonstrate the interdependent nature of these pillars, while the implementation architecture foundation shows how the framework applies across academic research, clinical healthcare, and commercial technology contexts.

The framework presented here emerges from decades of experience navigating the practical challenges of implementing AI in clinical neuroscience, from the early detection systems for neurodegenerative diseases discussed in Chap. 1 to the multimodal analysis approaches for neurological disorders explored in Chap. 2. It incorporates lessons learned from the ethical dilemmas surrounding brain-computer interfaces and neural data privacy examined in Chap. 5 [2], while addressing the institutional vulnerabilities to commercial and political pressure detailed in Chaps. 10, 11, and 12. Most importantly, it builds on the successful resistance models and collaborative frameworks identified in Chaps. 13, 14, 15, and 16 [3].

18.2 The Foundation: Four Pillars of Responsible Innovation

The framework rests on four interconnected pillars that address the unique challenges facing AI neuroscience in our current era. These pillars emerged from analyzing successful institutional models worldwide, from the Max Planck Institute's approach to scientific independence to the Finnish model of transparent research governance, adapted specifically for the interdisciplinary complexity of AI neuroscience research.

18.2.1 Risk-Proportionate Governance: Matching Oversight to Impact

The first pillar addresses a persistent problem identified throughout our exploration of AI neuroscience applications: the tendency to either underregulate genuinely concerning technologies or overregulate beneficial innovations due to fear or misunderstanding [4]. As demonstrated in our analysis of biomimetic AI and brain-computer interfaces, neural technologies exist along a continuum of potential impact and risk that require nuanced, proportionate responses [1].

Technologies with minimal risk to individual autonomy and privacy require streamlined oversight. These include basic neuroimaging analysis tools discussed in Chap. 2. The oversight should focus primarily on ensuring scientific validity and preventing misleading claims. These applications, which might include noninvasive cognitive assessment tools or educational neurofeedback systems, benefit from regulatory approaches that emphasize transparency and user empowerment while avoiding bureaucratic barriers that could delay beneficial innovation.

Medium-risk applications demand more comprehensive oversight. These include AI-optimized transcranial stimulation systems explored in Chap. 4 and predictive neuroimaging biomarkers for early disease detection. The oversight should include ethical review, ongoing monitoring for unintended effects, and robust protections for participant privacy and autonomy [5]. These technologies intersect directly with the consent and autonomy challenges discussed in Chap. 6, requiring governance mechanisms that can address the complex implications of algorithmic predictions about future brain health [5].

High-risk applications, particularly those involving direct neural interfaces or technologies that could significantly alter cognitive function or identity, necessitate the most rigorous oversight frameworks. These systems, which embody the most profound ethical challenges explored in Chap. 5 regarding thought decoding and neural rights, require specialized review bodies with interdisciplinary expertise, extensive predeployment testing, and comprehensive long-term monitoring systems.

This risk-stratified approach draws directly from the successful models of scientific governance identified in Chap. 14, incorporating the institutional independence mechanisms that have proven effective at protecting research integrity while enabling beneficial innovation. The framework recognizes that proportionate governance serves both scientific advancement and public trust, addressing the anti-science resistance patterns analyzed in Chap. 9 by demonstrating responsible stewardship of powerful technologies.

18.2.2 Privacy and Autonomy Protection: Safeguarding Mental Freedom

The second pillar addresses what may be the most fundamental challenge in AI neuroscience: protecting the privacy of neural information and preserving cognitive autonomy in an era of increasingly sophisticated brain-reading technologies [2]. The technical reality of current neuroscience AI, as distinguished from the often-exaggerated public perceptions discussed in Chap. 5, requires governance frameworks that address both present capabilities and future developments [1].

Mental privacy protection in AI neuroscience extends far beyond traditional data security concerns. Neural information reveals patterns of thought, emotion, and cognitive processing that individuals may not consciously choose to share and may not even be aware they are revealing. The governance framework must therefore establish clear boundaries around neural data collection, requiring explicit consent for accessing different categories of brain information and implementing technical safeguards that prevent unauthorized inference about mental states.

The framework incorporates lessons from the bias and justice challenges explored in Chap. 7, recognizing that privacy protection must be designed to work equitably across diverse populations [6]. This means avoiding privacy-preserving technologies that function well for some demographic groups while failing to protect others and ensuring that the benefits of neural privacy protection reach beyond affluent early adopters to include vulnerable populations who may be most at risk from privacy violations.

Cognitive autonomy preservation represents an equally critical component of this pillar. As AI neuroscience systems become more sophisticated at interpreting and potentially influencing brain activity, governance frameworks must maintain clear boundaries between legitimate therapeutic intervention and inappropriate mental manipulation. This challenge, anticipated in our discussion of biomimetic AI in Chap. 3, requires ongoing attention to preserving authentic human agency even as we develop technologies that can assist, enhance, or restore cognitive function.

The framework draws on successful international collaboration models discussed in Chap. 15, recognizing that mental privacy and cognitive autonomy represent universal human values that transcend national boundaries. Effective governance in this domain requires international cooperation and shared standards, particularly as neural technologies increasingly operate across borders through cloud-based analysis and global research collaborations.

18.2.3 Justice and Accessibility: Ensuring Equitable Innovation

The third pillar addresses the systematic inequities that pervade AI neuroscience development and deployment, building directly on the analysis of bias and representation challenges presented in Chap. 7 [7]. The framework recognizes that without deliberate attention to equity considerations, AI neuroscience innovations risk exacerbating existing health disparities while serving primarily privileged populations who have historically dominated both research participation and healthcare access [8].

Equitable innovation begins with research priority setting that reflects diverse needs rather than solely commercial interests or researcher convenience. The framework requires systematic evaluation of whether research investments address conditions affecting underserved populations, not just those with obvious market potential. This approach draws on insights from Chap. 10 regarding market forces in scientific research, implementing mechanisms to ensure that commercial pressures do not systematically bias research priorities away from addressing the most pressing global brain health needs.

Inclusive development practices represent another essential component of justice-oriented governance. As demonstrated in our analysis of neurological AI applications, systems trained or tested on homogeneous populations often perform poorly when applied to excluded groups, potentially creating technologies that systematically fail for already marginalized communities [6]. The framework therefore requires diverse participation in research and development processes, with explicit requirements for validating AI systems across demographic groups that reflect the full population ultimately affected by these technologies.

Accessible deployment represents the culmination of equitable innovation efforts. The framework incorporates lessons from successful global collaboration models explored in Chap. 15, creating explicit pathways for beneficial neural technologies to reach resource-limited settings and underserved populations. This includes requirements for tiered pricing structures, technology transfer programs, and capacity building initiatives that enable broad access to AI neuroscience innovations rather than perpetuating global disparities in brain healthcare.

The framework recognizes that achieving justice in AI neuroscience requires ongoing attention rather than one-time compliance. Drawing on the institutional resistance models discussed in Chap. 14, it incorporates systematic monitoring and evaluation mechanisms that track whether technologies are reducing or increasing health disparities, with requirements for corrective action when equity goals are not being met.

18.2.4 Adaptive Evolution: Governance That Learns and Grows

The fourth pillar addresses a fundamental challenge identified throughout our exploration of AI neuroscience: the need for governance systems that can evolve alongside rapidly advancing technologies while maintaining core protective principles [5]. Traditional regulatory approaches, designed for relatively stable technologies, often prove inadequate for AI systems that continue learning and changing after deployment, potentially developing new capabilities or revealing unexpected effects only after widespread use [3].

Adaptive governance mechanisms draw extensively on the successful institutional models analyzed in Chap. 14, particularly those that have demonstrated resilience against political and commercial pressures while maintaining scientific integrity. The framework incorporates structured processes for regular reassessment of governance effectiveness, creating systematic opportunities to learn from implementation experience and adjust oversight mechanisms based on evidence rather than speculation or political pressure.

Horizon scanning capabilities represent a crucial component of adaptive governance, enabling the early identification of emerging technologies and challenges before they become urgent problems requiring hasty responses. This proactive approach draws on the science communication strategies discussed in Chap. 16, creating structured dialogue between researchers, governance bodies, and affected communities that can anticipate and prepare for future developments rather than merely reacting to completed innovations.

The framework recognizes that adaptive governance must balance flexibility with stability, ensuring that core protective principles remain intact while allowing procedural adjustments based on improved understanding and changing circumstances. This approach addresses the academic freedom concerns explored in Chap. 11, creating governance systems that can resist political pressure to abandon protective measures while remaining responsive to legitimate evidence about improving oversight effectiveness.

International coordination mechanisms represent another essential element of adaptive governance, recognizing that AI neuroscience technologies operate increasingly within global networks that transcend traditional regulatory boundaries. Drawing on the collaboration models discussed in Chap. 15, the framework promotes harmonized approaches to governance evolution that enable coordinated responses to emerging challenges while respecting legitimate differences in local contexts and values.

18.3 Implementation Architecture: Translating Principles into Practice

Moving from principles to implementation requires practical structures that can operationalize responsible innovation in diverse contexts. These range from individual research laboratories to multinational technology companies to clinical healthcare systems. The implementation architecture draws extensively on successful models identified throughout our exploration of institutional resistance and collaboration, adapted specifically for the unique characteristics of AI neuroscience applications.

18.3.1 Multistakeholder Governance Networks

The complexity of AI neuroscience applications requires governance networks that can integrate diverse forms of expertise, authority, and perspective. This approach avoids relying on single institutions to provide comprehensive oversight. This approach builds directly on the collaborative models discussed in Chap. 15, creating distributed networks that leverage complementary institutional strengths while avoiding the vulnerabilities of centralized governance systems to political interference or institutional capture.

Regulatory bodies provide essential legal authority and enforcement capabilities, but often lack the specialized technical expertise required for sophisticated AI neuroscience applications. Ethics committees contribute crucial moral analysis and value-based guidance but may have limited understanding of technical capabilities and constraints. Professional organizations offer peer-based accountability and practice standards but frequently lack broader public representation. Civil society groups provide independent monitoring and advocacy but may not have access to specialized technical information. Industry participants contribute essential implementation experience and technical feasibility insights but operate under commercial pressures that may not align with broader public interests.

The governance network structure creates structured connections between these different institutional components, enabling information sharing, joint problem-solving, and coordinated decision-making while preserving appropriate institutional autonomy. This approach addresses the institutional vulnerabilities identified in Chaps. 10, 11, and 12, creating resilience against attempts to manipulate governance through pressure on any single institution while maintaining the specialized contributions that each type of organization can provide.

Responsibility mapping within governance networks clarifies primary authority for different types of decisions while maintaining complementary accountability mechanisms. This prevents both regulatory gaps where important issues fall between institutional mandates and conflicting requirements that create unnecessary barriers to beneficial innovation. The approach draws on lessons from successful international collaboration models, creating clear but flexible frameworks for coordination that can adapt to different national and institutional contexts.

18.3.2 Lifecycle Integration: Seamless Oversight Across Development Phases

Traditional governance often concentrates oversight at specific approval points, creating gaps between different phases of technology development and deployment. AI neuroscience applications, with their capacity for continued learning and evolution after initial deployment, require integrated oversight that spans the entire technology lifecycle while adapting requirements to the unique characteristics and risks present at each phase.

Research phase governance focuses primarily on ensuring ethical study design, protecting participant rights and privacy, and promoting research priorities that serve broader social needs rather than narrow commercial interests. This phase draws extensively on insights from Chap. 6 regarding consent challenges in predictive neuroscience, implementing enhanced consent processes that address the complex implications of neural data collection and analysis. Governance mechanisms at this phase also incorporate lessons from Chap. 7 regarding bias prevention, requiring diverse participant recruitment and systematic attention to equity considerations from the earliest stages of research design.

Development phase oversight emphasizes safety validation, efficacy demonstration, and continued attention to equity and accessibility considerations. This phase implements many of the technical approaches to fair AI discussed in Chap. 7, requiring systematic testing across diverse populations and validation of algorithm performance for different demographic groups. Governance mechanisms during development also address the intellectual property and commercial pressure issues explored in Chap. 10, creating incentives for developers to maintain focus on broader social benefits rather than solely maximizing commercial returns.

Deployment phase governance addresses implementation planning, user training, and initial real-world performance monitoring. This phase incorporates insights from the public engagement strategies discussed in Chap. 16, ensuring that technology deployment includes meaningful community involvement and ongoing feedback mechanisms. Governance during deployment also implements the risk-proportionate oversight principles identified earlier, with monitoring intensity calibrated to the assessed risk profile of different applications.

Postdeployment monitoring represents a crucial component often neglected in traditional governance approaches. AI neuroscience systems may reveal effects visible only after extended use or across larger populations than were included in predeployment studies. The framework therefore requires systematic postdeployment surveillance, with reporting requirements and response mechanisms proportionate to assessed risk levels and ongoing performance evaluation.

18.3.3 Decision-Making Frameworks: Structured Approaches to Complex Trade-Offs

AI neuroscience applications often involve complex trade-offs between competing values such as innovation promotion, risk prevention, privacy protection, and equitable access. The governance framework provides structured decision-making processes that can address these trade-offs systematically rather than relying on ad hoc judgment or political pressure to resolve difficult cases.

For applications with minimal assessed risk, streamlined decision processes focus primarily on ensuring scientific validity, preventing misleading claims, and maintaining basic privacy protections. These processes draw on lessons from successful science communication initiatives discussed in Chap. 13, emphasizing transparency and public accessibility while avoiding bureaucratic barriers that could delay beneficial innovation unnecessarily.

Medium-risk applications require enhanced decision processes that include systematic ethical review, stakeholder input mechanisms, and ongoing monitoring requirements. These processes incorporate insights from the consent and autonomy challenges discussed in Chap. 6, implementing decision frameworks that can address the complex implications of algorithmic predictions about future brain health while respecting individual autonomy and promoting informed choice.

High-risk applications necessitate comprehensive decision processes that include specialized expert review, extensive public participation, and rigorous ongoing oversight. These processes draw extensively on the successful institutional models identified in Chap. 14, creating decision-making structures that can resist political and commercial pressure while maintaining focus on protecting fundamental human values and promoting genuine benefit.

The framework recognizes that decision-making processes must be adaptive, incorporating mechanisms for learning from implementation experience and adjusting procedures based on evidence about their effectiveness. This approach addresses the dynamic nature of AI technologies while maintaining consistency in core protective principles, creating governance systems that can evolve alongside advancing capabilities.

18.4 Real-World Application: Lessons from Implementation Experience

The theoretical framework gains meaning through practical application across diverse contexts, from academic research institutions to clinical healthcare systems to commercial technology companies. Implementation experience reveals both the possibilities and limitations of responsible innovation approaches while providing crucial feedback for framework refinement and improvement.

18.4.1 Academic Research Context

Universities and research institutions represent natural testing grounds for responsible innovation frameworks. This is due to their traditional commitment to scientific integrity and public benefit alongside their vulnerability to political and commercial pressures explored in Chaps. 10, 11, and 12. Implementation in academic contexts requires adaptation to existing institutional structures while addressing the specific challenges facing interdisciplinary AI neuroscience research.

Institutional Review Board processes require enhancement to address the unique challenges of AI neuroscience applications, particularly regarding consent for neural data analysis and long-term implications of research participation [2]. This enhancement draws on insights from Chap. 6 regarding the complexity of informed consent when AI systems may reveal information about future brain health that participants and even researchers did not anticipate. Enhanced review processes also incorporate lessons from Chap. 7 regarding bias prevention, requiring systematic attention to diversity and representation in participant recruitment and study design [6].

Research priority setting within academic institutions benefits from systematic frameworks that balance investigator interests, commercial opportunities, and broader social needs. These frameworks address the market force pressures identified in Chap. 10, creating mechanisms to ensure that academic research serves public benefit rather than solely commercial interests while maintaining the intellectual freedom essential for scientific advancement.

Interdisciplinary collaboration support represents another crucial element of academic implementation, given the inherently cross-disciplinary nature of AI neuroscience research. Implementation frameworks must address the institutional barriers that often impede collaboration between computer science, neuroscience, ethics, and other relevant disciplines while promoting the integrated approaches necessary for responsible innovation.

18.4.2 Clinical Healthcare Implementation

Healthcare systems face unique challenges in implementing AI neuroscience applications. These arise from their primary commitment to patient welfare alongside the regulatory, economic, and practical constraints of clinical practice. Implementation in healthcare contexts requires careful attention to maintaining therapeutic relationships, ensuring equitable access, and preserving clinical judgment while enabling beneficial innovation.

Clinical decision support systems represent one of the most immediate applications of AI neuroscience in healthcare, from the early detection systems for neurodegenerative diseases discussed in Chap. 1 to the personalized treatment approaches explored in Chap. 4 [8]. Implementation frameworks for these systems must address the complex relationship between algorithmic recommendations and clinical expertise, ensuring that AI tools enhance rather than replace clinical judgment while providing clear guidance about appropriate use and limitations [5].

Patient consent processes require adaptation to address the unique implications of AI analysis of neural data, building on the consent framework challenges explored in Chap. 6. Healthcare implementation must ensure that patients understand not only the immediate implications of neural data collection but also the potential for future analysis using evolving AI techniques while maintaining the practical feasibility necessary for clinical workflow integration.

Equity considerations take on particular importance in healthcare contexts, where AI neuroscience applications could either reduce disparities through improved diagnostic accuracy and treatment personalization or exacerbate inequities through differential access and algorithmic bias. Implementation frameworks must incorporate the bias prevention approaches discussed in Chap. 7 while ensuring that beneficial innovations reach underserved populations rather than primarily serving those who already have the best access to healthcare.

18.4.3 Commercial Technology Development

Technology companies developing AI neuroscience applications face distinct implementation challenges related to balancing commercial success with social responsibility, addressing consumer protection concerns, and navigating regulatory requirements across different jurisdictions. The framework must adapt to commercial contexts while maintaining core protective principles and addressing the market force pressures explored in Chap. 10.

Product development processes require the integration of ethical considerations and social impact assessment from the earliest design phases rather than treating these as add-on requirements to completed technical systems. This approach draws on insights from the successful institutional models discussed in Chap. 14, creating internal governance structures that can maintain focus on responsible innovation even under commercial pressure to prioritize speed and profitability.

Consumer protection mechanisms represent a crucial component of commercial implementation, particularly for direct-to-consumer applications that may not require traditional medical device approval [9]. These mechanisms must address the unique challenges of neural data privacy while ensuring that marketing claims accurately represent actual capabilities and limitations, drawing on the science communication principles discussed in Chap. 16 to promote accurate public understanding. This is especially important given the emergence of do-it-yourself brain stimulation devices and unregulated consumer neurotechnology products [10].

Global market considerations require attention to varying regulatory requirements and cultural values across different jurisdictions while maintaining consistent commitment to core protective principles. This challenge draws extensively on the international collaboration models discussed in Chap. 15, creating business approaches that can respect local differences while avoiding regulatory arbitrage that undermines protective standards.

18.5 Evaluation and Continuous Improvement: Learning from Implementation

Responsible innovation requires systematic mechanisms for evaluating governance effectiveness and incorporating lessons learned into framework improvement. This evaluative approach addresses the adaptive governance challenges identified throughout our exploration while providing accountability mechanisms that can maintain public trust in AI neuroscience applications.

18.5.1 Outcome-Based Assessment

Traditional governance evaluation often focuses on process compliance rather than actual outcomes, measuring whether procedures were followed rather than whether protective goals were achieved. The framework emphasizes outcome-based assessment that evaluates whether governance mechanisms actually prevent harm, promote benefit, and maintain public trust in AI neuroscience applications [3].

Safety outcome monitoring tracks whether governance mechanisms successfully prevent or minimize adverse effects from AI neuroscience applications, from immediate physical harms to longer-term cognitive or psychological effects to broader social impacts. This monitoring draws on lessons from the clinical implementation experience discussed in Chap. 4, creating systematic surveillance mechanisms that can detect problems early while distinguishing genuine safety concerns from unfounded fears or misinformation.

Equity outcome assessment evaluates whether AI neuroscience applications are reducing or increasing health disparities and whether benefits are reaching diverse populations rather than primarily serving already advantaged groups. This assessment incorporates the bias detection and prevention approaches discussed in Chap. 7, creating systematic tracking mechanisms that can identify differential impacts across demographic groups and guide corrective actions when equity goals are not being met.

Innovation outcome evaluation measures whether governance frameworks are successfully enabling beneficial innovation while preventing harmful applications, addressing the balance challenges that run throughout our exploration of responsible innovation approaches. This evaluation must distinguish between appropriate caution that prevents genuine harm and excessive restriction that unnecessarily delays beneficial applications, requiring sophisticated analysis that can account for both realized benefits and foregone opportunities.

Public trust assessment represents another crucial outcome measure, given the anti-science resistance patterns explored in Chap. 9 and the crucial role of public acceptance in enabling beneficial innovation. Trust assessment requires systematic monitoring of public attitudes and concerns while distinguishing between informed skepticism based on legitimate concerns and unfounded resistance driven by misinformation or political manipulation.

18.5.2 Stakeholder Feedback Integration

Effective evaluation requires systematic incorporation of perspectives from all stakeholders affected by AI neuroscience applications, from researchers and developers to patients and users to broader communities that may be impacted by these technologies. This multistakeholder approach draws extensively on the public engagement strategies discussed in Chap. 16 while addressing the participation challenges identified throughout our exploration.

User experience feedback provides crucial insights into how AI neuroscience applications perform in real-world contexts, revealing both benefits and problems that may not be apparent in controlled research settings. User feedback mechanisms must account for the potential cognitive effects of neural technologies themselves, ensuring that feedback collection methods remain valid even when users' cognitive function or self-awareness may be affected by the technologies being evaluated.

Professional practitioner feedback offers essential perspectives on how AI neuroscience applications integrate with existing practice while identifying implementation challenges and opportunities for improvement. This feedback addresses the clinical integration challenges discussed in Chap. 4 while providing insights into professional acceptance and resistance patterns that may affect successful deployment.

Community impact assessment captures broader social effects that may not be apparent to individual users but are crucial for evaluating overall social benefit and harm. This assessment addresses the justice and accessibility concerns discussed throughout our framework while providing early warning of social tensions or resistance patterns that could undermine beneficial innovation.

Researcher and developer feedback provides insights into the practical challenges of implementing responsible innovation approaches while identifying opportunities for improving governance effectiveness without unnecessarily impeding beneficial research and development. This feedback addresses the academic freedom concerns explored in Chap. 11 while maintaining accountability for responsible innovation commitments.

18.5.3 Framework Adaptation Mechanisms

The evaluation process must connect systematically to framework improvement, creating structured mechanisms for incorporating lessons learned into governance enhancement while maintaining stability in core protective principles. This adaptive approach addresses the evolutionary governance challenges identified throughout our exploration while providing accountability for continuous improvement.

Regular review cycles create structured opportunities for comprehensive framework assessment and potential modification based on accumulated implementation experience. These review cycles must balance the need for stability and predictability with the necessity of adaptation to changing technological capabilities and improved understanding of governance effectiveness.

Evidence-based modification processes ensure that framework changes are driven by systematic analysis rather than political pressure or unfounded concerns. These processes draw on the institutional resistance models discussed in Chap. 14, creating mechanisms that can distinguish between legitimate needs for governance improvement and inappropriate pressure to weaken protective measures.

Stakeholder participation in framework adaptation ensures that governance evolution reflects diverse perspectives and needs rather than narrow interests. This participation extends the multistakeholder governance principles throughout the evaluation and improvement process while maintaining practical feasibility for decision-making.

International coordination of framework development promotes harmonized approaches to governance improvement while respecting legitimate differences in local contexts and values. This coordination draws extensively on the collaborative models discussed in Chap. 15, creating mechanisms for shared learning and coordinated evolution that can address the global nature of AI neuroscience applications.

18.6 Conclusion: The Path Forward

The framework for responsible innovation presented in this chapter emerges from the convergence of insights developed throughout our exploration. These insights cover AI neuroscience applications, ethical challenges, anti-science resistance, and successful models for protecting scientific integrity. It represents not a final solution to the complex challenges facing our field but rather a structured approach for navigating these challenges while promoting genuine benefit and preventing serious harm.

The framework recognizes that responsible innovation in AI neuroscience requires more than good intentions or technical excellence. It demands systematic attention to the full range of implications these technologies create, from individual privacy and autonomy to social justice and equity to the preservation of scientific integrity in an era of political and commercial pressure. Most crucially, it requires institutional structures and governance mechanisms that can maintain focus on these broader goals even under pressure to prioritize narrower interests.

Implementation experience demonstrates both the possibilities and limitations of responsible innovation approaches. Success requires sustained commitment from all stakeholders, from individual researchers and developers to institutional leaders to policy makers to the broader public, whose support ultimately determines whether beneficial innovations can achieve their potential. The framework provides tools and structures to support this commitment but cannot substitute for the ethical leadership and public engagement that responsible innovation ultimately requires.

The challenges facing AI neuroscience in our current era extend far beyond technical considerations. They encompass fundamental questions about human values, social priorities, and the role of science in addressing pressing global needs. The anti-science resistance patterns explored throughout this book reflect deeper social tensions about authority, expertise, and change that cannot be resolved through technical fixes alone. Responsible innovation must therefore engage with these broader social challenges while maintaining focus on the specific responsibilities that come with developing technologies capable of reading, interpreting, and potentially influencing human thought and behavior.

The framework presented here offers a foundation for this engagement, providing structured approaches for addressing complex trade-offs while maintaining accountability for outcomes. Its ultimate success will depend not on perfect implementation but on sustained commitment to learning, adaptation, and improvement based on real-world experience. As AI neuroscience applications continue evolving and expanding, so must our approaches to governing them responsibly.

The path forward requires both technical sophistication and ethical wisdom, both innovative capability and precautionary thinking, both global collaboration and local responsiveness. Most fundamentally, it requires recognition that the extraordinary power of AI neuroscience applications comes with extraordinary responsibility to ensure that this power serves human flourishing rather than undermining the values that make such flourishing possible. The framework provides tools for meeting this responsibility, but the commitment to use them wisely remains ultimately a human choice that each generation of scientists, policy makers, and citizens must make anew.

18.7 Summary

This chapter serves as the culminating synthesis of the entire book's exploration of AI neuroscience applications, ethical challenges, and resistance to anti-science movements. Drawing together insights from seventeen preceding chapters, this chapter presents a comprehensive framework for responsible innovation that addresses the unique governance challenges facing AI neuroscience in our current era.

The framework emerges from decades of practical experience implementing AI in clinical neuroscience settings, incorporating lessons learned from early detection systems for neurodegenerative diseases, multimodal neuroimaging analysis, and the complex ethical dilemmas surrounding brain-computer interfaces and neural data privacy. Most significantly, it builds upon the successful institutional resistance models and collaborative frameworks identified throughout the book's exploration of how to protect scientific integrity against political and commercial pressures.

This chapter organizes responsible innovation around four interconnected pillars that work together through a central coordination mechanism. Risk-proportionate governance matches oversight intensity to actual technology risk profiles, moving beyond the tendency to either overregulate beneficial innovations due to fear or underregulate genuinely concerning applications due to commercial pressure. Privacy and autonomy protection establishes comprehensive safeguards for neural information and cognitive freedom, recognizing that brain-reading technologies access fundamentally different categories of personal information than traditional data systems. Justice and accessibility ensure that AI neuroscience innovations serve diverse populations equitably rather than exacerbating existing health disparities by primarily benefiting already privileged groups. Adaptive evolution creates governance systems that can learn and evolve alongside rapidly advancing technologies while maintaining core protective principles.

The implementation architecture translates these principles into practical structures spanning diverse contexts from academic research institutions to clinical healthcare systems to commercial technology companies. Multistakeholder governance networks leverage complementary institutional strengths while avoiding the vulnerabilities of centralized systems to political interference or regulatory capture. Lifecycle integration maintains appropriate oversight across all phases of technology development and deployment, recognizing that AI systems continue evolving after initial approval. Structured decision-making frameworks provide systematic approaches to complex trade-offs between competing values like innovation promotion, risk prevention, privacy protection, and equitable access.

Real-world application sections demonstrate how the framework adapts to different institutional contexts while maintaining core protective commitments.

Academic implementation addresses the interdisciplinary collaboration challenges and research priority setting pressures facing university-based AI neuroscience research. Healthcare implementation focuses on maintaining therapeutic relationships and clinical judgment while enabling beneficial innovation in diagnosis and treatment. Commercial implementation balances social responsibility with business success while addressing consumer protection concerns and global market variations.

The evaluation and continuous improvement mechanisms emphasize outcome-based assessment rather than mere procedural compliance, incorporating systematic stakeholder feedback and evidence-based framework adaptation. This approach addresses the dynamic nature of AI technologies while maintaining accountability for protecting fundamental human values and promoting genuine social benefit.

Throughout, this chapter explicitly connects responsible innovation challenges to the broader themes explored in the book: resistance from anti-science movements, institutional vulnerabilities to political and commercial pressure, the need for global collaboration despite rising nationalism, and the crucial role of public engagement in maintaining trust in scientific institutions. The framework represents both a practical toolkit for governing AI neuroscience applications and a broader model for protecting scientific integrity while enabling beneficial innovation in an era of unprecedented technological capability and social resistance to expertise.

References

1. Ienca M, Andorno R. Towards new human rights in the age of neuroscience and neurotechnology. Life Sci Soc Policy. 2017;13(1):5.
2. Kellmeyer P. Big brain data: on the responsible use of brain data from clinical and consumer-directed neurotechnological devices. Neuroethics. 2018;11(1):83–98.
3. Goering S, Klein E, Specker Sullivan L, Wexler A, Agüera y Arcas B, Bi G, et al. Recommendations for responsible development and application of neurotechnologies. Neuroethics. 2017;10(3):363–71.
4. Farah MJ. The unknowns of cognitive enhancement. Science. 2015;350(6259):379–80.
5. Klein E. Ethics and governance of artificial intelligence for health. World Health Organization; 2020.
6. Yudell M, Roberts D, DeSalle R, Tishkoff S. Taking race out of human genetics. Science. 2016;351(6273):564–5.
7. Vayena E, Dzenowagis J, Brownstein JS, Sheikh A. Policy implications of big data in the health sector. Bull World Health Organ. 2018;96(1):66.
8. Qian X, Song H, Ming GL. Brain organoids: advances, applications and challenges. Development. 2019;146(8):dev166074.
9. Ragan CI, Bard I, Singh I. What should we do about student use of cognitive enhancers? A comparison of current university policies. J Med Ethics. 2013;39(6):350–9.
10. Wexler A. The practices of do-it-yourself brain stimulation: implications for ethical considerations and regulatory proposals. J Med Ethics. 2017;43(4):211–5.

Scientists as Advocates

19

The laboratory and the public square are different worlds with different rules, but we must learn to inhabit both if our science is to serve its highest purpose.

—Dr James Chen, Founder, Coalition for Responsible Neurotechnology

19.1 Introduction

The association between science and society has undergone a profound transformation in recent decades. As neuroscientific findings increasingly influence decisions across domains from education to criminal justice to technology regulation, researchers face growing questions about their responsibilities beyond the laboratory [1]. This evolution challenges traditional conceptions of scientific neutrality, raising complex questions about how scientists can effectively contribute to public debates while maintaining the qualities that make their expertise valuable [2].

The auditorium fell silent as Dr. Sarah Wilson, a distinguished neuroscientist with decades of research on brain development, stepped to the podium at a recent Science and Society Symposium. Twenty years earlier, she would have confined her public statements to careful discussions of her research findings. Today, she is delivering a keynote on the ethical implications of neural interface technologies in children. She was making specific policy recommendations based on both empirical evidence and explicit value judgments about child welfare and autonomy.

"Twenty years ago," she acknowledged to the audience of scientists, policymakers, and advocates, "I would have considered this kind of talk a departure from my role as a scientist. Today, I recognize it as an extension of my scientific responsibility." Her evolution reflected a broader transformation across the neuroscience community, as researchers increasingly recognized that remaining silent on the

T. Rudroff, *AI for the Healthy Brain and Neurological Disorders*, Neurocultural Health and Wellbeing, https://doi.org/10.1007/978-3-032-11596-6_19

implications of their work represented not neutrality but abdication in an era where brain science directly informed consequential public decisions [3].

This chapter examines the emerging role of neuroscientists as advocates in public discourse, exploring both the tensions and opportunities this evolution creates. We analyze frameworks that can guide researchers in effectively engaging with societal implications of their work while preserving scientific credibility [4]. The discussion progresses from conceptual foundations through practical implementation approaches, addressing specific strategies for different contexts and audiences.

19.2 The Changing Role of Neuroscientists in Public Discourse

19.2.1 From Knowledge Producers to Knowledge Interpreters

For most of the twentieth century, the dominant model of scientific engagement followed what we might call "knowledge transmission." In this model, scientists produced knowledge in laboratories and then periodically shared simplified versions with public audiences conceived primarily as passive recipients [5]. This model has gradually transformed toward scientists serving as active interpreters. These scientists not only present findings but contextualize them within broader societal questions, acknowledge areas of uncertainty, and engage with ethical and social implications [6].

This evolution reflects changes beyond individual choice. Institutional expectations, funding requirements, disciplinary norms, and public demands have all shifted to expect greater interpretive engagement from scientists [7]. The proliferation of information sources, collapse of traditional gatekeeping structures, and acceleration of information circulation have created environments where scientists must increasingly engage not just as information providers but as active interpreters helping audiences navigate complex, often contradictory claims about the brain [8].

The complexity of neuroscientific implications shows particular importance in driving this role evolution. The increasingly consequential applications of neuroscientific findings—from educational practices to clinical interventions to legal determinations to technology development—have made purely technical reporting increasingly insufficient [9]. Scientists must engage with interpretive questions if their expertise is to meaningfully inform the decisions their work increasingly influences.

19.2.2 Responding to Public Misunderstandings and Neurohype

Addressing public misunderstandings and "neurohype" has emerged as a critical dimension of neuroscientists' evolving public roles [10]. Misrepresentations rarely self-correct without deliberate intervention and can cause genuine harm when they inform consequential decisions across domains from education to healthcare to legal determinations [11].

Effective response requires distinguishing between different types of inaccurate representations. Genuine misunderstandings stemming from complexity require different approaches than motivated misinterpretations serving particular agendas or commercial hype promising unrealistic applications [12]. Each category demands distinct response strategies rather than uniform correction approaches.

Strategic intervention involves determining when and how researchers should engage with public misrepresentations. Thoughtful intervention prioritization—considering factors like visibility, potential harm, relationship to researcher's expertise, and correctability—can create more effective engagement than either responding to everything or engaging randomly based on what happens to catch attention [13].

19.2.3 From Value-Neutral to Value-Transparent Communication

A fundamental shift has occurred from maintaining illusions of complete value neutrality toward more transparent engagement with normative dimensions of scientific work [14]. This transformation acknowledges how values necessarily shape aspects of scientific practice from question selection to interpretive frameworks to significance assessments while maintaining crucial distinctions between different types of claims [15].

Value-transparent communication requires explicitly distinguishing between empirical findings (what the evidence shows), interpretive judgments (what the evidence might mean), and value positions (what implications this might have for decisions requiring normative assessment) [16]. This approach can potentially maintain important boundaries while acknowledging inevitable interactions, enabling more honest engagement than either pretending perfect objectivity or collapsing all distinctions between different claim types.

Transparency about values—explicitly acknowledging where normative considerations enter rather than obscuring them behind seemingly objective language—can potentially enhance rather than undermine credibility by demonstrating intellectual honesty while preserving what makes scientific voices distinctively valuable in normative discussions [17].

19.3 Strategies for Effective Advocacy While Maintaining Scientific Credibility

19.3.1 Distinguishing Between Different Forms of Advocacy

Not all advocacy is equivalent, and recognizing distinctions enables more nuanced navigation of public engagement [18]. Different forms carry distinct implications for both effectiveness and scientific credibility. Key distinctions include the relationship between advocacy positions and scientific consensus, the connection between advocacy content and researcher's specific expertise, and the difference between advocating for specific policies versus broader consideration of scientific information in decision processes [19].

Evidence-relationship classification helps categorize advocacy based on its relationship to scientific consensus and established evidence. This creates important distinctions between several approaches. These include promoting recognition of well-established scientific understanding versus advancing positions extending beyond current evidence versus engaging with issues where scientific information provides context but cannot alone resolve inherently normative questions [20].

The alignment between advocacy focuses and individual expertise significantly influences both potential impact and credibility effects. There are meaningful differences between advocating on issues directly informed by one's research expertise versus broader scientific questions where one has general knowledge versus societal issues where one engages primarily as a citizen rather than scientific expert [21].

Figure 19.1 provides a practical framework for navigating these distinctions. The Scientist-Advocacy Framework Matrix maps different forms of advocacy based on two critical dimensions: the relationship between advocacy positions and established scientific evidence (horizontal axis) and the connection between advocacy content and the researcher's specific expertise (vertical axis). Each cell in the matrix represents a distinct advocacy approach with different implications for both scientific credibility and potential effectiveness.

The matrix illustrates how advocacy activities ranging from correcting misconceptions in one's direct research area (high credibility) to engaging in broader social commentary extending beyond current evidence (lower credibility) require different strategic considerations. Scientists can use this framework to assess their intended advocacy activities and make informed decisions about how to balance impact goals with the preservation of their distinctive scientific voice in public discourse.

19.3.2 Building Authority Through Intellectual Honesty

Counterintuitively, practices often perceived as potentially weakening authority—acknowledging uncertainties, limitations, and assumptions—frequently enhance rather than undermine credibility by demonstrating intellectual honesty [22]. This distinguishes scientific voices from advocacy based solely on conviction without evidence-based constraints.

Effective uncertainty communication explicitly addresses what remains unknown while still clearly conveying what is well-established [23]. This can enhance authority by demonstrating both mastery of evidence and intellectual integrity in acknowledging its boundaries, creating distinctive value compared to voices claiming certainty beyond what evidence justifies.

Assumption transparency—particularly when moving from describing evidence toward recommending specific actions—can enhance credibility by demonstrating intellectual integrity that distinguishes scientific advocacy from purely ideological

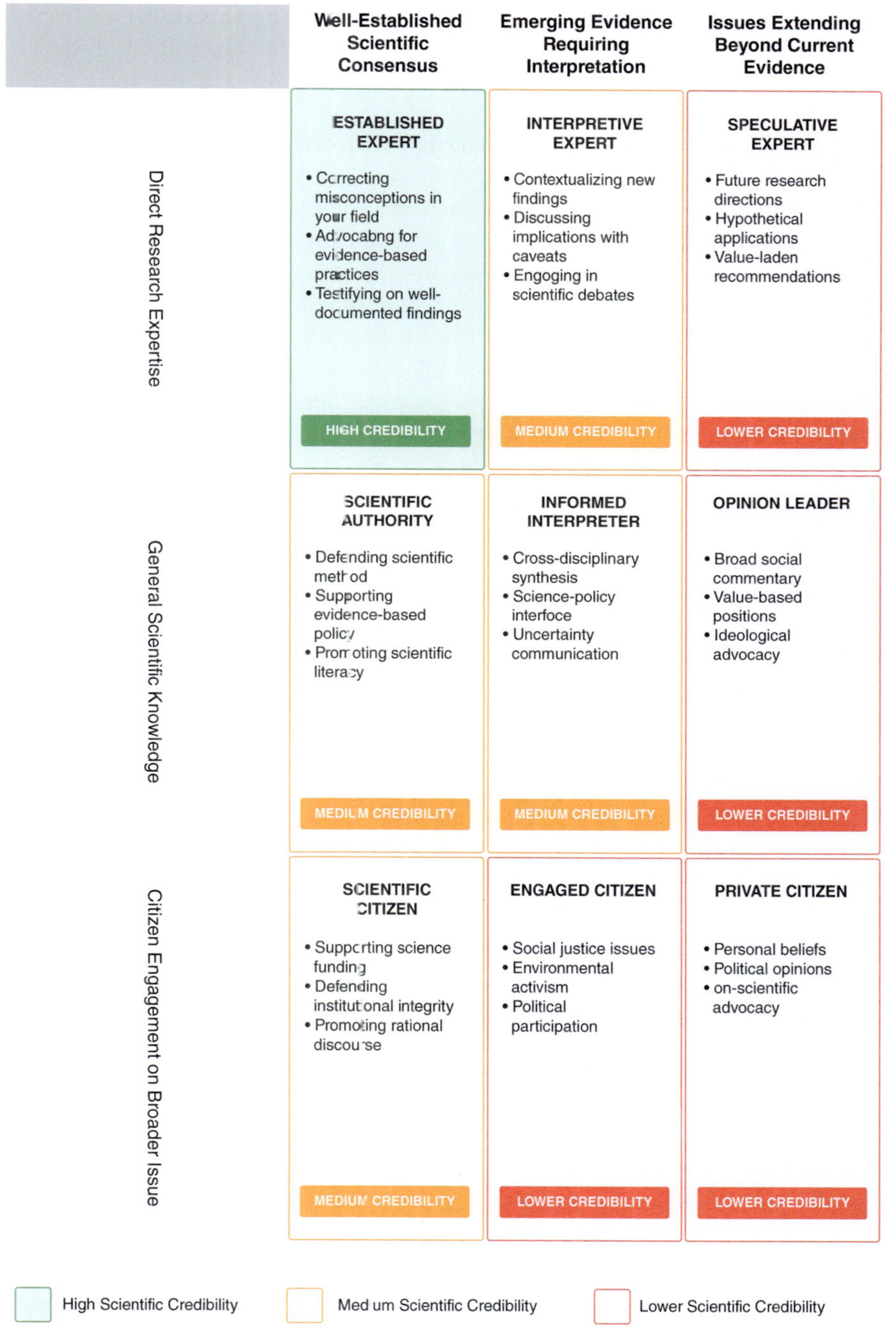

Fig. 19.1 Scientist-Advocacy Framework Matrix

positions, obscuring their underlying value assumptions behind seemingly objective claims [24].

19.3.3 Strategic Communication for Different Audiences

Effective advocacy requires moving beyond one-size-fits-all approaches toward recognition that different stakeholders have fundamentally different information needs, concerns, contexts, and existing knowledge [25]. This requires thoughtful calibration rather than standardized messaging or mere simplification.

Successful audience analysis identifies what truly matters about different stakeholders—including their existing knowledge levels, primary concerns, decision responsibilities, information processing patterns, and value priorities [26]. This enables communication tailored to specific characteristics rather than relying on generic assumptions about "public audiences."

Value resonance involves connecting scientific content with what specific audiences already care about rather than assuming universal appreciation for knowledge regardless of perceived relevance [27]. This can enhance engagement without requiring content distortion, creating meaningful connections between evidence and existing priorities.

19.4 Building Coalitions Across Disciplines

19.4.1 Finding Common Ground with Diverse Partners

Meaningful partnerships across disciplinary boundaries rarely emerge from comprehensive agreement on all aspects of complex issues [28]. Instead, they require sophisticated approaches focusing on shared outcome priorities while acknowledging legitimate differences. This enables effective collaboration on specific objectives despite ongoing diversity in conceptual frameworks and professional perspectives.

Successful coalition development involves identifying shared priorities across different professional contexts. This often requires moving beyond surface terminology differences to identify substantive value connections—recognizing when seemingly different professional objectives actually reflect shared underlying priorities expressed through different vocabularies and frameworks [29].

Disciplinary respect practices acknowledge the legitimate expertise and perspectives each field brings to complex challenges [30]. Genuine respect recognizes that complex problems require multiple knowledge domains rather than privileging single disciplines, creating collaborative environments where different fields enhance rather than undermine each other's contributions.

19.4.2 Translating Across Professional Languages

Effective cross-disciplinary communication extends beyond simplification toward genuine translation, maintaining substantive content while expressing it through conceptual frameworks and vocabulary meaningful within different professional contexts [31]. This requires developing a sufficient understanding of other fields to reformulate knowledge through diverse disciplinary lenses.

Disciplinary framework mapping involves identifying how different fields conceptualize related issues through distinctive frameworks reflecting their particular concerns, methodologies, and objectives. This enables more sophisticated translation that reconfigures information through alternative structures rather than simply substituting terminology while maintaining neuroscience-centric conceptual organization [32].

Concept transformation involves expressing core ideas through different disciplinary lenses without distorting essential content. This can potentially communicate content more effectively than either technical precision that ignores recipient frameworks or oversimplification that loses critical substance [33].

19.4.3 Navigating Power Dynamics and Status Differences

Meaningful collaboration requires explicitly addressing rather than ignoring the reality that different disciplines and sectors carry varying levels of institutional prestige, resource access, and cultural authority [34]. These power dynamics inevitably shape interactions unless deliberately managed through approaches acknowledging differentials while establishing practices preventing them from undermining equitable partnership.

Status recognition involves explicitly acknowledging hierarchical realities rather than pretending they do not exist. This creates space for open discussion of how different disciplines, credentials, and institutional affiliations carry varying prestige and authority across different contexts [35].

Power-sharing mechanisms create governance structures, decision processes, and resource allocation approaches that ensure authority extends beyond those with the greatest formal status [36]. This can potentially transform collaborations toward more equitable partnerships, enabling contributions based on relevant expertise rather than institutional position.

19.5 Engaging with Policy Makers Effectively

19.5.1 Understanding Policy-Making Realities

Effective policy engagement requires a genuine understanding of the actual environments within which policy decisions occur. This involves recognizing constraints, timelines, competing priorities, and decision processes that determine how scientific information might be used [37]. This enables more sophisticated approaches based on how policy actually functions rather than idealized models.

Decision-maker context assessment involves understanding the specific situations of individual policymakers. Policymakers represent not a homogeneous group but diverse roles. These roles have vastly different authorities, constraints, timeframes, and priorities.

Policy process mapping involves understanding the specific sequences through which decisions develop from initial consideration to final determination [38]. This can identify optimal intervention points where scientific input might prove most influential, enabling strategic engagement when opportunities actually exist.

19.5.2 Strategic Timing and Opportunity Recognition

Effective policy influence often depends less on developing novel scientific arguments than on recognizing specific moments when decision contexts particularly align with potential research contributions [39]. This requires both attention to when conditions favor evidence consideration and the capability for rapid mobilization during sometimes brief windows.

Policy cycle awareness involves understanding cyclical patterns in governance processes including budget cycles, legislative sessions, regulatory review periods, and program evaluation timelines [40]. These create recurring windows when decisions necessarily occur, enabling strategic engagement aligned with established rhythms.

External event monitoring tracks developments that temporarily reshape decision environments—media coverage, public attention patterns, leadership changes, and crisis responses—that may briefly create receptivity unlikely during normal periods [41].

19.5.3 Building Relationships Beyond Transactional Encounters

Effective policy influence often emerges not from isolated communications during specific debates but from established trust and understanding developed through ongoing engagement [42]. This creates relationships where credibility, communication channels, and mutual understanding accumulate over time.

Reciprocal value approaches create genuine benefits for both research and policy communities rather than one-way information provision [43]. Policymakers receive valuable insights while researchers gain improved understanding of policy challenges, implementation considerations, and potential research applications.

Trust development occurs through consistent reliability, appropriate boundary maintenance between evidence and advocacy, explicit acknowledgment of limitations, and follow-through on commitments [44]. This can transform relationships from a skeptical reception toward earned credibility.

19.5.4 Balancing Evidence Presentation with Clear Implications

Effective policy engagement requires navigating beyond binary choices between "just the facts" or comprehensive policy advocacy [45]. This involves finding nuanced approaches that provide decision-relevant information addressing policymakers' actual needs while maintaining appropriate boundaries preserving scientific credibility.

Implication derivation involves creating structured processes for identifying what specific findings might reasonably suggest for decision consideration while explicitly recognizing where implications represent reasonable extensions rather than direct empirical conclusions [46].

Option presentation involves explicitly identifying different potential approaches consistent with existing evidence while noting their respective advantages, limitations, and assumptions [47]. This can provide more decision-useful information than singular recommendations, implying scientific consensus on inherently value-laden choices.

19.6 Personal Reflections on Navigating Scientist-Advocacy

19.6.1 Finding Your Authentic Voice

Effective public engagement often emerges not from adopting standardized "scientist personas" but from developing authentic voices integrating scientific expertise with individual values, communication styles, and personal characteristics [48]. This creates distinctive contributions impossible through generic approaches attempting to separate professional from personal identity.

Value alignment examination involves exploring how scientific pursuits reflect and connect with personal values rather than maintaining artificial separation [49]. This enables more coherent advocacy emerging from genuine rather than compartmentalized motivations.

Communication style integration involves identifying and developing distinctive approaches aligned with individual characteristics and strengths rather than attempting to conform to standardized scientific presentation models [50].

19.6.2 Navigating Tensions Between Research and Advocacy

Effective navigation requires moving beyond viewing research and advocacy as inherently competing commitments toward more nuanced recognition of both tensions and potential complementarities [51]. This involves acknowledging real constraints while identifying possible synergies.

Boundary-setting practices establish appropriate distinctions between different professional activities without artificial separation, creating a clear demarcation between roles while maintaining awareness of their inevitable interactions [52].

Time allocation frameworks involve developing explicit approaches for determining how much time and energy different activities warrant during particular career phases rather than allowing immediate demands to dictate distribution without strategic consideration [53].

19.6.3 Learning from Setbacks and Adjusting Approaches

Sustainable impact requires moving beyond viewing setbacks as either random misfortunes or evidence of fundamental futility [54]. Instead, sophisticated learning approaches extract valuable insights from disappointing experiences, enabling continuous improvement through what might otherwise become discouraging failures.

Failure analysis involves examining fundamental assumptions, approach limitations, and knowledge gaps revealed through disappointing results rather than attributing setbacks primarily to external circumstances [55].

Approach adaptation requires creating explicit processes for translating experiential learning into strategic modifications rather than either persisting with initial approaches regardless of results or making reactive changes without systematic consideration [56].

19.6.4 Maintaining Hope and Perspective on Long-Term Change

Sustainable advocacy requires frameworks extending beyond immediate results to encompass appreciation for participation in essential transformations that typically unfold across extended timeframes [57]. This creates more resilient engagement through realistic rather than idealized conceptions of how scientific knowledge ultimately translates into societal impact.

Realistic expectation calibration involves developing mental models that recognize how structural factors naturally create resistance to change regardless of evidence quality or advocacy effectiveness [58]. This enables continued engagement despite opposition by understanding it as normal within change processes.

The historical perspective involves examining how scientific knowledge has ultimately influenced societal practices across extended timeframes despite often encountering similar resistance patterns [59]. This can transform present challenges from seemingly exceptional barriers into recognizable features of documented influence processes.

19.7 Conclusion: Toward Engaged Neuroscience in Society

As neuroscience becomes increasingly central to consequential societal decisions, the models guiding researchers' public engagement will significantly shape both the quality of public discourse and the ultimate societal impact of neuroscientific knowledge [60]. Several key principles emerge from examining effective scientist-advocacy approaches.

First, authenticity appears increasingly essential for both effectiveness and sustainability, requiring approaches that integrate scientific expertise with individual values and characteristics rather than attempting artificial separation between professional and personal identity. Second, boundary awareness emerges as fundamental for maintaining credibility while engaging with societal implications—creating clear distinctions between empirical findings, reasonable implications, and acknowledgment where normative judgments extend beyond what evidence alone establishes. Third, relationship investment shows particular importance for meaningful impact—recognizing that effective influence typically emerges from established trust and understanding developed through ongoing engagement rather than isolated communications.

The evolving landscape creates both opportunities and challenges for neuroscientists navigating public roles. The most profound questions facing engaged neuroscientists are not whether to participate in public discourse but how to do so effectively. The challenge is preserving the qualities that make scientific voices worth hearing while effectively contributing to societal conversations where values and evidence necessarily interact [61].

Perhaps most fundamentally, effective scientist-advocacy requires ongoing learning and adaptation rather than fixed approaches [62]. The most successful engaged scientists demonstrate continuous refinement through both successes and failures—creating more sustainable and ultimately impactful engagement through willingness to evolve approaches based on experience rather than defending initial strategies regardless of evidence suggesting need for reconsideration.

This perspective highlights the inherently experimental nature of public engagement requiring the same commitment to evidence-based refinement that characterizes good science itself. Responsible advocacy demands not just initial skill development but ongoing evolution reflecting accumulated understanding of how scientific knowledge most effectively contributes to the complex social processes through which societies navigate questions where values and evidence necessarily interact in determining how neuroscientific advances ultimately serve human flourishing [63].

19.8 Summary

This chapter explores the evolving landscape of neuroscientist advocacy, examining how researchers navigate the complex terrain between maintaining scientific integrity and engaging meaningfully with public discourse. This chapter opens with the transformation of scientists from passive knowledge producers to active interpreters who contextualize their findings within broader societal implications, acknowledging that remaining silent on consequential issues represents abdication rather than neutrality.

The discussion traces the shift from traditional "knowledge transmission" models toward more sophisticated approaches where scientists address public misunderstandings, combat "neurohype," and move from value-neutral to value-transparent communication. Rather than maintaining artificial separation between facts and values, effective scientist-advocates learn to distinguish explicitly between empirical findings, interpretive judgments, and normative positions while preserving their distinctive credibility.

This chapter provides practical frameworks for effective advocacy, emphasizing that not all forms of public engagement carry the same implications for scientific credibility. It explores how intellectual honesty, including acknowledgment of uncertainties and limitations—paradoxically enhances rather than undermines authority, and how strategic communication requires tailoring messages to different audiences' specific needs and contexts.

Building successful coalitions across disciplines emerges as a critical skill, requiring scientists to find common ground with diverse partners, translate complex concepts across professional languages, and navigate power dynamics that inevitably arise when different fields collaborate. This chapter dedicates substantial attention to policy engagement, emphasizing the importance of understanding policy-making realities, recognizing strategic timing opportunities, and building relationships that extend beyond transactional encounters.

The final sections offer personal reflections on the scientist-advocacy journey, exploring how researchers can find their authentic voice while navigating tensions between research and advocacy roles. This chapter concludes by acknowledging that effective scientist-advocacy requires ongoing learning and adaptation, treating public engagement as an inherently experimental process requiring the same evidence-based refinement that characterizes good science itself.

References

1. Jasanoff S. Technologies of humility: citizen participation in governing science. Minerva. 2003;41(3):223–44.
2. Pielke RA Jr. The honest broker: making sense of science in policy and politics. Cambridge University Press; 2007.
3. Illes J, Bird SJ. Neuroethics: A modern context for ethics in neuroscience. Trends Neurosci. 2006;29(9):511–7.
4. Fischhoff B. The sciences of science communication. Proc Natl Acad Sci. 2013;110(Supplement 3):14033–9.
5. Wynne B. Public engagement as a means of restoring public trust in science–hitting the notes, but missing the music? Community Genet. 2006;9(3):211–20.
6. Collins H, Evans R. The third wave of science studies: studies of expertise and experience. Soc Stud Sci. 2002;32(2):235–96.
7. Nowotny H, Scott P, Gibbons M. Re-thinking science: knowledge and the public in an age of uncertainty. Polity Press; 2001.
8. Davies SR. Constructing communication: talking to scientists about talking to the public. Sci Commun. 2008;29(4):413–34.
9. Reardon S. 'Brain doping' may improve athletes' performance. Nature. 2016;531(7594):283–4.
10. Weisberg DS, Keil FC, Goodstein J, Rawson E, Gray JR. The seductive allure of neuroscience explanations. J Cogn Neurosci. 2008;20(3):470–7.
11. McCabe DP, Castel AD. Seeing is believing: the effect of brain images on judgments of scientific reasoning. Cognition. 2008;107(1):343–52.
12. Racine E, Bar-Ilan O, Illes J. fMRI in the public eye. Nat Rev Neurosci. 2005;6(2):159–64.
13. O'Connor C, Rees G, Joffe H. Neuroscience in the public sphere. Neuron. 2012;74(2):220–6.
14. Longino HE. Science as social knowledge: values and objectivity in scientific inquiry. Princeton University Press; 1990.
15. Douglas HE. Science, policy, and the value-free ideal. University of Pittsburgh Press; 2009.

16. Kitcher P. Science, truth, and democracy. Oxford University Press; 2001.
17. Steel D. Epistemic values and the argument from inductive risk. Philos Sci. 2010;77(1):14–34.
18. Mehta G, Hopf H, Krief A, Matlin SA. Realigning science, society and policy in uncertain times. R Soc Open Sci. 2020;7(5):200554.
19. Sarewitz D. How science makes environmental controversies worse. Environ Sci Pol. 2004;7(5):385–403.
20. Kahan DM. Climate-science communication and the measurement problem. Polit Psychol. 2015;36:1–43.
21. Gieryn TF. Cultural boundaries of science: credibility on the line. University of Chicago Press; 1999.
22. Shapin S. The scientific life: a moral history of a late modern vocation. University of Chicago Press; 2008.
23. Fischhoff B. Evaluating science communication. Proc Natl Acad Sci. 2019;116(16):7670–5.
24. Elliott KC. A tapestry of values: an introduction to values in science. Oxford University Press; 2017.
25. Bucchi, M., & Trench, B. (Eds.). (2021). Routledge handbook of public communication of science and technology. Routledge.
26. Weingart P, Guenther L. Science communication and the issue of trust. J Sci Commun. 2016;15(5):C01.
27. Corner A, Whitmarsh L, Xenias D. Uncertainty, scepticism and attitudes towards climate change: biased assimilation and attitude polarisation. Clim Chang. 2012;114(3–4):463–78.
28. Star SL, Griesemer JR. Institutional ecology, 'translations' and boundary objects: amateurs and professionals in Berkeley's Museum of Vertebrate Zoology, 1907–39. Soc Stud Sci. 1989;19(3):387–420.
29. Galison P. Image and logic: a material culture of microphysics. University of Chicago Press; 1997.
30. Klein JT. Typologies of interdisciplinarity: the boundary work of definition. In: Frodeman R, editor. The Oxford handbook of interdisciplinarity. Oxford University Press; 2017. p. 21–34.
31. Thompson Klein J. Communication and collaboration in interdisciplinary research. In: O'Rourke M, Crowley S, Eigenbrode SD, Wulfhorst JD, editors. Enhancing communication & collaboration in interdisciplinary research. SAGE Publications; 2014. p. 11–30.
32. Kockelmans JJ. Why interdisciplinarity? In: Kockelmans JJ, editor. Interdisciplinarity and higher education. Pennsylvania State University Press; 1979. p. 123–60.
33. Lattuca LR. Creating interdisciplinarity: interdisciplinary research and teaching among college and university faculty. Vanderbilt University Press; 2001.
34. Bourdieu P. The specificity of the scientific field and the social conditions of the progress of reason. Soc Sci Inf. 1975;14(6):19–47.
35. Abbott A. The system of professions: an essay on the division of expert labor. University of Chicago Press; 1988.
36. Frickel S, Moore K, editors. The new political sociology of science: institutions, networks, and power. University of Wisconsin Press; 2006.
37. Cash DW, Clark WC, Alcock F, Dickson NM, Eckley N, Guston DH, et al. Knowledge systems for sustainable development. Proc Natl Acad Sci. 2003;100(14):8086–91.
38. Kingdon JW. Agendas, alternatives, and public policies. Pearson; 2014.
39. Sabatier, P. A. (Ed.). (2007). Theories of the policy process. Westview Press.
40. Jones BD, Baumgartner FR. The politics of attention: how government prioritizes problems. University of Chicago Press; 2005.
41. Birkland TA. After disaster: agenda setting, public policy, and focusing events. Georgetown University Press; 1997.
42. Heclo H. Issue networks and the executive establishment. New Am Polit Syst. 1978;1:87–124.
43. Weiss CH. The many meanings of research utilization. Public Adm Rev. 1979;39(5):426–31.
44. Head BW. Toward more "evidence-informed" policy making? Public Adm Rev. 2016;76(3):472–84.

45. Nutley SM, Walter I, Davies HT. Using evidence: how research can inform public services. Policy Press; 2007.
46. Sanderson I. Evaluation, policy learning and evidence-based policy making. Public Adm. 2002;80(1):1–22.
47. Cartwright N, Hardie J. Evidence-based policy: a practical guide to doing it better. Oxford University Press; 2012.
48. Treise D, Weigold MF. Advancing science communication: a survey of science communicators. Sci Commun. 2002;23(3):310–22.
49. Poliakoff E, Webb TL. What factors predict scientists' intentions to participate in public engagement of science activities? Sci Commun. 2007;29(2):242–63.
50. Davies SR. Constituting public engagement: meanings and genealogies of PEST in two UK studies. Sci Commun. 2013;35(6):687–707.
51. Felt U, Fochler M. The bottom-up meanings of the concept of public participation in science and technology. Sci Public Policy. 2008;35(7):489–99.
52. Horlick-Jones T, Walls J, Rowe G, Pidgeon N, Poortinga W, Murdock G, O'Riordan T. The GM Science Review Panel: the construction of scientific advice? Soc Stud Sci. 2007;37(5):723–48.
53. Callon M, Lascoumes P, Barthe Y. Acting in an uncertain world: an essay on technical democracy. MIT Press; 2009.
54. Stirling A. "Opening up" and "closing down": power, participation, and pluralism in the social appraisal of technology. Sci Technol Hum Values. 2008;33(2):262–94.
55. Rowe G, Frewer LJ. A typology of public engagement mechanisms. Sci Technol Hum Values. 2005;30(2):251–90.
56. Wilsdon J, Willis R. See-through science: why public engagement needs to move upstream. Demos; 2004.
57. Gross AG, Harmon JE, Reidy M. Communicating science: the scientific article from the 17th century to the present. Oxford University Press; 2002.
58. Gregory J, Miller S. Science in public: communication, culture, and credibility. Perseus Publishing; 1998.
59. Irwin A. Constructing the scientific citizen: science and democracy in the biosciences. Public Underst Sci. 2001;10(1):1–18.
60. Rip A. Constructing expertise: in a third wave of science studies? Soc Stud Sci. 2003;33(3):419–34.
61. Jasanoff S. Designs on nature: science and democracy in Europe and the United States. Princeton University Press; 2005.
62. Ezrahi Y. The descent of Icarus: science and the transformation of contemporary democracy. Harvard University Press; 1990.
63. Winner L. Upon opening the black box and finding it empty: social constructivism and the philosophy of technology. Sci Technol Hum Values. 1993;18(3):362–78.

Epilogue: Charting the Path Forward 20

> *The best way to predict the future is to invent it, but the best way to invent a good future is to understand the present clearly and act with wisdom proportional to the power of our tools.*
>
> *—Adapted from Alan Kay*

In a research symposium held virtually in late 2024, neuroscientists from six continents gathered to discuss a peculiar paradox. Despite unprecedented computational power to analyze brain function, machine learning algorithms that can decode neural signals with remarkable precision, and growing investment in neurotechnology development, public trust in neuroscience research appeared to be declining in many regions. This paradox mirrors what Peter Hotez and Michael Mann documented in *Science Under Siege* and what Hotez further explored in *The Deadly Rise of Anti-Science*, showing how extraordinary technical capabilities can coincide with growing skepticism of scientific expertise. The disconnect between technical capability and social acceptance had become a defining challenge for the field.

This paradox encapsulates the central tension explored throughout this book. We possess extraordinary sophistication for understanding and treating disorders of the brain and mind, yet we operate within institutional and social contexts that often seem unprepared for the implications of these capabilities.

The convergence of artificial intelligence and neuroscience has created possibilities that would have seemed impossible just years ago. However, it has also generated challenges that extend far beyond technical considerations to encompass fundamental questions about privacy, equity, governance, and the role of expertise in democratic societies. The journey through 19 chapters has revealed a field caught between extraordinary technical potential and extraordinary social complexity. The early chapters documented genuine achievements. These include multimodal neuroimaging analysis that reveals hidden patterns of disease, biomimetic AI systems that learn from the brain's own computational principles, and clinical applications

T. Rudroff, *AI for the Healthy Brain and Neurological Disorders*, Neurocultural Health and Wellbeing, https://doi.org/10.1007/978-3-032-11596-6_20

that are beginning to transform patient care. These advances demonstrate real potential to revolutionize how we understand and treat neurological conditions. Applications range from predicting Alzheimer's disease years before symptoms appear to developing personalized treatments based on individual neural characteristics. Yet each technical advance has been accompanied by challenges that traditional frameworks struggle to address. The capacity to decode thoughts raises unprecedented questions about mental privacy. The ability to predict future disease states complicates established approaches to informed consent. The potential for algorithmic bias threatens to exacerbate rather than reduce health disparities. The interdisciplinary nature of AI neuroscience challenges institutional boundaries designed for more traditional research domains.

Perhaps most fundamentally, the power of these technologies has attracted attention from commercial and political actors whose interests may not align with scientific integrity or public benefit. Anti-science movements exploit public fears about neurotechnology to undermine research that could benefit millions of patients—a pattern that Hotez and Mann analyzed in *Science Under Siege* and that Hotez documented extensively in *The Deadly Rise of Anti-Science*, showing how similar dynamics during the COVID-19 pandemic led to over 200,000 preventable deaths among unvaccinated Americans despite widespread vaccine availability. Commercial pressures create incentives for the premature translation of findings and the publication of results that favor narrow interests over scientific truth. Political forces increasingly view neuroscience research through partisan lenses, threatening the independence essential for rigorous inquiry.

The vulnerability of scientific institutions to these pressures has emerged as a persistent theme throughout our exploration. As Hotez and Mann detailed in *Science under Siege*, this vulnerability spans multiple scientific domains, while Hotez's personal experience documented in *The Deadly Rise of Anti-Science*—facing death threats and harassment while advocating for vaccines—illustrates how individual scientists become targets when expertise conflicts with political narratives. Researchers engage in self-censorship to avoid controversial topics. This dynamic, which Hotez experienced firsthand as he became "one of the main targets of anti-science rhetoric" during the pandemic, demonstrates how external pressures can distort research priorities even within ostensibly independent institutions. Institutions implement risk-averse policies that may protect against criticism while limiting beneficial research. Funding agencies shape research priorities through mechanisms that reflect political considerations alongside scientific merit. The analysis of successful institutional models revealed that protecting scientific integrity requires sophisticated governance structures, distributed accountability mechanisms, and sustained investment in building resilience against various forms of pressure.

One of the most significant insights from this exploration is that technical excellence alone is insufficient for beneficial innovation. The systematic analysis of bias revealed how seemingly neutral technical choices can embed discriminatory patterns that disadvantage already marginalized populations. Without deliberate attention to equity and justice considerations, AI neuroscience risks perpetuating and amplifying existing disparities in healthcare access and quality. This challenge extends beyond technical fixes to require fundamental reconsideration of how we conceptualize neurological diversity, conduct research, and distribute benefits from innovation.

The persistent gap between scientific reality and public perception has created vulnerabilities that can be exploited to undermine beneficial research. Hotez and Mann's analysis reveals how this gap becomes a strategic target—anti-science movements deliberately exploit public unfamiliarity with complex topics to sow doubt about expertise itself, transforming legitimate questions into the wholesale rejection of scientific authority. The exploration of anti-science movements revealed how this gap creates openings for those who seek to discredit expertise itself, while the analysis of public engagement strategies demonstrated that addressing these challenges requires moving beyond traditional deficit models of science communication toward genuine dialogue and collaborative approaches. Artificial intelligence presents both opportunities and challenges in this domain, with the potential to personalize educational content and detect misinformation at scale while also risking oversimplification of complex concepts or inadvertent amplification of existing biases.

The challenge of developing governance frameworks adequate to the pace and complexity of AI neuroscience development has proven particularly complex. Traditional regulatory approaches, designed for more stable technologies, prove insufficient for systems that continue learning after deployment and applications that span multiple domains from healthcare to enhancement to research. The framework for responsible innovation presented in the previous chapter represents one approach to this challenge, emphasizing risk-proportionate oversight, privacy protection, justice considerations, and adaptive mechanisms. However, the success of any governance framework ultimately depends on the commitment of multiple stakeholders to implement and refine these approaches based on real-world experience.

20.1 The Path Forward

Looking ahead, the transformation of brain health research and care depends on the actions of diverse stakeholders whose collaboration is essential for navigating these challenges successfully. Researchers and scientists bear primary responsibility for maintaining the integrity and social relevance of AI neuroscience research. This responsibility extends beyond conducting rigorous research to encompass thoughtful engagement with ethical implications, proactive attention to bias and representation issues, and meaningful communication with diverse publics. They must develop algorithmic literacy that encompasses understanding of bias sources, fairness metrics, and interpretability approaches. They must also implement inclusive research practices that prioritize diverse participation and validation across demographic groups.

Healthcare professionals serve as crucial intermediaries between AI neuroscience innovations and patient benefit, bringing essential expertise about clinical workflows, patient needs, and healthcare delivery systems. Their engagement in evaluating AI-based clinical decision support tools, advocating for transparency in algorithmic systems, and championing equity considerations in deployment will significantly influence whether technological advances translate into improved care or additional barriers. The integration of AI capabilities with clinical judgment requires developing protocols that preserve the essential human elements of therapeutic relationships while leveraging computational insights effectively.

Policymakers and regulators play critical roles in creating frameworks that enable beneficial innovation while preventing harm and protecting public interests. This requires developing expertise in rapidly evolving technical domains while maintaining focus on broader societal values and democratic accountability. Investment in regulatory capacity for oversight of AI applications in neuroscience must include both technical expertise and ethical analysis capabilities, supported by adaptive regulatory frameworks that can evolve alongside technological developments while maintaining core protective principles.

Institutional leaders in universities, research institutes, and healthcare organizations provide the contexts within which AI neuroscience research and application occur. Their policies and practices significantly influence whether these activities serve a broad social benefit while maintaining scientific integrity. Strengthening institutional review processes to address the unique challenges of AI neuroscience research, investing in interdisciplinary collaboration infrastructure, and creating incentive structures that reward attention to social implications alongside technical achievement represent immediate opportunities for positive impact.

Commercial organizations developing AI neuroscience applications bear significant responsibility for ensuring that market incentives align with social benefit. This requires implementing ethical design principles from the earliest stages of product development, investing in diverse and representative development teams, prioritizing transparency and interpretability in AI systems, and engaging meaningfully with affected communities during design and testing processes. Long-term sustainability requires building business models that can succeed while serving diverse populations equitably and maintaining accountability to broader social values alongside commercial objectives.

Community organizations and advocacy groups represent essential voices in ensuring that AI neuroscience development serves the interests of those most affected by neurological conditions. Their participation in governance processes, advocacy for inclusive research practices, monitoring of deployment for bias or accessibility barriers, and support for science communication efforts that acknowledge legitimate concerns provide crucial counterbalances to commercial and professional interests.

The realization of AI neuroscience's transformative potential requires sustained collaboration across these different domains of expertise and stakeholder communities, even when their immediate interests may diverge. Building and maintaining such coalitions requires ongoing investment in relationship-building, mutual understanding, and shared commitment to outcomes that serve broad human benefit.

20.2 Vision and Principles

The vision that emerges from this exploration is one of a fundamentally transformed landscape of brain health research and care, shaped by the responsible development and deployment of AI neuroscience applications. In this future, early detection systems routinely identify neurological conditions years before symptoms appear. This enables interventions when they can be most effective. Personalized treatment approaches, based on individual neural characteristics and predictive models, replace one-size-fits-all protocols with precision interventions tailored to each person's unique brain profile. Brain-computer interfaces seamlessly restore communication and mobility for those who have lost these capabilities. Meanwhile, cognitive enhancement technologies expand human potential in ways that are both safe and equitably accessible. This technological transformation occurs within governance frameworks that successfully balance innovation with protection, ensuring that powerful new capabilities serve human flourishing rather than exacerbating existing inequities. Research institutions maintain independence from inappropriate commercial and political pressures while remaining accountable to the communities they serve. International collaboration flourishes despite geopolitical tensions, enabling global coordination on challenges that transcend national boundaries.

Realizing this vision requires sustained commitment to evidence-based humility that maintains realistic assessments of current capabilities while remaining open to future possibilities. It demands inclusive innovation that ensures AI neuroscience development serves diverse populations equitably rather than primarily benefiting those who are already privileged. Adaptive governance must create institutional frameworks that can evolve alongside technological developments while maintaining core protective principles, supported by democratic engagement that maintains meaningful public participation in decisions about how AI neuroscience applications are developed, deployed, and governed.

Scientific integrity must be protected through institutional resilience against inappropriate pressures while maintaining accountability for social relevance. Global collaboration must be fostered on challenges that require coordinated responses while respecting legitimate differences in values and priorities across different contexts. The complexity of both brain function and social systems means that linear progress is unlikely, and successful navigation will require tolerance for ambiguity alongside commitment to core values of reducing human suffering, expanding human potential, and building more just and sustainable societies.

The enduring foundation of scientific integrity represents more than adherence to methodological protocols—it encompasses honest reporting of findings, transparency about limitations and uncertainties, and commitment to truth over convenience or commercial advantage. Maintaining these commitments in the face of anti-science movements, commercial pressures, and political interference requires both individual resolve and collective action. Individual scientists must continue pursuing rigorous research while engaging thoughtfully with social implications. Institutions must create structures that protect scientific autonomy while ensuring appropriate accountability. Society must maintain support for independent research while exercising legitimate oversight over applications with broad social implications.

Resilience in the scientific enterprise comes not from uniformity but from maintaining multiple pathways for important research to advance even when particular channels face obstacles. This principle applies across scales, from individual career strategies that maintain multiple options to international collaborations that distribute capabilities across different jurisdictions. Building such diversity requires deliberate investment and ongoing attention, as natural tendencies toward efficiency and standardization can create vulnerabilities that become apparent only during periods of stress.

The challenges facing AI neuroscience are substantial but not unprecedented. Throughout history, powerful new technologies have created analogous tensions between potential benefits and legitimate concerns about risks and social implications. The most successful resolutions have generally involved sustained dialogue among diverse stakeholders, adaptive governance mechanisms, and commitment to values that transcend immediate interests. The current moment may seem particularly challenging due to the pace of technological development and the complexity of global social and political dynamics, yet these same characteristics create opportunities for more inclusive and thoughtful approaches to innovation that can learn from past mistakes while building on accumulated wisdom about governing powerful technologies responsibly.

The framework for responsible innovation should be understood not as a fixed solution but as a starting point for ongoing evolution. As AI neuroscience capabilities continue advancing and social contexts continue changing, governance approaches must adapt while maintaining core commitments to human dignity, equity, and flourishing. This evolutionary process requires mechanisms for learning from experience, incorporating diverse perspectives, and adjusting approaches based on evidence about their effectiveness.

The convergence of artificial intelligence and neuroscience represents one of the most significant technological developments of our time, with implications that extend far beyond technical advancement to encompass fundamental questions about human nature, social organization, and the kind of future we want to create. The choices we make today about how to develop, govern, and deploy these technologies will shape possibilities for generations to come.

Realizing the benefits of AI neuroscience while avoiding its potential harms requires sustained attention to ethical implications, inclusive development processes, robust governance frameworks, and meaningful public engagement. Most fundamentally, it requires recognition that the extraordinary power of these technologies comes with extraordinary responsibility to ensure they serve values that make human flourishing possible.

The path forward will require collaboration across disciplinary boundaries, sustained dialogue across different communities and interests, and ongoing commitment to learning and adaptation as both technologies and social contexts evolve. It will demand both technical sophistication and ethical wisdom, both innovative capability and institutional resilience, both global coordination and local responsiveness. The potential to understand and treat conditions that cause immense human suffering, to expand possibilities for human flourishing, and to build more just and sustainable societies represents one of the most important opportunities available to our generation.

By engaging thoughtfully with both the promise and the challenges of AI neuroscience, we can work toward beneficial futures. These futures would honor both the extraordinary capabilities of human intelligence and the values that make its exercise worthwhile. The conversation that began with technical questions about machine learning algorithms and brain imaging data ultimately leads to fundamental questions about who we are, who we want to become, and how we can build technologies that serve our deepest aspirations rather than undermining them.

The future of AI in neuroscience remains unwritten, awaiting the contributions of all who will be shaped by its development. The time for thoughtful engagement is now, while the trajectory of development can still be influenced by values and commitments that prioritize genuine human benefit. In this essential task, we all have roles to play.

MIX
Papier aus verantwortungsvollen Quellen
Paper from responsible sources
FSC® C105338

If you have any concerns about our products,
you can contact us on
ProductSafety@springernature.com

In case Publisher is established outside the EU,
the EU authorized representative is:
Springer Nature Customer Service Center GmbH
Europaplatz 3, 69115 Heidelberg, Germany

Printed by Libri Plureos GmbH
in Hamburg, Germany